D0945385

Conservation
Biological
Control

Conservation Biological Control

Edited by
Pedro Barbosa

Department of Entymology
University of Maryland
College Park, Maryland

Academic Press

San Diego   London   Boston   New York   Sydney   Tokyo   Toronto

This book is printed on acid-free paper. ∞

Academic Press
a division of Harcourt Brace & Company
525 B Street, Suite 1900, San Diego, California 92101-4495, USA
http://www.apnet.com

Academic Press Limited
24-28 Oval Road, London NW1 7DX, UK
http://www.hbuk.co.uk/ap/

Library of Congress Catalog Card Number: 97-80315

International Standard Book Number: 0-12-078147-6

PRINTED IN THE UNITED STATES OF AMERICA
98 99 00 01 02 03 BB 9 8 7 6 5 4 3 2 1

CONTENTS

CHAPTER **1**

CONSERVATION BIOLOGICAL CONTROL:
PAST, PRESENT, AND FUTURE

L. E. Ehler

CHAPTER **2**

CONSERVATION BIOLOGY: LESSONS FOR CONSERVING
NATURAL ENEMIES

Deborah K. Letourneau

CHAPTER **3**

AGROECOSYSTEMS AND CONSERVATION
BIOLOGICAL CONTROL

P. Barbosa

CHAPTER **4**

THE INFLUENCE OF PLANTS ON INSECT PARASITOIDS:
IMPLICATIONS FOR CONSERVATION BIOLOGICAL CONTROL

Pedro Barbosa and Betty Benrey

CHAPTER **5**

INFLUENCE OF PLANTS ON INVERTEBRATE PREDATORS:
IMPLICATIONS TO CONSERVATION BIOLOGICAL CONTROL

P. Barbosa and S. D. Wratten

CHAPTER **6**

ECOLOGICAL CONSIDERATIONS IN THE CONSERVATION OF
EFFECTIVE PARASITOID COMMUNITIES IN
AGRICULTURAL SYSTEMS

D. A. Landis and F. D. Menalled

CHAPTER **7**

HABITAT ENHANCEMENT AND CONSERVATION OF NATURAL
ENEMIES OF INSECTS

David N. Ferro and Jeremy N. McNeil

CHAPTER **8**

SOWN WEED STRIPS: ARTIFICIAL ECOLOGICAL COMPENSATION
AREAS AS AN IMPORTANT TOOL IN CONSERVATION
BIOLOGICAL CONTROL

Wolfgang Nentwig, Thomas Frank, and Christa Lethmayer

CHAPTER **11**

PESTICIDES AND CONSERVATION OF NATURAL ENEMIES

John R. Ruberson, Hisashi Nemoto, and Yoshimi Hirose

CHAPTER **12**

CONSERVATION BIOLOGICAL CONTROL OF MOBILE PESTS:
PROBLEMS AND TACTICS

Yoshimi Hirose

CHAPTER **13**

A CONSERVATION APPROACH TO USING ENTOMOPATHOGENIC
NEMATODES IN TURF AND LANDSCAPES

Edwin E. Lewis, James F. Campbell, and Randy Gaugler

CHAPTER **14**

ENVIRONMENTAL MANIPULATION FOR MICROBIAL CONTROL
OF INSECTS

James R. Fuxa

CHAPTER **15**

DEPLOYMENT OF THE PREDACEOUS ANTS AND THEIR
CONSERVATION IN AGROECOSYSTEMS

Ivette Perfecto and Antonio Castiñeiras

CHAPTER **16**

CONSERVATION OF APHIDOPHAGA IN PECAN ORCHARDS

James D. Dutcher

CHAPTER **17**

CHAPTER **18**

CHAPTER **19**

BIOLOGICAL CONTROL OF SOIL-BORNE PATHOGENS WITH
RESIDENT VERSUS INTRODUCED ANTAGONISTS: SHOULD
DIVERGING APPROACHES BECOME STRATEGIC CONVERGENCE?

Philippe Lucas and Alain Sarniguet

CHAPTER **20**

CONSERVATION STRATEGIES FOR THE BIOLOGICAL
CONTROL OF WEEDS

Raymond M. Newman, David C. Thompson, and David B. Richman

CONTRIBUTORS[1]

Pedro Barbosa Department of Entomology, University of Maryland College Park, Maryland 20742 (3,4,5)

Betty Benrey Centro de Ecologia, Universidad Nacional Autonoma de Mexico, Ciudad Universitaria, Mexico D.F., 04510 Mexico (4)

James F. Campbell Department of Entomology, University of California, Davis, California 95616 (13)

Antonio Castiñeiras University of Florida, Tropical Research and Education Center, Homestead, Florida 33031 (15)

James Dutcher Department of Entomology, University of Georgia, Coastal Plain Experiment Station, Tifton, Georgia 31793 (16)

Les Ehler Department of Entomology, University of California, Davis, California 95616-8584 (1)

Greg English-Loeb Department of Entomology, NYS Agricultural Exp. Station, Geneva, New York, 14456 (17)

Jennifer Feldman Nature Mark, Boise, Idaho 83706 (10)

David Ferro Department of Entomology, University of Massachusetts, Amherst, Massachusetts 01003 (7)

[1]Numbers in parentheses indicate chapter number.

Thomas Frank   Zoological Institute, University of Berne, CH-3012,
                        Berne, Switzerland (8)

James Fuxa   Department of Entomology, Louisiana State
                        University Baton Rouge, Louisiana 70803 (14)

Randy Gaugler   Department of Entomology, Rutgers University, New
                        Brunswick, New Jersey 08903 (13)

Fred Gould   Department of Entomology, North Carolina State
                        University, Raleigh, North Carolina 27695 (10)

G. M. Gurr   Orange Agricultural College, The University of
                        Sydney, Orange, New South Wales, 2800
                        Australia (9)

Yoshimi Hirose   Institute of Biological Control, Kyushi University,
                        Fukuoka 812-81, Japan (11,12)

Casey Hoy   Department of Entomology, The Ohio State
                        University, Ohio Agricultural Research and
                        Development Center, Wooster, Ohio 44691 (10)

George Kennedy   Department of Entomology, North Carolina State
                        University, Raleigh, North Carolina 44691 (10)

Douglas Landis   Department of Entomology, Michigan State
                        University, East Lansing, Michigan 48824 (6)

Christa Lethmayer   Federal Office and Research Centre of Agriculture, A-
                        1226 Vienna, Austria (8)

Deborah Letourneau College Eight, University of California, Santa Cruz,
                        California 95064 (2)

Edwin E. Lewis   Department of Entomology, University of Maryland,
                        College Park, Maryland 20742 (13)

Philippe Lucas   INRA, Station de Pathologie Végétale, BP 29, Cedex,
                        France (19)

Jeremy N. McNeil   Department of Biology, Laval University, Quebec PQ,
                        Canada G2K7P4 (7)

Fabian D. Menalled Department of Entomology, Michigan State
                        University East Lansing, Michigan 48824 (6)

Hisashi Nemoto   Saitama Horticultural, Experiment Station, Saitama
                        346, Japan (11)

Wolfgang Nentwig   Zoological Institute, University of Berne, CH-3012
                        Berne, Switzerland (8)

Raymond M. Newman   Department of Fisheries and Wildlife, University of
                        Minnesota, St. Paul, Minnesota 55108 (20)

Jan Nyrop   Department of Entomology, NY State Agricultural
                        Experimental Station, Geneva, New York
                        14456 (17)

Ivette Perfecto   School of Natural Resources and Environment,
                             University of Michigan, Ann Arbor,
                             Michigan 48109-1115 (15)

Gary Reed   Hermiston Agricultural Research and Education
                             Center, Oregon State University, Hermiston,
                             Oregon 97838 (10)

David B. Richman   Department of Entomology, Plant Pathology, and
                             Weed Science, New Mexico State University,
                             Las Cruces, New Mexico 88003 (20)

A. Roda   Department of Entomology, NY State Agricultural
                             Experimental Station, Geneva, New York
                             14456 (17)

John Ruberson   Department of Entomology, NY State
                             Agricultural Experimental Station, Geneva,
                             New York 14456 (17)

Alain Sarniguet   INRA, Station de Pathologie Végétale, BP 29, Cedex,
                             France (19)

David C. Thompson   Department of Entomology, Plant Pathology, and
                             Weed Science, New Mexico State University,
                             Las Cruces, New Mexico 88003 (20)

Charles L. Wilson   USDA, ARS, Appalachian Fruit Research Station,
                             Kearneysville, West Virginia 25430 (18)

H. F. van Emden   Department of Horticulture and Landscape, The
                             University of Reading, Plant Science
                             Laboratories, Whiteknights, Reading, RG6
                             6AS, England (9)

Steve D. Wratten   Department of Entomology and Animal Ecology,
                             Lincoln University, Canterbury, New Zealand
                             (5,9)

Jeff Wyman   Department of Entomology, University of Wisconsin,
                             Madison, Wisconsin 53706 (10)

PREFACE

Agriculture and many of the tactics currently used for pest management, although resulting in high productivity, may be too perilous for the environment, too risky for the consumer, and too problematic for the farmer. A variety of circumstances have combined to produce a pending national and global crisis. This crisis stems largely from the widespread reliance on chemicals to control pests. some of the negative consequences of the use of pesticides are due to the fact that broad-spectrum chemical pesticides are often toxic to nontarget beneficial and endangered species, to wildlife, and to humans. In addition, improper or excessive pesticide use has led to the development of resistance in pests and created new pest problems, while threatening groundwater quality. Finally, many chemical pesticides are no loner available due to registration revocations: Thus, alternate approaches are urgently needed to restore and maintain the balance between the human need for food and a need to maintain the ecological integrity of the environment. In recent years, in the U.S.A. and other parts of the world, there has been increasing interest in approaches which promote sustainable agroecosystems. Biological control, and in particular Conservation Biological Control, can be an extremely important alternative to widespread pesticide use as well as a significant component of a sustainable agriculture.

Conservation biological control involves the use of tactics and approaches that involve the manipulation of the environment (i.e., the habitat) of natural enemies so as to enhance their survival, and/or physiological and behavioral performance, and resulting in enhanced effectiveness. This approach to biological control can be applied to exotic (i.e., introduced) natural enemies, used as part of classical or augmentative biological control programs, as well as to indigenous (native) natural enemies. In general, habitat manipulations may entail the elimination or mitigation of detrimental conditions, or the enhancement or induction of favorable factors that are lacking in the habitat, or present at inadequate levels.

Historically, conservation biological control has entailed the use of selective insecticides that effectively control the target pest without detrimentally affecting natural enemies, or the use of non-persistent pesticides. The ideal was presumed to be the use of insecticides such as microbial formulations that could be used (sprayed) like synthetic insecticides, e.g., Bt. Alternatively, when selective insecticides were unavailable (and few commercially available insecticides were benign to natural enemies) the recommended conservation biological control tactic usually involved the "appropriate" timing of insecticide spraying, or limited distribution or placement of insecticides, so as to minimize their detrimental effects. However, as one would expect such appropriate timing is frequently circumstance specific, difficult to translate into a general protocol, and thus impractical to apply in a routine manner in many integrated pest management schemes. In classical biological control programs, if released biological control agents have any hope of surviving and establishing populations, the elimination or reduction of pesticide spraying is a bare minimum requirement.

As is the case with all other insects, microclimatic factors (i.e., temperature, moisture levels, light intensity and quality, etc.) have significant effects on biological control agents. Although it is difficult, if not impossible, to control ambient conditions in the field, edaphic factors can be moderated and/or mitigated to favor natural enemies. The alteration of cultivation practices (such as disc harrowing, tilling, burning of crop residue, irrigation, pruning, etc.) not only can have a direct impact on natural enemy survival, but can change microclimates to favor the development, survival and behavior of natural enemies. In many of the latter examples the appropriate use of the cultural practices enhance physical and/or biological diversity of the habitat and thus enhance natural enemy performance. Still another recommendation that historically has been suggested to conserve natural enemies has been the direct enhancement of vegetational diversity of agroecosystems. Such diversification is presumed to provide alternate food sources, favorable microclimate, or alternate hosts which would maintain natural enemy populations at densities that would be effective against target pest species. Still other approaches have involved maintaining food sources such as flowers, spraying sugar or yeast-sugar solutions for parasitoids and invertebrate predators, or providing nest boxes for vertebrate or invertebrate predators.

In spite of the rigorous and important work of researcher in the past, for a variety of reasons, the implementation of conservation biological control has received a great deal of lip service but little serious consideration. However, changes in the pest management arena, described above, have provided the impetus for a serious analysis and reappraisal of conservation biological control. Indeed, over the last several decades more and more fascinating and important research on the ecological underpinnings and the practical implementation of tactics and methods of conserving natural enemies has been conducted. That research is highlighted in this book.

It is my goal to highlight current research on the conservation of natural enemies to demonstrate its potential and utility is pest management scenarios. The research and experiences discussed in the chapters will not only synthesize the data available (nationally and internationally) on each topic but will also generate working hypotheses or broad generalizations that may guide the work of future researchers. Finally, by including discussions of the conservation of biological control species effective against arthropods, plant pathogens, nematodes, and weeds, I hope to convey that conservation biological control is not only relevant but a comprehensively applicable approach to the management of many different types of pests.

CHAPTER
1

CONSERVATION BIOLOGICAL CONTROL: PAST,
PRESENT, AND FUTURE

L. E. Ehler

I. INTRODUCTION

The modern concept of biological pest control has been developed primarily by entomologists and in practice is normally taken to mean the use of living natural enemies to control pest species. This can be accomplished either through (1) importation of exotic enemies against either exotic or native pests (i.e., classical biological control) or (2) conservation and augmentation of enemies that are already in place or are readily available. Most authors define conservation as actions that preserve or protect natural enemies and augmentation as actions that increase the populations of natural enemies (e.g., Rabb *et al.,* 1976; Gross, 1987; Hoy, 1988; Nordlund, 1996). However, DeBach (1964a) considered conservation to mean environmental modification to protect and enhance natural enemies, a definition that is consistent with the major theme of this book. These definitions are, however, compatible if we consider conservation and augmentation as two points on a continuum. At one extreme is conservation in the form of pesticide selectivity or selective use of pesticides, and at the other extreme is augmentation through inoculative or inundative releases.

Many environmental modifications are designed to both preserve and enhance natural enemies and thus lie at an intermediate point on this continuum. Although this approach to biological control has received some attention in the literature (e.g., van den Bosch and Telford, 1964; Coppel, 1986; Pickett and Bugg, in preparation), this pales in comparison with that received by classical biological control (e.g., DeBach, 1964b; Hagen and Franz, 1973; Huffaker and Messenger, 1976; Clausen 1978; Hokkanen, 1985; Greathead and Greathead, 1992) and augmentative releases (e.g., Ridgway and Vinson, 1977; Knipling, 1992; Parrella *et al.,* 1992; Hunter, 1994,

Ridgway *et al.,* in press). In addition, most of the work on environmental modification has dealt with arthropod pests only. Clearly, the time has come for a reappraisal of this approach to biological control, not only for insects but for other classes of pests as well. Although other pest disciplines (e.g., plant pathology) may view biological control from a different conceptual framework (see Barbosa and Braxton, 1993; Nordlund, 1996) the underlying principles should be similar.

In certain situations biological control of insect pests through environmental modification has inherent advantages over either classical biological control or augmentative releases. Although classical biological control should be the priority for introduced pests, there will be cases where conflicting interests preclude importation (e.g., against native pests) or where the candidate natural enemy lacks the degree of host specificity required for environmental safety. The latter problem is common in temporary or highly disturbed agroecosystems where mobile, opportunistic pest species predominate and natural enemies with similar ecological strategies are often required for effective biological control. These natural enemies tend to have three attributes in common: (1) colonizing ability to allow the enemy to keep pace with the spatial and temporal disruption of the habitat, (2) temporal persistence, especially in the absence of the target pest, and (3) opportunistic feeding habits to allow for persistence and permit the enemy to rapidly exploit the pest population (Ehler, 1990). Polyphagous or general predators (including omnivorous species) commonly fit this pattern. However, intentional introduction of such agents poses considerable environmental risk and is not likely to be approved under current regulatory guidelines. In such situations, conservation biological control *(sensu latu)* should receive priority, especially if augmentative release of large numbers of insectary produced agents is not economical. Thus, even if polyphagous enemies are enhanced, the effect will be local and can be relaxed as needed. Conservation biological control also relies on naturally occurring enemies that are presumably well adapted to the target system. This is a major advantage over augmentative release of insectary reared agents whose fitness (vis a vis the target system) may well have been reduced through selection, inbreeding, and genetic drift while in culture. These advantages can, of course, be offset by problems in implementation of conservation biological control; these will be addressed in this chapter.

II. HISTORICAL DEVELOPMENT

Conservation of natural enemies is probably the oldest form of biological control of insect pests. As early as 900 AD, Chinese citrus growers placed nests of the predaceous ant *Oecophylla smaragdina* F. in mandarin orange trees to reduce populations of foliage feeding insects (Sweetman, 1958; Doutt, 1964; Simmonds *et al.,* 1976). In an early review of the subject, Sweetman (1958) described some additional methods that had been suggested over the years, ranging from preserving enemies to interplanting of insectary crops; but then came to the following conclusion, "The data supporting many of the statements, however, are inadequate to justify much of the enthusiasm that has been exhibited. Further evidence of this is shown by the abandonment of most of the suggested procedures for conserving the desirable species." In the first major review of biological control through environmental modification, van den Bosch and Telford (1964) recognized the following techniques: building artificial structures (e.g., for nesting), providing supplementary food (e.g., honeydew), supplying alternative hosts, improving pest-enemy synchrony, control of ants, and modification of adverse production practices. However, no assessment of implementation in production agriculture was made. In the next major review of the subject, Rabb *et al.* (1976) observed that "Most of the techniques ... are of potential rather than realized value in pest management." Similar concerns were expressed by Coppel (1986) and Gross (1987).

One of the best examples of conservation biological control to emerge since van den Bosch and Telford's (1964) seminal review is the practice of strip-harvesting hay alfalfa in California (Stern *et al.,* 1964, 1976; van den Bosch and Stern, 1969). When an entire field of alfalfa is mowed (= solid cut) during hot weather, the native lygus bug *(Lygus hesperus* Knight) migrates within 24 hours, often to cotton where it is a key pest. However, when fields were harvested in alternating strips up to 400 ft. wide (=strip cut), lygus bugs moved from the cut strips to the remaining strips (usually half grown) rather than migrating to cotton. This cultural practice can conserve natural enemies in cotton (due to reduced chemical control of lygus) and in hay alfalfa where mobile natural enemies can disperse from cut strips to half grown strips. Another method for conserving natural enemies in cotton is to interplant alfalfa (a preferred host of lygus) at regular intervals to hold lygus bugs and prevent them from dispersing into the adjacent cotton (Stern, 1969). Although both strip cutting and interplanting can be highly effective management tools, neither practice has been widely adopted by growers. As van den Bosch and Stern (1969) noted, strip cutting poses operational problems and is more expensive than solid cutting. Interplanting alfalfa in cotton poses similar difficulties. This California example seems typical of the history of conservation biological control, at least in developed countries, where economics and integration with production practices have precluded the implementation of many ecologically sound methods of environmental modification.

III. CURRENT SITUATION

A recent report from the U.S. Congressional Office of Technology Assessment (OTA) notes that, from the 1950's to about 1980, use of conventional pesticides in the U.S.A. grew dramatically, doubling between 1964 and 1978 (Office of Technology Assessment, 1995). Pesticide use since 1980 has stabilized, and the report concludes that pesticides now pervade all aspects of pest management in the U.S.A. Similar patterns are evident in global pesticide use (Waage, 1997). Biological control is, of course, one of the major alternatives to chemical pesticides and, with respect to conservation biological control, OTA reports that "the approach is rarely used as a major and deliberate component of pest management." The situation in lesser developed countries may be more favorable.

Clearly, there is a gap between research and implementation in conservation biological control. Although similar gaps exist with respect to other biologically based technologies (Office of Technology Assessment, 1995), the one for conservation biological control (excluding use of selective chemical insecticides) is one of the greatest. How such a gap or "valley of death," as OTA puts it, can be overcome is a challenging and highly complicated matter that must be addressed in the future.

IV. CHALLENGES FOR THE FUTURE

As we approach the millennium, crop protection specialists throughout the world are being challenged to reduce pesticide usage and adopt integrated pest management (IPM). Two recent initiatives illustrate the nature of the challenge. Europe, Sweden, Denmark, and The Netherlands have passed legislation that mandates a reduction of 50% or more in the use of agricultural pesticides by the year 2000 (Matteson, 1995). In North America, the U.S. Department of Agriculture, the Environmental Protection Agency, and the Food and Drug Administration have called for the development and implementation of IPM on 75% of U.S.A. crop acreage by the year 2000 (Office of Technology Assessment, 1995). In these and related initiatives, biological control can be expected to play an important role in achieving the stated objectives. In principle, conservation biological control seems well suited for the challenge at hand; however, its future remains uncertain.

Conservation biological control will likely face stiff competition from an array of competing management tactics, including other approaches to biological control. Classical biological control has an advantage in that, when successful, it is "self- implementing." Augmentative biological control has an attendant industry (i.e., producers and suppliers of natural enemies) that has a vested interest in both the science and practice of augmentative release. In contrast, conservation biological control is not self- implementing and has no industry to turn to for financial or political support. Moreover, some industry analysts predict that the appeal of all biological-based tactics will diminish when new types of "Goldilocks" compounds reach the market (Office of Technology Assessment, 1995). Such compounds are touted as not too hard (i.e., on the environment), not too soft (i.e., on the target pest), but just right. Biotechnology will provide yet another challenge, particularly in the form of genetically engineered crop plants that possess pesticidal properties.

Perhaps the biggest challenge for conservation biological control will be to shed its image of not being implementable in the production agriculture of developed countries. This will require renewed commitment in two key areas: integration of conservation biological control into IPM programs and training of a cadre of specialists who have the holistic outlook required to achieve this goal.

Prokopy et al. (1990, 1994) recognized four levels of integration in IPM (see also Kogan, 1988). First-level IPM integrates management tactics for a single class of pests (e.g., insects), whereas second-level IPM integrates multiple management tactics across all classes of pests. Third-level IPM integrates management tactics with the entire system of crop production. Fourth-level IPM involves social, cultural, and political realms and integrates the concerns of everyone with an interest in IPM. Under this definition of IPM, there are very few complete IPM programs and the U.S.A. initiative of 75% IPM by 2000 would seem unattainable. Conceptually, most procedures in conservation biological control of arthropod pests have not progressed beyond first-level IPM. Failure to address the remaining levels of integration is probably the main reason why these techniques were not implemented. Strip-cutting of hay alfalfa in California is a good example. As Norris (1986) noted, entomologists concerned with conservation biological control tend to emphasize interactions between arthropod pests and plants, thereby overlooking changes in disease, nematode, or vertebrate problems: in addition to increasing the complexity of weed management. If conservation biological control is to play a key role in future IPM programs, broader participation of crop protection specialists will be required.

In the U. S. A., the land-grant colleges of agriculture (LGCA) that have been so critical to the development of modern agriculture are facing increased pressures for institutional change (Meyer, 1993). This includes a reduction in focus on production agriculture which does not bode well for conservation biological control. With the decline in traditional state support for LGCA, many researchers find themselves increasingly concerned with generating revenue and addressing trendy issues as opposed to conducting the long-term field research that is necessary for implementation of both conservation biological control and IPM. At the same time, there is concern that the broad training at the graduate level that encourages a holistic approach to agroecosystem management is losing its appeal. In entomology, for example, there is considerable pressure at many LGCA to dispense with the traditional core curriculum in favor of a largely research degree, i.e., "Entomology Lite," as it were. Conservation biological control and IPM are holistic, not reductionist, disciplines! We simply cannot afford to train a cadre of narrow reductionists and ask them to implement holistic approaches to agroecosystem management. New funding initiatives will be required to encourage long-term, holistic research that has an implementation component. Business as usual (i.e., exploiting pest problems for publications) will not get the job done.

V. CONCLUDING REMARKS

Conservation biological control in developed countries is at a crossroads. The choice is between continuing down the same path that has yielded management techniques that are essentially of academic interest or taking a new path that leads to the implementation of these techniques in production agriculture. Only then will conservation biological control assume its proper place as an equal partner with both classical and augmentative biological control. Let us hope that the chapters that follow will stimulate movement in this direction.

REFERENCES

Barbosa, P., and Braxton, S. (1993). A proposed definition of biological control and its relationship to related control approaches. *In* "Pest Management: Biologically Based Technologies." (R. D. Lumsden, and J. L. Vaughn, eds.), pp. 21-27. Amer. Chem. Soc. Washington, DC.

Clausen, C. P. (1978). "Introduced Parasites and Predators of Arthropod Pests and Weeds: A World Review. " U.S.D.A. Agriculture Handbook No. 480. Washington, DC.

Coppel, H. C. (1986). Environmental management for furthering entomophagous arthropods. *In* "Biological Plant and Health Protection." (J. M. Franz, ed.), pp. 57-73. Gustav Fischer Verlag. Stuttgart, Germany.

DeBach, P. (1964a). The scope of biological control. *In* "Biological Control of Insect Pests and Weeds." (P. DeBach, ed.), pp. 3-20. Chapman and Hall. London, U.K.

DeBach, P. (1964b). "Biological Control of Insect Pests and Weeds." Chapman and Hall. London, U.K.

Doutt, R. L. (1964). The historical development of biological control. *In* "Biological Control of Insect Pests and Weeds." (P. DeBach, ed.), pp. 21-42. Chapman and Hall. London, U.K.

Ehler, L. E. (1990). Introduction strategies in biological control of insects. *In* "Critical Issues in Biological Control." (M. Mackauer, L. E. Ehler, and J. Roland, eds.), pp. 111-134. Intercept. Andover, Hants, U.K.

Greathead, D. J., and Greathead, A. H. (1992). Biological control of insect pests by insect parasitoids and predators: the BIOCAT database. *Biocont. News Inform.* 13, 61N-68N.

Gross, H. R., Jr. (1987). Conservation and enhancement of entomophagous insects--a perspective. *J. EntomoL Sci.* 22, 97-105.

Hagen, K. S., and Franz, J. M. (1973). A history of biological control. *In* "History of Entomology." (R. F. Smith, T. E. Mittler, and C. N. Smith, eds.), pp. 433-476. Annual Reviews Inc. Palo Alto, CA.

Hokkanen, H. M. T. (1985). Success in classical biological control. *CRC Critical Rev. Plant Scl.* 3, 35-72.

Hoy, M. A. (1988). Biological control of arthropod pests: traditional and emerging technologies. *Amer. J. Altern. Agric.* 3, 63-68.

Huffaker, C. B., and Messenger, P. S. (1976). "Theory and Practice of Biological Control." Academic Press. New York, NY.

Hunter, C. D. (1994). "Suppliers of Beneficial Organisms in North America." Calif. Environ. Protection Agency. Sacramento, CA.

Knipling, E. F. (1992). "Principles of Insect Parasitism Analyzed from New Perspectives: Practical Implications for Regulating Insect Populations by Biological Means." U.S.D.A. Agriculture Handbook No. 693. Washington, DC.

Kogan, M. (1988). Integrated pest management theory and practice. *Entomol. Exp. Appl.* 49, 59-70.

Matteson, P. C. (1995). The "50% pesticide cuts" in Europe: a glimpse of our future? *Amer. Entomol.* 41, 210-220.

Meyer, J. H. (1993). The stalemate in food and agricultural research, teaching, and extension. *Science* 260, 881 and 1007.

Nordlund, D. A. (1996). Biological control, integrated pest management and conceptual models. *Biocont. News Inform.* 17, 35N-44N.

Norris, R. F. (1986). Weeds and integrated pest management systems. *Hort. Sci.* 21, 402-410.

Office of Technology Assessment (U. S. Congress). (1995). "Biologically Based Technologies for Pest Control." OTA-ENV-636. Washington, DC.

Parrella, M. P., Heinz, K. M., and Nunney, L. (1992). Biological control through augmentative releases of natural enemies: a strategy whose time has come. *Amer. Entomol.* 38, 172-179.

Pickett, C. H., and Bugg, R. L. "Enhancing Biological Control: Habitat Management to Promote Natural Enemies of Agricultural Pests." University of California Press. Berkeley, CA (in preparation).

Prokopy, R. J., Johnson, S. A., and O'Brien, M. T. (1990). Second-stage integrated management of apple arthropod pests. *Entomol. Exp. Appl.* 54, 9-19.

Prokopy, R. J., Cooley, D. R., Autio, W. R., and Coli, W. M. (1994). Second-level integrated pest management in commercial apple orchards. *Amer. J. Altern. Agric.* 9, 148-156.

Rabb, R. L., Stinner, R. E., and van den Bosch, R. (1976). Conservation and augmentation of natural enemies. *In* "Theory and Practice of Biological Control." (C. B. Huffaker, and P. S. Messenger, eds.), pp. 233-254. Academic Press. New York, NY.

Ridgway, R. L., and Vinson, S. B. (1977). 'Biological Control by Augmentation of Natural Enemies." Plenum Press. New York, NY.

Ridgway, R. L., Hoffmann, M. P., Inscoe, M. N., and Glenister, C. S. "Mass-reared Natural Enemies: Application, Regulation, and Needs." Thomas Say Publ. Entomol., Entomol. Soc. Amer. Lanham, MD. (in press).

Simmonds, F. J., Franz, J. M., and Sailer, R. I. (1976). History of biological control. *In* "Theory and Practice of Biological Control." (C. B. Huffaker, and P. S. Messenger, eds.), pp. 17-39. Academic Press. New York, NY.

Stern, V. M. (1969). Interplanting alfalfa in cotton to control lygus bugs and other insect pests. *In* "Proc. Tall Timbers Conf. On Ecol. Anim. Cont. By Habitat Manag." (R. Komarek, ed.), 1, 55-69.

Stern, V. M., van den Bosch, R., and Leigh, T. F. (1964). Strip cutting alfalfa for lygus bug control. *Calif. Agric.* 18, 4-6.

Stern, V. M., Adkisson, P. L., Beingolea, G. 0., and Viktorov, G. A. (1976). Cultural controls. *In* "Theory and Practice of Biological Control." (C. B. Huffaker, and P. S. Messenger, eds.), pp. 593-613. Academic Press. New York.

Sweetman, H. L. (1958). "The Principles of Biological Control." Wm. C. Brown Company. Dubuque, IA.

van den Bosch, R., and Stern, V. M. (1969). The effect of harvesting practices on insect populations in alfalfa. *In* "Proc. Tall Timbers Conf. On Ecol. Anim. Contr. By Habitat Manag." (R. Komarek, ed.), 1, 47-54.

van den Bosch, R., and Telford, A. D. (1964). Environmental modification and biological control. *In* "Biological Control of Insect Pests and Weeds." (P. DeBach, ed.), pp. 459-488. Chapman and Hall. London, U.K.

Waage, J. (1997). "Yes, but does it work in the field?" The challenge of transferring biological control technology. *Entomophaga* 41, 315-332.

CONSERVATION BIOLOGY: LESSONS FOR CONSERVING NATURAL ENEMIES

Deborah K. Letourneau

I. INTRODUCTION

Ecological theory has been used as a basis to manage populations of organisms in agroecosystems and natural systems. Since the 1970s applied ecology in agricultural systems has developed as the subdiscipline agroecology, whereas a prominent subdiscipline of applied ecology in natural systems has developed as conservation biology. Both subdisciplines are concerned with managing species populations in their habitats. However, the fields of agroecology and conservation biology have advanced separately, with few attempts to adapt lines of inquiry laterally from one field to the other. In this chapter I explore how ecological theory aimed at the conservation of threatened or endangered species can be transferred to foster new concepts in the conservation of natural enemies in agroecosystems.

To set the stage, I will develop two arguments. First, landscape trends in natural and agricultural lands have parallels with respect to loss of habitat. Whether target animal species are predatory birds or primary parasitoids, some species are declining because of changes in the amount and quality of critical habitat. Second, the aims and objectives in conservation biology are comparable, in many ways, to those in conservation biological control. Strategies to mitigate species extinction are directly applicable to biological control if extinction is defined in terms of the minimum population size necessary to regulate pests in the agroecosystem. These sections are followed by a series of critical reviews to elucidate ideas and lessons developed in conservation biology that may be useful conceptually or practically in the conservation of natural enemies in agroecosystems. I present these narratives with the caveat that theoretical ecology, conservation biology, and agroecology are dynamic fields of study full of controversy and debate. Thus, for every lesson or principle mentioned in this chapter, there are challenges and exceptions. Challenges to these operating principles produce a constellation of corollary lessons that may also apply to the conservation of natural enemies in agroecosystems. Therefore, I have attempted to account for major debates and remaining questions rather than to transfer wholesale current dogma from conservation biology to agroecology.

II. LESSONS FROM CONSERVATION BIOLOGY

A.   Habitat Loss, Fragmentation, Isolation, and Degradation and Species Extinction

Landscape level patterns of habitat loss, fragmentation, isolation and degradation are primary concerns for conservation biologists attempting to slow the tide of species extinction (Primack, 1993; Doak, 1995; Wiens, 1997). These phenomena can be dramatic. For example, in the U.S. Pacific northwest, clear felling of ancient forests that support northern spotted owl *(Strix occidentalis caurinus)* populations radically decreases the resource base required by this species and results in remnant forest fragments only large enough to support a small population (Bart, 1995; Spies and Franklin, 1996). The more subtle process of habitat degradation from optimal to suboptimal to unsuitable for the maintenance of a species or group of species is more difficult to measure, yet can strongly affect population growth rates. For example, Doak (1995) warns that subtle, steady degradation of Yellowstone grizzly bear *(Ursus arctos horribilis)* habitat (say, 1% conversion of good to poor habitat per year) can trigger long-term population decline long before bear surveys can detect any response. Parallel processes occur for arthropods in general (Samways, 1994). The reduction of suitable habitat in the landscape is also a common cause of natural enemy depletion in agroecosystems (van den Bosch and Telford, 1964; Altieri and Letourneau, 1982; Hill *et al.,* 1995; Chapter 6).

Natural habitats in the vicinity of crop lands can be quite important in providing alternative foods and refugia for natural enemy populations (see Chapters 8 and 9). However, trends in agriculture over the past several decades include decreasing landscape heterogeneity, increasing agrichemical inputs, increasing mechanization, and decreasing genetic diversity (USDA, 1973; Bottrell, 1979; Whitham, 1983; Altieri and Anderson, 1986; Risch, 1987). Further, economies of scale that dominate the agricultural sector in developed countries encourage farmers to reduce unit production costs by increasing farm size and becoming more specialized. Thus, in conservation biology jargon, the landscape mosaic, especially in developed countries, has been transformed over time from one that featured "disturbance patches" (i.e., agricultural fields in a sea of natural habitats) to one that features "remnant patches" (i.e., remnants of natural habitats in a sea of agricultural enterprises).

The loss and degradation of natural habitat creates serious impediments to the conservation of natural enemies that rely on resources not provided in modern agricultural habitats. In particular, natural enemies require hosts or prey, sources of nectar and/or free water, and refugia. Most of these components increase, on average, in habitats with greater vegetational diversity, temporal stability, and sheltering capacities. However, large-scale monocultures resulting from agricultural expansion and commodification of agricultural enterprises (that is, the institutional and socioeconomic changes associated with a shift from subsistence production to local and global markets) are characterized by homogeneous vegetation, frequent disruptions due to harvest and fallow cycles, and a lack of shelter from biocides harmful to natural enemies and their alternative hosts. Thus, in the same way that protected habitats preserve viable populations of threatened and endangered species, natural lands such as riparian belts, non-crop vegetation such as hedgerows, or even unsprayed crop habitats can be crucial for maintaining and fostering population growth of predators and parasitoids in agricultural landscapes (see Chapters 8 and 9).

B. Locally and Ecologically Extinct Species

Legally, a species is defined as endangered if it is in imminent danger of extinction throughout all or a significant portion of its range. A threatened species is one that is likely to become endangered in the foreseeable future. The terms imply that the species is the functional unit of concern and that extinction is a threat to be avoided. The Endangered Species Act provides statutory means whereby the ecosystems upon which endangered species and threatened species depend may be conserved and provides a program for the conservation of target species. Species are tagged for inclusion in these categories based upon their population sizes in different parts of their geographic distribution. Species extinction, however, is not the focus of conservation biological control. Anecdotal evidence exists for some species extinctions of natural enemies in agroecosystems, such as an aphidophagous *Hippodamia* sp. (Coleoptera: Coccinellidae) which was once common in California's central valley but dependent on overwintering habitat in native bunch grasses (K. Hagen, pers. comm.). Two other categories of extinction are relevant to a wider range of natural enemies, local extinction and ecological extinction.

Local extinction occurs in response to habitat alteration, either as a gradual process or a dramatic disturbance. Even once abundant species may become extinct locally. Examples of such local extinctions in conservation biology range from the middle spotted woodpecker *(Dendrocopos medius)*, which declined to zero as regional habitat fragmentation occurred in Sweden (Meffe and Carroll, 1994) to the California grizzly bear *(Ursus arctos)*, whose habitat has also seen massive alterations, but may still have been sufficient to support the populations that ultimately were forced to extinction through hunting pressures (Dasmann, 1964; S. Minta, pers. comm.). For species with high vagility and/or genetic resources for adaptation to environmental change, local populations can be replenished by colonization from source pools outside the region. Theoretically, these populations, if relieved of adverse pressures through measures such as habitat restoration or hunting restriction, can reestablish and flourish. Neve *et al.* (1996) show how dispersal is critical to the conservation of *Proclossiana eunomia* (Esper) (Lepidoptera: Nymphalidae) in Europe. Arthropod natural enemies also tend toward high vagility and reasonable levels of adaptability. Thus, recovery from local extinction is often feasible.

However, persistence of a species *per se* is necessary but not sufficient as a goal for the conservation of natural enemies. Perhaps most appropriate for conservation biological control is the concept of "ecological extinction" (Primack, 1993). A population is "ecologically extinct" if the number of individuals in the population is below the number needed to carry out its ecological roles in the ecosystem. To illustrate, Primack (1993) uses the example of tigers: "So few tigers remain in the wild that their impact on prey populations is insignificant." Thus, if we were to define the ecological role of an arthropod predator or parasitoid as the ability to regulate pest populations (or add the functional component of regulation below an economic threshold level), then ecological extinction would occur if natural enemies population levels are too low for effective control of pests in the system. It is precisely this problem that conservation biological control most commonly addresses, yet the idea of ecological extinction, and the concomitant lessons from conservation biology that address the problem have not been fully considered by its research community.

By applying the concept of ecological extinction to functionally low levels of natural enemies, the goals of conservation biological control become parallel to those of species conservation in natural systems (halting extinction, preserving species, maintaining viable populations), and many of the tenets of conservation biology can be aptly transferred. For example, biological control practitioners could devise monitoring schemes in agroecosystems which would indicate whether or not key natural enemy species are thriving, ecologically extinct, or locally extinct. Such a measure could be used as a criterion of success for conservation biological control tactics. Findings of "endangerment" (i.e., some likelihood of extinction), coupled with an assessment of their host densities, would drive management plans for species enhancement or recovery, just as it would for endangered plants or animals in the wild. Tractable measurements of natural enemy densities have already been devised in some systems, such as the sequential sampling schemes for egg parasitoids of tomato pests in IPM protocols implemented widely in California (Division of Agriculture and Natural Resources, 1985).

C. Disturbance and Biodiversity: Influence of Scale, Intensity, and the Frequency of Disturbance Regimes

In conservation biology, attention to disturbance regimes has become more critical as theoretical and empirical interest has shifted from equilibrium paradigms and purely deterministic viewpoints toward the roles of various kinds of patch dynamics and other stochastic processes. Stochastic processes are now regarded as major influences on the probabilities of extinction and colonization and result in landscape level changes in biodiversity (Wu and Loucks, 1995). Subsequently, conservationists have recognized the value of a moderate level of disturbance as a viable alternative to preserving an ecosystem as static and unchanging (e.g., Meffe and Carroll, 1994). In concert with this notion is the "minimum dynamic area" rule for reserve design, which Pickett and Thompson (1978) defined as the smallest area with a complete, natural disturbance regime. Intermediate levels of disturbance, either within a habitat patch or over the landscape mosaic, should promote the persistence and coexistence of the greatest number of species: those that are disturbance adapted and those that are competitively dominant, and would otherwise exclude species from the habitat.

In many kinds of agroecosystems, particularly annual systems, large-scale disturbance is the rule and the intensity and frequency of disturbance are far greater than occur in natural habitats (see Hobbs and Huenneke, 1996). Therefore, species diversity in conventional agroccosystems (especially with annual crops) is expected to be relatively low, favoring those species that can survive high levels of disturbance. Cyclic patterns of cropping, harvest, and fallow serve to remove habitat for natural enemies before several generations have been achieved. Soil fumigation, herbicides, and tillage practices disrupt soil fauna directly (Baker et al., 1985) and indirectly by removing noncrop flora which may provide critical resources for natural enemies. Populations of natural enemies can be reduced to ecological extinction and biological control disrupted by the application of broadly toxic insecticides to the crop habitat (Ridgway et al., 1976; Riehl et al., 1980; Hill et al., 1995).

The minimum dynamic area concept would require the incorporation of uncultivated habitats into the landscape. The greater permanency and more subtle disturbance regimes of these habitats would complement and compensate for highly disturbed crop areas. This approach to the conservation of natural enemies in agroecosystems relies on diversifying the landscape to include all successional habitats within a reasonable proximity to crop production operations. The resulting mosaic would foster high species diversity and provide temporary refugia for natural enemies whose habitats are subject to extreme disturbance levels (Samways, 1979; Nazzi *et al.,* 1989; Chapters 8 and 9).

A complementary approach would entail the modification of agricultural practices to reduce the rate, intensity, and size as well as the temporal or spatial autocorrelation among individual disturbance events (Moloney and Levin, 1996) within agricultural fields. Indeed, slight modifications in cultural practices for sowing, maintaining, and harvesting annual crops can effect substantial changes in natural enemy populations, which bring species richness and abundance nearer to those observed in less disturbed, perennial counterparts (Barfield and Gerber, 1979; Arkin and Taylor, 1981; Risch and Carroll, 1982; Blumberg and Crossley, 1983; Altieri and Schmidt, 1984; Herzog and Funderburk, 1985; Andow and Hidaka, 1989; Stinner *et al.,* 1989).

D. Spatial Fragmentation, Species Richness, and the Fate of Species in a Habitat

1. Island biogeography theory

MacArthur and Wilson's (1967) equilibrium model of island biogeography offered predictions about the number of species present on an island. Species richness on a given island results from a dynamic equilibrium established as a function of continuous immigration from mainland source pools and extinction rates on islands. Because migration rates and extinction rates vary according to distance from a source of immigrants and the size of the island, species richness would be greatest on islands that are closer to the mainland and relatively larger in size. These ideas were rapidly applied to the size and design of nature preserves as insular habitats in a matrix of inhospitable or unsuitable terrain (Terborgh, 1974; Willis, 1974; Wilson and Willis, 1975; Diamond and May, 1976). The empirical relationship showing increased species richness as island size increases and the theoretical explanations for this tendency prompted several notions about biodiversity reserves. Reserves should be as large as possible; the attributes of a single large reserve are fundamentally different than those of a collection of small reserves. Long, thin reserves may preserve environmental heterogeneity while compact reserves maximize the area to circumference ratio and thus preserve core species. Theoretical and empirical arguments have been made for each of these design options (e.g., Wilson and Willis, 1975; Simberloff and Abele, 1976; MacGuire, 1986; Robinson and Quinn, 1992) for the conservation of biodiversity.

Soon after the theory of island biogeography and species-area concepts were applied in the design of nature reserves, the insular nature of cultivated areas themselves motivated analogies regarding crops as islands available for colonization by arthropods (Strong, 1974; Strong *et al.,* 1977; Simberloff, 1986; Price and Waldbauer, 1994). The critical question for conservation biological control might be, would it be possible to achieve a diverse natural enemy community by simply increasing the size of a crop island? One condition may be that the habitat must persist for a sufficient amount of time to achieve a balance between colonization and extinction. The assumption of equilibrium is often inappropriate given that plants are supplied to many agricultural systems, or "reset," at certain intervals (Levins and Wilson, 1980). Nevertheless, natural enemy species richness may be expected to increase with increasing crop island size if passive sampling processes occur for islands with larger perimeter to area ratios (Connor and McCoy, 1979).

However, conservation biologists clearly recognize that the area component of species-area curves for islands of different sizes is confounded by habitat diversity on those islands (He and Legendre, 1996). Simberloff (1991) suggested that an increase in the number of habitats, on average, with increased size accounted for species-area relationships on all but a few sets of sites studied. Therefore, species number in a homogeneous monoculture may quickly reach an asymptotic value and an increase in the size of the homogeneous crop island may not be expected to result in a dramatic increase in species richness. Thus, the applicability and desirability of island biogeography theory as a guide for crop field design remains a complex question (Simberloff, 1986). Fragments of noncrop habitat, however, are likely to exhibit typical species-area curves because for these taxonomically and structurally diverse habitats heterogeneity is likely to increase with size.

Thus, to the extent that noncrop habitats act as insular refugia for natural enemies which then colonize or visit agricultural fields and inflict mortality on target pests, the size, design, abundance, and location of noncrop habitats are relevant to conservation biological control. Indeed, the predicted relationship between arthropod species richness and habitat patch or fragment size of noncrop vegetation is supported by many data (Ryszkowski 1993; Didham *et al.,* 1996). Strong (1979) found that larger patches of British trees were host to more species of lepidopteran herbivores than were small patches. Opler (1974) found the same trend on oak trees, with respect to both oak leaf miners and miner parasitoids, in patches of different sizes. Similarly, the number of parasitoid species reared from vetch seed pods increased with patch size in naturally occurring meadows ranging from 300 to 800,000 m^2. The increase in parasitoid species richness detected in seed pod samples from larger meadow habitats was associated with an increase in the parasitism rate of a seed-feeding weevil within the vetch pods. Weevil parasitism rates ranged from 40% in the smallest meadow habitat islands to 83% in the largest meadows (Kruess and Tscharntke, 1994).

2. The importance of species richness

In agricultural habitats, does increased species richness of natural enemies result in increased mortality of target pests? This question strikes at the heart of long-debated practices in classical biological control concerning single or multiple introductions and to community-level aspects of conservation biological control (Ehler, 1996; Heinz and Nelson, 1996). MacArthur and Wilson's (1967) models treat all members of a species source pool as equivalent colonizers. However, trophic structure and interspecific relationships among natural enemies and their prey/hosts cannot be ignored when the goal is prophylactic pest control. The primary concern is intraguild predation or adverse types of competition among natural enemies, which may reduce the overall efficiency of natural control of pests in the crop habitat (Rosenheim *et al.*, 1993). However, synergistic or additive effects of natural enemies are known from many different systems and have been demonstrated experimentally in recent years (Provencher and Riechert, 1994; Losey and Denno, in press; Letourneau, unpubl. data). Finally, competition among parasitoids can be quite severe, yet lead to an over-all improvement in biological control of the pest in different regions (Murdoch *et al.*, 1996).

What if the equilibrium number of species is high, but the majority of the species that become established are not beneficial? Pests also may proliferate in noncrop environments, depending on plant species composition (Altieri and Letourneau, 1984; Collins and Johnson, 1985; Levine, 1985; Slosser *et al.*, 1985; Lasack and Pedigo, 1986) and migrate to vulnerable crop habitats. However, the presence of low levels of pest populations and/or alternate hosts may be necessary to maintain natural enemies in the area. Thus, the composition of the arthropod community is a better indicator of its capacity for conservation biological control than is species richness (Becker, 1992; also see Chapter 8, Section IV and Chapter 9, Section III, B). Liss *et al.* (1986) present a modification of the MacArthur and Wilson's (1967) model that incorporates colonizer source composition and changes in "island" habitats over time. The authors called for further research toward understanding the organization of species pools as sources of arthropod communities in agroecosystems.

Reserve design rules for threatened and endangered species have been challenged in parallel ways (Simberloff and Abele, 1976; Margules *et al.,* 1982; Simberloff and Gotelli, 1984). Doak and Mills (1994) noted that although theory could make strong and accurate predictions about species numbers and extinction rates it did not address the identity of the species that persist or the sequence of colonization and extinction on islands of different sizes. They proposed the use of nested subset analysis (Patterson, 1987; Worthen *et al.,* 1996) to elucidate patterns of the sequence of extinction on shrinking habitat islands. In the absence of detailed natural history data on particular species in the habitat, the technique is applied in conservation biology to identify suites of extinction prone species that require special attention. Similarly, data collected for the analysis of species-area relationships in natural vegetation could be used as a first step in identifying natural enemy species that are relatively more sensitive to patch size. Such information would guide the design of agroecosystems, including the size of particular habitats and refugia needed to enhance the species richness and abundance of natural enemies.

The relationships between species richness and locations or shapes of "donor" reserves for natural enemies have not been studied directly. The fact that noncrop refugia are meant as sources of natural enemies for nearby crop habitats and not as reserves that retain core species within their boundaries means that habitat design goals differ fundamentally from those for conservation biology of many endangered species. It would be prudent, then, to investigate reserve designs that both maintain abundant and diverse populations and allow for optimum rates of movement of natural enemies between reserves and agricultural lands (Duelli *et al.,* 1990). For example, the SLOSS (single large or several small reserves) debate in conservation biology would be viewed not only from the point of view of conserving biodiversity or even conserving high quality species but also with respect to critical movement capacities into adjacent habitats. There may be tradeoffs, then, as there are in reserves for threatened species, with respect to size, shape, and edge to core ratios of noncrop vegetational reserves (Altieri, 1991).

E.   Functional Populations and Communities: Maintenance of
     Subpopulations and Habitat Patches as Source Pools
     for Recolonization

1.   Metapopulation theory and conservation biological control

A metapopulation is a set of populations distributed over an array of spatially arranged patches that are connected, to different degrees, by dispersal movements

(Levins, 1969; Hanski and Gilpin, 1991; Harrison, 1991). In different localities population trends may proceed simultaneously in opposite directions. Some ideas from island biogeography can apply when habitat patches act as small islands, unable to support populations indefinitely without recolonization events (e.g., de Vries and den Boer, 1990). The classical metapopulation scenario (Levins, 1969) is based on

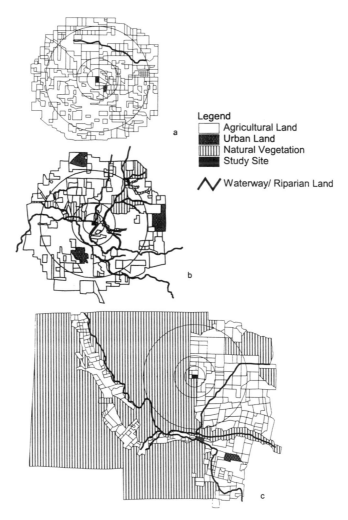

Figure 1. Geographic information system (GIS) depictions of selected tomato fields sampled for natural enemy species richness by D. K. Letourneau (see Drinkwater *et al.,* 1995) and their environs in the Sacramento valley region of California; circles depict 1 km, 2 km, and 5 km radii around the center of the field (black) and show how fields a, b, and c are situated within dramatically different matrices of agricultural (field outlines showing different crop categories), urban, and noncrop vegetation (maps digitized and produced using Arc-Info, by B. Goldstein).

a large network of small habitat patches inhabited by an array of local populations with a substantial risk of stochastic, local extinction (Hanski *et al.,* 1996). Single species metapopulation models that focus on the balance between local extinction and recolonization of patches have strongly influenced conservation strategies for threatened and endangered species but have not been actively used for conceptualizing the conservation of natural enemies in agroecosystems. However, the agricultural landscape can be viewed as a mosaic of habitat patches for natural enemies with spatiotemporal sources and sinks (Fig. 1).

The application of the metapopulation concepts in conservation biology before their application in agroecology is particularly interesting because Levins (1969) proposed these models first in the context of pest control in crops. He suggested that local pest eradication could be approached by either reducing the migration rate or increasing the extinction rate in the patch. Hess (1996) showed how conservation biologists have inverted the reasoning that extends from Levins' (1969) model, expressed as:

$$dp / dt = mp (1 - p) - xp,$$

where p is the proportion of patches occupied, m is the migration rate per unit time, and x is the random, local extinction rate per unit time. Here, the single, nonzero equilibrium occurs at $p^* = 1 - (x / m)$ when the local extinction rate is lower than the migration rate. To prevent metapopulation extinction, conservation biologists focus on ways to increase migration rates (Hess, 1996). Like the approach taken by conservation biologists, it may be useful for biological control specialists to apply such an inversion of Levins' (1969) model to a conceptual model of the agricultural mosaic.

The stochastic nature of extinction in a patch is gaining importance in conservation biology; catastrophic events can figure prominently in the local extinction of a species even if abundance prior to the event is comfortably high (Tscharntke, 1992; Mangel and Tier, 1994). The most likely cause of repeated catastrophic mortality of nontarget species occurs through treatment of pest populations with insecticides. The metapopulation paradigm could be extremely appropriate when the following three conditions are met: (1) broad-spectrum insecticide applications cause local or ecological extinction of natural enemies; (2) insecticide treatments are applied differentially, either in time or space, across the agricultural mosaic; and (3) natural enemies migrate from one patch to another within the mosaic. In this scenario, untreated fields or natural areas would act as a source of dispersing natural enemies that may rescue *(sensu* Brown and Kodric-Brown, 1977; Harrison and Taylor, 1996) local populations from ecological extinction. Sinks are expected in a metapopulation scenario. Optimally, patches with low or extinct subpopulations of natural enemies will contain a relatively abundant supply of prey/hosts for new colonists.

If prey scarcity due to insecticide applications is an important factor in indirectly increasing the mortality of natural enemies, then metapopulation dynamics may be used conceptually for promoting the persistence of extinction prone interactions (Karieva, 1990; Tilman, 1994; Tilman *et al.,* 1994). For example, Holyoak and Lawler (1996) demonstrated in experimental habitat subdivisions that metapopulation dynamics allowed increased persistence times of predators and prey and that the ability of predators to disperse among local populations decreased the local persistence of prey, while increasing the local persistence of predators.

2. The design and management of agroecosystems

How distant can natural enemy source pools be from sites of dwindling populations such that recolonization can exceed local extinction rates? Dispersal ranges of natural enemies are likely to vary greatly. However, even distant sites can be colonized by individuals migrating on high air currents along paths of turbulent convection. Weak flying insects can disperse over long distances and across wide areas by exploiting the ephemeral but very structured nature of air movement (Wellington, 1983). For example, robust hosts and minute parasitoids can exhibit coupled displacement in long distance migration, as shown by the Australian plague locust *Chortoicetes terminifera* Walker (Orthoptera: Acrididae) and its egg parasitoid *Scelio fulgidus* Crawford (Hymenoptera: Scelionidae), which disperse independently on wind currents to the same location (Farrow, 1981).

Cumulative numbers over a growing season, however, may be irrelevant if immigration rates of natural enemies are very slow in relation to rising levels of the pest (Doutt and Nakata, 1973; Price, 1976; Letourneau and Altieri, 1983; Williams, 1984). Accurate descriptions of the species composition of colonizer source pools, phenology, and flight patterns, then, are relevant to the design and management of regional scale agroecosystems for optimal biological control (see Haas, 1995). Flight capacity studies and mathematical models describing movement patterns based on continuous diffusion or discrete random walk equations have focused on predicting dispersal and migration of herbivores (Okubo, 1980; Stinner *et al.,* 1983, 1986). Detailed biological information on flight capacities of natural enemies (e.g., Duelli, 1980) and behaviors in habitat mosaics (Duelli *et al.,* 1990) coupled with predictive models of natural enemy movement will aid in predicting synchrony. In some cases, synchronies are difficult to achieve because local species are adapted to exploit conditions of prey or host phenologies typical of unmanaged habitats as opposed to those imposed, or novel, conditions in agroecosystems. For example, coccinellid beetles in California aestivate during times of high prey availability. Irrigated crops provide a continuous food supply that was not available in the valleys before agricultural expansion had occurred (Hagen, 1962).

Conservation biologists have examined the use of habitat corridors designed to effectively reduce interhabitat distance and link patches of high quality habitat (sources). Networks of wildlife corridors are being advocated as a key component for the conservation of biodiversity, even though there are still few data on their effectiveness (Lindenmayer and Nix, 1993). They recognize that corridors can facilitate movement among subpopulations, but that different species use them disproportionately (Noss, 1991). Also, whereas habitat corridors can encourage gene flow among different subpopulations, facilitate escape from insecticide treatments, harvest, or tillage, and allow for recolonization after local extinction (Soulé, 1991) they can also provide a conduit for disease, pests, or fires.

To my knowledge, movement patterns of arthropod natural enemies through corridors has not been documented. Yet, corridors may be extremely useful in fostering synchronies of natural enemy colonization with pest build-up to achieve conservation biological control (Thacker, 1996). Frequently disturbed monocultures often favor the rapid colonization and growth of herbivore populations. Initial conditions of enemy-free space and high abundance of pests further reduces the ability of natural enemies to regulate them (Price, 1981). In many farming situations these negative factors can be minimized or eliminated by providing continuity of vegetation (and the associated food and shelter) in time and space, thus aiding natural enemies.

The scale of effective habitat corridors for arthropod natural enemies could be conveniently small. Whereas larger animals such as the Florida panther (*Felis concolor coryi)* or the red wolf *(Canis lyceon)* require relatively wide habitat corridors, parasitoids or arthropod predators may be accommodated by riparian vegetation strips, weedy patches along roads, or even single beds of cover crop vegetation left uncultivated until crops are established in the field. In the face of large-scale disruption of habitat in annual agriculture, established vegetation corridors could provide an innovative way to incorporate continuous vegetation that allows natural enemies to move into areas of target pest populations as they develop early in the season.

F.   Conservation by Preservation and Enhancement of Biodiversity:
        Active Programs of Ecological Restoration of Habitat Quality

1.   Habitat restoration

Species loss and species imbalance are common consequences of habitat degradation. The developing subfield of conservation biology known as restoration ecology focuses on degraded habitats and is contributing an array of experimental approaches for revitalizing habitat on a variety of scales. When possible, natural environments in the immediate area are selected as comparison model ecosystems. Detailed methodologies have been devised for comparative sampling and monitoring the suitability of restored habitats for attracting and retaining a diverse fauna.

The goal of restoration ecology programs, i.e., to create habitats that compare to natural models, may be only loosely related to the goal of agroecosystem management as a prophylactic control strategy (sensu Vandermeer and Andow, 1986). However, as a conceptual model, ecological restoration may provide insights on crucial elements of ecosystem health. Also, some of the monitoring schemes may be particularly useful for determining indicators of habitat suitability for key parasitoids or predators. The application of strict restoration ecology concepts and methods in the conservation of natural enemies in agroecosystems, then, suggests two approaches: (1) intentional alteration of a habitat to mimic a defined, indigenous, historic ecosystem; and (2) the application of methodologies to monitor and compare indicators of ecosystem integrity, including structure, function, diversity, and dynamics in restored habitats, degraded habitats, and if possible, relevant natural habitats.

Efforts to use natural ecosystems as models for agroecosystems are relatively rare. Notable research programs include perennial grain cropping systems designed to mimic prairie vegetation (Jackson and Piper, 1989) and successional vegetation models of tropical crop systems proposed by Hart (1980) and tested by Ewel (1986). These schemes, like ecological restoration programs, were designed to promote as many natural processes and feedback mechanisms as possible and increase the ecological sustainability of the crop production system. Evaluative monitoring techniques from restoration ecology could be helpful in assessing the success of these crop habitats in restoring ecosystem level processes. Many of the indicator taxa used by restoration ecologists to monitor biotic integrity are also appropriate for agroecosystems. Whereas early assessments of restoration projects were based on plant surveys and mammal counts, multimetric measures of soil microorganisms, arthropods, and birds have become more prominent (Holloway, 1980; Majer, 1984; Kremen et al., 1993; Oliver and Beattie, 1993; Holl, 1996). Williams (1993) has used relative abundance measures of functional groups (i.e., pollinators, herbivores, predators, parasitoids, and detritivores) to compare restored habitat to a natural model system. She also monitored a group of prey items designated to represent the food resources of the least Bell's vireo, for which the restored habitat was created. Because species identification can be unwieldy in arthropod communities, techniques to monitor morphospecies have been used successfully by restorationists (e.g., diversity indices based on sequential runs of morphospecies devised by Cairns and Dickson, 1971).

Measurements of the diversity and relative abundance of functional groups or indicator taxa have been used in some comparative studies of alternative agricultural practices. These data show that the species richness of arthropod communities and specific taxa commonly increases in response to habitat restoration efforts such as hedgerows, organic farming methods, and low-input methods (Hendrix *et al.*, 1986; Drinkwater *et al.*, 1995; Marino and Landis, 1996). Recent efforts to design methods for sampling indicators of sustainability would seem to be a rich area for exchange between conservation biologists and agroecologists. Indeed, conservation of biodiversity in agricultural and agroforestry systems has emerged recently as an alternative to viewing agricultural landscapes as the "sea of inhospitable terrain" around insular reserves of natural habitat (Roth *et al.*, 1996; Gajaseni *et al.*, 1996).

Practical guidelines for agroecosystem design, in the absence of a natural ecosystem model, can be based on accumulated knowledge of resource needs and habitat preferences of natural enemies. Thus, the concept of ecological rehabilitation *(sensu* Aronson *et al.*, 1993), which concentrates on repairing damaged or blocked ecosystem functions, may be applicable to agroecosystems that are prone to pest outbreaks due to imbalances in natural enemy communities. Application of the rehabilitation concept in the conservation of natural enemies in agroecosystems could consist of introducing critical elements to degraded sites that provide resources and increase the attractiveness of the habitat. For example, successful ecological restoration or rehabilitation of terrestrial habitats requires the selection of appropriate component plant species (Cottam, 1990; Howell and Jordan, 1991; Samways, 1994). The vegetational composition of an agroecosystem (including the crops, weeds, and natural vegetation areas) may be just as important in determining the fauna of a crop field, yet tillage and cropping decisions are often made for a variety of reasons unrelated to the concept of habitat design for natural enemies. We know that environmental properties of crop fields can be manipulated to provide critical resources and preferred habitats for natural enemies (e.g., van den Bosch and Telford, 1964; van Emden and Williams, 1974; Thresh, 1981; Altieri and Letourneau, 1982, 1984; Price and Waldbauer, 1994; Letourneau and Altieri, 1983; Risch *et al.*, 1983; Herzog and Funderburk, 1985; van Emden, 1988; Andow, 1991; Altieri *et al.*, 1993). Many of these studies involve intercropping, a cropping method that tends to increase the diversity of resources and the structural complexity of the habitat. Crop mixtures, in time and space, can improve conditions for natural enemies that require multiple food resources or refugia (but see Andow and Risch, 1985; Nafus and Schreiner, 1986).

Noncrop plants within and around fields can also be used to restore agricultural fields to predisturbance levels of biological control agents (Altieri and Whitcomb,

1979a,b; Barney *et al.*, 1984; Norris, 1986; Bugg and Wilson, 1989; see Chapters 6 to 9). For example, crop fields with dense weed cover and high diversity usually have more predaceous arthropods than do weed-free fields (Pimentel, 1961; Garcia, 1991). Carabid beetles (Dempster, 1969; Speight and Lawton, 1976; Thiele, 1977), syrphid flies (Pollard, 1971; Smith, 1976; Cowgill, 1989), and coccinellid beetles (Bombosch, 1966; Perrin, 1975) are abundant in weed-diversified systems. Rapidly colonizing, fast growing plants offer many important requisites for natural enemies such as alternate prey or hosts, pollen, or nectar as well as microhabitats that are not available in weed- free monocultures (van Emden, 1965; Doutt and Nakata, 1973; Cerutti *et al.*, 1989; Rosenheim *et al.*, 1993).

Clearly, vegetational diversity is a key factor in designing crop habitats to restore natural functions not present in degraded agroecosystems. However, the effect of any particular crop-crop or weed-crop assemblage on species richness, species composition, reproduction, survival, and efficacy of natural enemies is difficult to predict (Flaherty *et al.*, 1985; Andow, 1986; Powell *et al.*, 1986; Bugg *et al.*, 1991). Restoration ecologists face the same kind of uncertainties because of knowledge gaps about the resource needs of particular species, unpredictable dynamics among species, and the stochasticity of natural processes. Fine tuning of ecological restoration projects, then, requires periodic sampling schemes after vegetation is established. Minimally, the focal threatened or endangered species are monitored. Monitoring of target predators or parasitoids or community-level attributes such as species richness of natural enemies in agroecosystems would improve our management efforts to restore adequate biological control. In addition, these monitoring schemes could help to distinguish the kinds of interventions needed to enhance key natural enemies or enemy species richness in general. For example, some agroecosystems are likely to respond favorably to conservation tactics imposed within the crop system, and others may require noncrop habitat refugia (see Chapters 6 and 7).

2. Conservation of keystone species

Conservation biologists have acknowledged that all species in a community are not equal in their contributions to community structure and processes, and not all interactions are vital to the integrity of the ecosystem. Some species are particularly important because they interact with and affect many other species, process materials disproportionately to their numbers, or because their functions cannot be substituted or compensated for by other species in the community. This has led to more careful identification and focused efforts to preserve those keystone species in their communities. In natural systems, certain species can hold integral roles as seed dispersers and pollinators, providing essential resources during times of scarcity, as mobile links between distant populations, and as top predators that regulate populations; sometimes with cascading effects through multiple lower trophic levels.

Because of their trophic position, top predators are poised to play critical roles in natural communities. For example, Paine's (1974) classic study showed that the abundance, distribution, and diversity of sessile organisms in the rocky intertidal zone were determined by the abundance of a top predator. Furthermore, Power's (1990) predator inclusion/exclusion study showed that trophic cascades resulted when large, predatory fish reduced small predators, which, when abundant, control midge herbivores and increase algal biomass. Recent manipulative studies suggest that top-down trophic cascades operate among arthropods and plants in terrestrial ecosystems (Spiller and Schoener, 1990; Gomez and Zamora, 1994; Carter and Rypstra, 1995; Moran *et al.,* 1996; Letourneau and Dyer, 1997). It is precisely this sort of interaction that is central for conservation biological control: the maintenance or enhancement of predators that effectively reduce herbivores and have cascading effects on crop productivity.

Recent debates in conservation ecology concern the appropriate use of the keystone species as a focus for conservation biology efforts. Those who caution against its use argue that a combination of a few non-keystone species could have the same or greater effects as a single species considered to be a keystone and that conservation efforts geared to keystone species should not obviate the need for protecting other species of interest or the system at large (Mills *et al.,* 1993). This notion is particularly compatible with the "rivet" theory of species richness and ecosystem function, which suggests that all species make a contribution to ecosystems processes, and the loss of any group of species, like the loss of rivets in a complex machine, can cause systemic impairments (Ehrlich and Ehrlich, 1981; Lawton, 1997). Promoters of the keystone species concept maintain, however, that community level interactions are the focus of keystone species concepts, not single species, and that focusing on keystone species is preferable to haphazard management, treating all species within guilds or trophic levels as equals (Paine, 1995).

The latter argument, in favor of devoting disproportionate attention to the conservation of strong interacters, is probably the appropriate lesson for conservation biological control. At least proximally, the objective is to regulate pest densities and reduce damage to commodities rather than to conserve or enhance biodiversity. In simplified systems, the augmentation of specific predators or parasitoids may be an appropriate goal of conservation biological control. It is also possible to manipulate the system such that interaction strengths of natural enemies in the system (for example, the regulatory capacity of a parasitoid on its host population) are enhanced. For example, certain varieties of crop plants, because of their physical or chemical properties, are superior to others in attracting natural enemies or otherwise increasing their effects on pests (Read *et al.,* 1970; Shahjahan, 1974; Johnson and Hara, 1987; Gerard, 1989; Lewis and Gross, 1989; Martin *et al.,* 1990; see Chapter 4).

III.   CONCLUSION: CONSERVATION BIOLOGICAL CONTROL, POLICY, AND CHANGING PERSPECTIVES

The skills needed for effective conservation planning or species management entail more than knowledge of the theoretical and applied principles of population and community ecology. These principles, plus an understanding of relevant species' life histories and habitat relationships, are insufficient to conserve most of the species that are deemed threatened or endangered (Noon and Murphy, 1994). Instead, political forces, legal proceedings, policy decisions and economic pressures often dictate the level of success or failure in a management plan for preservation or recovery of species and their habitats. Although conservation biological control can have direct and dramatic economic consequences compared to many conservation efforts in natural systems, some of the same barriers exist. For example, regional cooperation on landscape level conservation efforts are extremely difficult to achieve. Rabb (1978) addressed these concerns when he criticized the propensity of single commodity, closed system approaches to pest management in research and decision making as deficient for problems which demand attention to "large unit ecosystem heterogeneity." Policy recommendations that encourage agricultural diversification and provide incentives for cooperative, multiple land-use options, however, are compatible with increasing interests in sustainability (Francis *et al.,* 1990; National Research Council, 1991; Goodland, 1995) and biodiversity (La Salle and Gauld, 1993).

Political and economic barriers apply to individual farming operations, as well. For example, the conservation of natural enemies and their ecological services in agroecosystems, as ecologically based solutions, are worth very little in the face of economic incentives and agricultural policies that reward simplified, agrichemical oriented management practices (Levins and Vandermeer, 1990). That is, current crop support systems focus on particular commodities and thus encourage both individual farm and regional specialization rather than diversified farming. If the definitions of crop acreage bases (the number of acres eligible for crop supports) were expanded, farmers would have more options to design their cropping systems to conserve natural enemies and other elements that lead to sustainable agriculture (Dahlberg, 1996).

Instead, the cost of replacing ecological services with chemical products is often subsidized by current taxation schemes.

However, conversations about incentives for sustainable farming methods in the process of deliberations for the 1996 Farm Bill began to create alliances among environmentalists and agriculturalists. These coalitions are fostering the rethinking of set-aside lands for conservation, not just price supports. Current research programs on strip farming options, vegetative buffer strips (designed to reduce agricultural chemical pollution off-farm), and intensive rotational management schemes, though designed for other objectives, could also provide the kinds of heterogeneity on farmscapes that foster natural enemy populations. Policy options, such as a pesticide mil tax, could generate funds to support research and provide compensation for lands taken out of production for conservation purposes. Recent movements to preserve traditional agricultural landscapes in Britain, along with their wildlife habitats, are part of new policy initiatives for multiple use land management on farms (Fry, 1991). Finally, alternatives to pesticide intensive management of crop pests may ultimately be supported because of the increasingly recognized, intimate linkages between human health, environmental health, and economic health (Bridgewater *et al.,* 1996). Just as conservation biologists benefit from coalitions with policy-makers, agroecologists interested in land-use options, cropping systems alternatives, and incentives for landscape-level design for conservation biological control must become familiar with economic and social incentives for change.

Acknowledgment

I thank P. Barbosa, C. Boggs, M. Fitzsimmons, P. Goldman, K. Holl, M. Kauffman, R. Krach, A. Lopez, M. Mangel, S. Minta, A. Shelton, and D. Thompson for contributing ideas that were helpful in the development of this chapter. Portions of this chapter were supported by a USDA-SARE grant (88-COOP-1-3525), NSF (DEB-931543) and faculty research grants from the Social Sciences Division and Academic Senate, UCSC.

REFERENCES

Altieri, M. A. (1991). Increasing biodiversity to improve insect pest management in agroecosystems. *In* "The Biodiversity of Microorganisms and Invertebrates: Its Role in Sustainable Agriculture." (D. L. Hawksworth, ed.), pp. 165-182. CAB International. Wallingford, U.K.

Altieri, M. A., and Anderson, M. K. (1986). An ecological basis for the development of alternative agricultural systems for small farmers in the third world. *Amer. J. Altern. Agric.* 1, 30-38.

Altieri, M. A., and Letourneau, D. K. (1982). Vegetation management and biological control in agroecosystems. *Crop Prot.* 1, 405-430.

Altieri, M. A., and Letourneau, D. K. (1984). Vegetation diversity and insect pest outbreaks. *CRC Critical Rev. Plant Sci.* 2,131-169.

Altieri, M. A., and Schmidt, L. L. (1984). Abundance patterns and foraging activity of ant communities in abandoned, organic and commercial apple orchards in northern California. *Agric., Ecosys., Environ.* 11, 3441-3452.

Altieri, M. A., and Whitcomb, W. H. (1979a). Manipulation of insect populations through seasonal disturbance of weed communities. *Prot. Ecol.* 1, 185-202.

Altieri, M. A., and Whitcomb, W. H. (1979b). The potential use of weeds in the manipulation of beneficial insects. *Hort. Sci.* 14, 12-18.

Altieri, M. A., Cure, J. R., and Garcia, M. A. (1993). The role and enhancement of parasitic Hymenoptera biodiversity in agroecosystems. *In* "Hymenoptera and Biodiversity." (J. LaSalle, and I. D. Gauld, eds.), pp. 257-275. CAB International. London, U.K.

Andow, D. A. (1986). Plant diversification and insect population control in agroecosystems. *In* "Some Aspects of Integrated Pest Management." (D. Pimentel, ed.), pp. 277-348. Dept. of Entomol., Cornell Univ. Ithaca, NY.

Andow, D. A. (1991). Vegetational diversity and arthropod population response. *Annu. Rev. Entomol.* 36, 561-586.

Andow, D. A., and Hidaka, K. (1989). Experimental natural history of sustainable agriculture: syndromes of production. *Agric., Ecosys., Environ.* 27, 447-462.

Andow, D. A., and Risch, S. J. (1985). Predation in diversified agroecosystems: relations between a coccinellid predator *Coleomegilla maculata* and its food. *J. Appl. Ecol.* 22, 357-372.

Arkin, G. F., and Taylor, H. M. (1981). "Modifying the Root Environment to Reduce Crop Stress." Amer. Soc. Agric. Engineer. St. Joseph, MI.

Aronson, J., Floret, C., Le Floch, E., Ovalle, C., and Pontanier, R. (1993). Restoration and rehabilitation of degraded ecosystems in arid and semi-arid land. II. Case studies in Southern Tunisia, Central Chile and Northern Cameroon. *Restor. Ecol.* 3, 168-187.

Baker, R. S., Laster, M. L., and Kitten, W. F. (1985). Effects of the herbicide monosodium methanearsonate on insect and spider populations in cotton fields. *J. Econ. Entomol.* 78, 1481-1484.

Barfield, B. J., and Gerber, J. F. (1979). "Modification of the Aerial Environment of Plants." Amer. Soc. Agric, Engineer. St. Joseph, MI.

Barney, R. J., Lamp, W. O., Ambrust, E. J., and Kapusta, G. (1984). Insect predator community and its response to weed management in spring-planted alfalfa. *Prot. Ecol.* 6, 23-33.

Bart, J. (1995). Amount of suitable habitat and viability of northern spotted owls. *Cons. Biol.* 9, 943-946.

Becker, P. (1992). Colonization of islands by carnivorous and herbivorous Heteroptera and Coleoptera: effects of island area, plant species richness, and "extinction" rates. *J. Biogeogr.* 19, 163-171.

Blumberg, A. Y., and Crossley, D. A., Jr. (1983). Comparison of soil surface: arthropod populations in conventional tillage, no-tillage and old field systems. *Agro-Ecosys.* 8, 247-253.

Bombosch, S. (1966). Occurrence of enemies on different weeds with aphids. *In* "Ecology of Aphidophagous Insects." (I. Hodek, ed.), pp. 177-179. Academic Printing House. Prague, Czechoslovakia.

Bottrell, D. R. (1979). "Integrated Pest Management." Council on Environmental Quality. U. S. Government Printing Office, Washington, DC. 120 pp.

Bridgewater, P., Walton, D. W., and Busby, J. R. (1996). Creating policy on landscape diversity. *In* "Biodiversity in Managed Landscapes: Theory and Practice." (R. C. Szaro, and D. W. Johnston, eds.), pp. 711-724. Oxford University Press, Oxford, U.K.

Brown, J. H., and Kodric-Brown, A. (1977). Turnover rates in insular biogeography: effect of immigration on extinction. *Ecology* 58, 445-449.

Bugg, R. L., and Wilson, L. T. (1989). *Ammi visnaga* (L.) Lamarck (Apiaceae): associated beneficial insects and implications for biological control, with emphasis on the bell-pepper agroecosystem. *Biol. Agric. Hort.* 6, 241-268.

Bugg, R. L., Wackers, F. L., Brunson, K. E., Dutcher, J. D., and Phatak, S. C. (1991). Cool-season cover crops relay intercropped with cantaloupe: influence on a generalist predator, *Geocrois punctipes* (Hemiptera: Lygaeidae). *J. Econ. Entomol.* 84, 408-416.

Cairns, J., Jr., and Dickson, K. L. (1971). A simple method for the biological assessment of the effects of waste discharges on aquatic bottom-dwelling organisms. *J. Water Pollution Cont. Fed.* 43, 755-771.

Carter, P. E., and Rypstra, A. L. (1995). Top-down effects in soybean agroecosystems: spider density affects herbivore damage. *Oikos* 72, 433-439.

Cerutti, F., Delucchi, V., Baumgartner, J., and Rubli, D. (1989). Research on the vineyard ecosystem in Tessin: II. Colonization of grapevines by the cicadellid *Empoasca vitis* Goethe (Hom., Cicadellidae, Typhlocybinae) and of its parasitoid *Anagrus atomus* Haliday (Hym., Mymaridae), and the importance of the surrounding flora. *Mitt. Schweiz. Entomol. Gesellschaft* 62, 253-267.

Collins, F. L., and Johnson, S. J. (1985). Reproductive response of caged adult velvetbean caterpillar and soybean looper to the presence of weeds. *Agric., Ecosys., Environ.* 14, 139-149.

Connor, E. F., and McCoy, E. D. (1979). The statistics and biology of the species-area relationship. *Amer. Nat.* 113, 791-833.

Cottam, G. (1990). Community dynamics on an artificial prairie. *In* "Restoration Ecology: Synthetic Approach to Ecological Research." (W. R. Jordan III, M.E. Gilpin, and J. D. Aber, eds.), pp. 257-270. Cambridge University Press. Cambridge, U.K.

Cowgill, S. (1989). The role of non-crop habitats on hoverfly (Diptera: Syrphidae) foraging on arable land. "Proc. Brighton Crop Prot. Conf., Weeds." Brighton, U.K., 20-23 Nov. 3, 1103-1108.

Dahlberg, K. A. (1996). Creating policy on community diversity. *In* "Biodiversity in Managed Landscapes: Theory and Practice." (R. C. Szaro, and D. W. Johnston, eds.), pp. 698-710. Oxford University Press. Oxford, U.K.

Dasmann, R. F. (1964). "Wildlife Biology." John Wiley and Sons. New York, NY.

Dempster, J. P. (1969). Some effects of weed control on the numbers of the small cabbage white (*Pieris rapae* L.) on brussels sprouts. *J. Appl. Ecol.* 6, 339-345.

de Vries, H. H., and den Boer, P. J. (1990). Survival of populations of *Agonum ericeti* Panz. (Col., Carabidae) in relation to fragmentation of habitats. *Neth. J. Zool.* 40, 484-498.

Diamond, J. M., and May, R. M. (1976). Island biogeography and the design of natural reserves. *In* "Theoretical Ecology: Principles and Applications." (R. M. May, ed.), pp. 163-186. W. B. Saunders. Philadelphia, PA.

Didham, R. K., Ghazoul, J., Stork, N. E., and Davis, A. J. (1996). Insects in fragmented forest: a functional approach. *Trends Ecol. Evol.* 11, 255-259.

Division of Agriculture and Natural Resources (1985). " Integrated Pest Management for Tomatoes." University of California. 2nd Edition. Berkeley, CA.

Doak, D. F. (1995). Source-sink models and the problem of habitat degradation: general models and applications to the Yellowstone grizzly. *Cons. Biol. 9,* 1370-1379.

Doak, D. F., and Mills, L. S. (1994). A useful role for theory in conservation. *Ecology* 75, 615-626.

Doutt, R. L., and Nakata, J. (1973). The *Rubus* leafhopper and its egg parasitoid: an endemic biotic system useful in grape pest management. *Environ. Entomol.* 2, 381-386.

Drinkwater, L. E., Letourneau, D. K., Workneh, F., van Bruggen, A. H. C., and Shennan, C. (1995). Fundamental differences between conventional and organic tomato agroecosystems in California. *Ecol. Appl.* 5, 1098-1112.

Duelli, P. (1980). Adaptive and appetitive flight in the green lacewing, *Chrysopa carnea. Ecol. Entomol.* 5, 213-220.

Duelli, P., Studer, M., Marchand, I., and Jakob, S. (1990). Population movements of arthropods between natural and cultivated areas. *Biol. Cons.* 54, 193-207.

Ehler, L. E. (1996). Structure and impact of natural enemy guilds in biological control of insect pests. *In* 'Food Webs: Integration of Patterns and Dynamics." (G. A. Polis, and K. 0. Winemiller, eds.), pp. 337-351, Chapman and Hall. New York, NY.

Ehrlich, P. R., and Ehrlich, A. H. (1981). "Extinction. The Causes and Consequences of the Disappearance of Species." Random House. New York, NY.

Ewel, J. J. (1986). Designing agricultural ecosystems for the humid tropics. *Annu. Rev. Ecol. Sys.* 17, 245-271.

Farrow, R. A. (1981). Aerial dispersal of *Scelio fulgidus* (Hymenoptera: Scellonidae), parasite of eggs of locusts and grasshoppers (Orthoptera: Acrididae). *Entomophaga* 26, 349-355.

Flaherty, D. L., Wilson, L. T., Stern, W. M., and Kido, H. (1985). Biological control in San Joaquin Valley vineyards. *In* "Biological Control in Agricultural IPM Systems." (M. A. Hoy, and D. C. Herzog, eds.), pp. 501-520. Academic Press. San Diego, CA.

Francis, G. A., Flora, D. B., and King, L. D. (1990). "Sustainable Agriculture in the Temperate Zones." John Wiley and Sons. New York, NY.

Fry, G. L. A. (1991). Conservation in agricultural ecosystems. *In* "The Scientific Management of Temperate Communities for Conservation." (I. F. Spellerberg, F. B. Goldsmith, and M. G. Morris, eds.), pp. 395-443. 31st Symp. Brit. Ecol. Soc., Southampton, U.K.

Gajaseni, J., Matta-Machado, R., and Jordan, C. F. (1996). Diversified agroforestry systems: buffers for biodiversity reserves, and land bridges for fragmented habitats in the tropics. *In* "Biodiversity in Managed Landscapes: Theory and Practice." (R. C. Szaro, and D. W. Johnston, eds.), pp. 506-513. Oxford University Press. Oxford, U.K.

Garcia, M.A. (1991). Arthropods in a tropical corn field: effects of weeds and insecticides on community composition. *In* "Plant-Animal Interactions: Evolutionary Ecology in Tropical and Temperate Regions." (P. W. Price, T. M. Lewinsohn, G. W. Fernandes, and W. W. Benson, eds.), pp. 619-634. John Wiley and Sons. New York, NY.

Gerard, P. J. (1989). Influence of egg depth in host plants on parasitism of *Scolypopa australis* (Homoptera: Ricaniidae) by *Centrodora scolypopae* (Hymenoptera: Aphelinidae). *New Zealand Entomol.* 12, 30-34.

Gomez, J. M., and Zamora, R. (1994). Top-down effects in a tritrophic system: parasitoids enhance plant fitness. *Ecology* 75, 1023-1030.

Goodland, R. (1995). The concept of environmental sustainability. *Annu. Rev. Ecol. Sys.* 26, 1-24.

Haas, C. A. (1995). Dispersal and use of corridors by birds in wooded patches on an agricultural landscape. *Cons. Biol.* 9, 845-850.

Hagen, K. S. (1962). Biology and ecology of predacious Coccinellidae. *Annu. Rev. Entomol.* 7, 289-326.

Hanksi, I., and Gilpin. M. E. (1991). Metapopulation dynamics: brief history and conceptual domain. *In* "Metapopulation Dynamics." (M. E. Gilpin, and I. Hanski, eds.), pp. 3-16. Academic Press, London, U.K.

Hanski, I., Moilanes, A, and Gyllenberg, M. (1996). Minimum viable metapopulation *size. Amer. Nat.* 147, 527-541.

Harrison, S. (1991). Local extinction in a metapopulation context: an empirical evaluation. *Biol. J. Linn. Soc.* 42, 73-88.

Harrison, S., and Taylor, A. D. (1996). Empirical evidence for metapopulation dynamics: a critical review. *In* "Metapopulation Biology: Ecology, Genetics, and Evolution." (I. Hanski, and M. E. Gilpin, eds.), pp. 27-42. Academic Press, San Diego, CA.

Hart R. D. (1980). A natural ecosystem analog approach to the design of a successional crop system for tropical forest environments. *Biotropica, Suppl. Tropical Succession,* 12, 73-95.

He, F., and Legendre, P. (1996). On species-area relations. *Amer. Nat.* 148, 719-737.

Heinz, K. M., and Nelson, J. M. (1996). Interspecific interactions among natural enemies of *Bemisia* in an inundative biological control program. *Biol. Cont.* 6, 384-393.

Hendrix, P. F., Parmelee, R. W., Crossley, Jr., D. A., Coleman, D. C., Odum, E. P., and Groffman, P. M. (1986). Detritus food webs in conventional and no-tillage agroecosystems. *Bioscience* 36, 374-380.

Herzog, D. C., and Funderburk, J. E. (1985). Plant resistance and cultural practice interactions with biological control. *In* "Biological Control in Agricultural IPM Systems." (M. A. Hoy, and D. C. Herzog, eds.), pp. 67-88. Academic Press, Orlando, FL.

Hess, G. R. (1996). Linking extinction to connectivity and habitat destruction in metapopulation models. *Amer. Nat.* 148, 226-236.

Hill, D. A., Andrews, J., Sotherton, N., and Hawkins, J. (1995). Farmland. *In* "Managing Habitats for Conservation." (W. J. Sutherland, and D. A. Hill, eds.), pp. 230-266. Cambridge University Press. Cambridge, U.K.

Hobbs, R. J., and Huenneke, L. F. (1996). Disturbance, diversity and invasions: implications for conservation. *In* "Ecosystem Management." (F. B. Samson, and F. L. Knopf, eds.), pp. 164-180. Springer-Verlag. New York, NY.

Holl, K. D. (1996). The effect of coal surface mine reclamation on diurnal lepidopteran conservation. *J. Appl. Ecol.* 33, 225-236.

Holloway, J. D. (1980). Insect surveys - an approach to environmental monitoring. *Atti XII Congr. Naz. Ital. Entomol.,* Roma, 239-261.

Holyoak, M., and Lawler, S. P. (1996). Persistence of an extinction-prone predator-prey interaction through metapopulation dynamics. *Ecology 77,* 1867-1879.

Howell, E. A., and Jordan, W. R. (1991). Tallgrass prairie restoration in the North American midwest. *In* "The Scientific Management of Temperate Communities for Conservation." (I. F. Spellerberg, F. B. Goldsmith, and M. G. Morris, eds.), pp. 395-414. 31st Symp. Brit. Ecol. Soc. Southampton, U.K.

Jackson, W., and Piper, J. (1989). The necessary marriage between ecology and agriculture. *Ecology* 70, 1591-1593.

Johnson, M. W., and Hara, A. H. (1987). Influence of host crop of parasitoids (Hymenoptera) of *Liriomyza* spp. (Diptera: Agromyzidae). *Environ. Entomol.* 16, 339-344.

Karieva, P. (1990). Population dynamics in spatially complex environments: theory and data. *Phil. Trans. Roy. Soc. Lond.,* Series B 330, 175-190.

Kremen, C., Colwell, R. K., Erwin, T. L., Murphy, D. D., Noss, R. F., and Sanjayan, M. A. (1993). Terrestrial arthropod assemblages: their use in conservation planning. *Cons. Biol.* 7, 796-808.

Kruess, A., and Tscharntke T. (1994). Habitat fragmentation, species loss, and biological control. *Science* 264, 1581-1584.

Lasack, P. M., and Pedigo, L. P. (1986). Movement of stalk borer larvae (Lepidoptera: Noctuidae) from noncrop areas into corn. *J. Econ. Entomol.* 79, 1697-1702.

LaSalle, J., and Gauld, I. D. (1993). "Hymenoptera and Biodiversity." CAB International, London, U.K.

Lawton, J. H. (1997). The role of species in ecosystems: aspects of ecological complexity and biological diversity. In "Biodiversity: An Ecological Perspective." (T. Abe, S. A. Levin, and M. Higashi, eds.), pp. 215-228. Springer-Verlag. New York, NY.

Letourneau, D. K., and Altieri, M. A. (1983). Abundance patterns of a predator *Orius tristicolor* (Hemiptera: Anthocoridae), and its prey, *Frankliniella occidentalis* (Thysanoptera: Thripidae): Habitat attraction in polycultures versus monocultures. *Environ. Entomol.* 122, 1464-1469.

Letourneau, D. K., and Dyer, L. E. (1997). Experimental test in lowland tropical forest shows top-down effects through four trophic levels. *Ecology* (in press).

Levine, E. (1985). Oviposition by the stalk borer, *Papaipema nebris* (Lepidoptera: Noctuidae), on weeds, plant debris, and cover crops in cage tests. *J. Econ. Entomol.* 78, 65-68.

Levins, R. (1969). Some demographic and genetic consequences of environmental heterogeneity for biological control, *Bull. Entomol. Soc. Amer.* 15, 237-240.

Levins, R., and Vandermeer, J. H. (1990). The agroecosystem embedded in a complex ecological community. *In* "Agroecology." (C. R. Carroll, J. H. Vandermeer, and P. M. Rosset, eds.), pp. 341-362. McGraw-Hill. New York, NY.

Levins, R., and Wilson, M. (1980). Ecological theory and pest management. *Annu. Rev. Entomol.* 25, 7-29.

Lewis, W. J., and Gross, H. R. (1989). Comparative studies on field performance of *Heliothis* larval parasitoids *Microplitis croceipes* and *Cardiochiles nigriceps* at varying densities and under selected host plant conditions. *Fla. Entomol.* 72, 6-14.

Lindenmayer, D. B., and Nix, H. A. (1993). Ecological principles for the design of wildlife corridors. *Cons. Biol.* 7, 627-630.

Liss, W. J., Gut, L. J., Westigard, P. H., and Warren, C. E. (1986). Perspectives on arthropod community structure, organization, and development in agricultural crops. *Annu. Rev. Entomol.* 31, 455-478.

Losey, J., and Denno R. F. Positive predator-predator interactions: Enhanced positive predation rates and synergistic suppression of aphid populations. Ecology (in press).

MacArthur, R. H., and Wilson, E. 0. (1967). "The Theory of Island Biogeography." Princeton University Press. Princeton, NJ.

MacGuire, L.A. (1986). Using decision analysis to manage endangered species populations. *J. Environ. Manag.* 22, 345-360.

Majer, J. D. (1984). Recolonization by ants in rehabilitated open-cut mines in northern Australia. *Reclam. Reveg. Res.* 2, 279-298.

Mangel, M., and Tier, C. (1994). Four facts every conservation biologist should know about persistence. *Ecology* 75, 607-614.

Margules, C. R., Higgs, A. J., and Rafe, R. W. (1982). Modern biogeographic theory: are there any lessons for nature reserve design? *Biol. Cons.* 24, 115-128.

Marino, P. C., and Landis, D. A. (1996). Effect of landscape structure on parasitoid diversity and parasitism in agroecosystems. *Ecol. Appl.* 6, 276-284.

Martin, Jr., W. R., Nordlund, D. A., and Nettles, W. C., Jr. (1990). Response of parasitoid *Eucelatoria bryani* to selected plant material in an olfactometer. *J. Chem. Ecol.* 16, 499-508.

Meffe, G. K., and Carroll, C. R. (1994). "Principles of Conservation Biology." Sinauer. Sunderland, MA.

Mills, L. S., Soulé, M. E., and Doak, D. F. (1993). The keystone-species concept in ecology and conservation. *BioScience* 43, 219-224.

Moloney, K. A., and Levin, S. A. (1996). The effects of disturbance architecture on landscape-level population dynamics. *Ecology* 77, 375-394.

Moran, M. D., Rooney, T. P., and Hurd, L. E. (1996). Top-down cascade from a bitrophic predator in an old-field community. *Ecology* 77, 2219-2227.

Murdoch, W. W., Briggs, S. J., and Nisbet, R. M. (1996). Competitive displacement and biological control in parasitoids: a model. *Amer. Nat.* 148, 807-826.

Nafus, D., and Schreiner, I. (1986). Intercropping maize and sweet potatoes. Effects on parasitization of *Ostrinia fumacalis* eggs by *Trichogramma chilonis*. *Agric., Ecosys., Environ.* 15, 189-200.

National Research Council (1991). "Sustainable Agriculture Research and Education in the Field." National Academy Press, Washington, DC.

Nazzi, F., Paoletti, M. G., and Lorenzoni, G. G. (1989). Soil invertebrate dynamics of soybean agroecosystems encircled by hedgerows or not in Friuli, Italy. First data. *Agric., Ecosys., Environ.* 27, 163-176.

Neve, G., Barascud, B., Hughes, R., Aubert, J., Descimon, H., Lebrun, P., and Baguette, M. (1996). Dispersal, colonization power and metapopulation structure in the vulnerable butterfly *Proclossiana eunomia* (Lepidoptera: Nymphalidae). *J. Appl. Ecol.* 33,14-22.

Noon, B. R., and Murphy, D. D. (1994). Management of the spotted owl: the interaction of science, policy, politics, and litigation. *In* "Principles of Conservation Biology." (G. K. Meffe, and C. R. Carroll, eds.), pp. 380-388. Sinauer. Sunderland, MA.

Norris, R. F. (1986). Weeds and integrated pest management systems. *Hort. Sci.* 21, 402-410.

Noss, R. F. (1991). Landscape connectivity: different functions at different scales. *In* "Landscape Linkages and Biodiversity." (W. E. Hudson, ed.), pp. 27-39. Island Press. Washington, DC.

Okubo, A. (1980). "Diffusion and Ecological Problems: Mathematical Models." Springer-Verlag. New York, NY.

Oliver, I., and Beattie, A. J. (1993). A possible method for the rapid assessment of biodiversity. *Cons. Biol.* 7, 562-568.

Opler, P. A. (1974). Oaks as evolutionary islands for leaf-mining insects. *Amer. Sci.* 62, 67-73.

Paine, R. T. (1974). Intertidal community structure: experimental studies on the relationship between a dominant competitor and its principal predator. *Oecologia* 15, 93 -120.

Paine, R. T. (1995). A conservation on refining the concept of keystone species. *Cons. Biol.* 9, 962-964.

Patterson, B. D. (1987). The principle of nested subsets and its implications for biological conservation. *Cons. Biol.* 1, 323-334.

Perrin, R. M. (1975). The role of the perennial stinging nettle *Urtica dioica,* as a reservoir of beneficial natural enemies. *Ann. Appl. Biol.* 81, 289-297.

Pickett, S. T. A., and Thompson, J. N. (1978). Patch dynamics and the design of nature reserves. *Biol. Cons.* 13, 27-37.

Pimentel, D. (1961). Species diversity and insect population outbreaks. *Ann. Entomol. Soc. Amer.* 54, 76-86.

Pollard, D. G. (1971). Hedges VI: Habitat diversity and crop pests - a study of *Brevicoryne brassicae* and its syrphid predators. *J. Appl. Ecol.* 8, 751-780.

Powell, W., Dean, G. J., and Wilding, N. (1986). The influence of weeds on aphid-specific natural enemies in winter wheat. *Crop Prot.* 4, 182-189.

Power, M. E. (1990). Effect of fish in river food webs. *Science* 250, 411-415.

Price, P. W. (1976). Colonization of crops by arthropods: non-equilibrium communities in soybean fields. *Environ. Entomol.* 5, 605-611.

Price, P. W. (1981). Relevance of ecological concepts to practical biological control. *In* "Biological Control in Crop Protection." (G. C. Papavizas, ed.), pp. 3-19. Beltsville Symp. Agric. Research. No. 5. Beltsville, MD.

Price, P. W., and Waldbauer, G. P. (1994). Ecological aspects of pest management. *In* "Introduction to Insect Pest Management." (R. L. Metcalf, and W. H. Luckmann, eds), pp. 33-65. 3rd Ed. Wiley-Interscience. New York, NY.

Primack, R. B. (1993). "Essentials of Conservation Biology." Sinauer Associates. Sunderland, MA.

Provencher, L., and S. E. Riechert. (1994). Model and field test of prey control effects by spider assemblages. *Environ. Entomol. 23,* 1-17.

Rabb, R. L. (1978). A sharp focus on insect populations and pest management from a wide-area view. *Bull. Entomol. Soc. Amer.* 24, 55-61.

Read, D. P., Feeny, P. P., and Root, R. B. (1970). Habitat selection by the aphid parasite *Diaeretiella rapae* (Hymenoptera: Braconidae) and hyperparasite *Charips brassicae* (Hymenoptera: Cynipidae). *Can. Entomol.* 102, 1567-1578.

Ridgway, R. L., King, E. G., and Carrillo, J. L. (1976). Augmentation of natural enemies for control of plant pests in the western hemisphere. *In* "Biological Control by Augmentation of Natural Enemies." (R. L. Ridgway, and S. B. Vinson, eds.), pp. 379-416. Plenum Press. New York, NY.

Riehl, L. A., Brooks, R. F., McCoy, C. W., Fisher, T. W., and Dean, H. A. (1980). Accomplishments toward improving integrated pest management for citrus. *In* "New Technology of Pest Control." (C. B. Huffaker, ed.), pp. 319-363. John Wiley and Sons. New York, NY.

Risch, S. (1987). Agricultural ecology and insect outbreaks. *In* "Insect Outbreaks." (P. Barbosa, and J. C. Schultz, eds.), pp. 217-238. Academic Press. New York, NY.

Risch, S. J., and Carroll, C. R. (1982). The ecological role of ants in two Mexican agroecosystems. *Oecologia 55,* 114-119.

Risch, S. J., Andow, D., and Altieri, M. A. (1983). Agroecosystem diversity and pest control: data, tentative conclusions, and new directions. *Environ. Entomol.* 12, 625-629.

Robinson, G. R., and Quinn, J. F. (1992). Habitat fragmentation, species diversity, extinction and the design of nature reserves. *In* "Viable Population for Conservation." (S. K. Jain, and L. W. Botsford, eds.), pp. 223-248. Cambridge University Press. Cambridge, U.K.

Rosenheim, J. A., Wilhoit, L. R., and Armer, C. A. (1993). Influence of intraguild predation among generalist insect predators on the suppression of an herbivore population. *Oecologia* 96, 439-449.

Roth, D. S., Perfecto, I., and Rathcke, B. (1996). The effects of management systems on ground-foraging ant diversity in Costa Rica. *In* "Ecosystem Management: Selected Readings." (F. B. Samson, and F. L. Knopf, eds.), pp. 313-330. Springer-Verlag. New York, NY.

Ryszkowski, L. (1993). Above-ground insect biomass in agricultural landscapes of Europe. *In* "Landscape Ecology and Agroecosystems." (R. G. H. Bunce, L. Ryszkowski, and M. G. Paoletti, eds.), pp.71-82, CRC Press/Lewis. Boca Raton, FL.

Samways, M. J. (1979). Immigration, population growth and mortality of insects and mites on cassava *(Manihot esculenta)* in Brazil. *Bull. Entomol. Res.* 69, 491-505.

Samways, M. J. (1994). "Insect Conservation Biology." Chapman and Hall. London, U.K.

Shahjahan, M. (*1974).* *Erigeron* flowers as food and attractive odor source for *Peristenus pseudopallipes,* a braconid parasitoid of the tarnished plant bug. *Environ. Entomol.* 3, 69-72.

Simberloff, D. (1986). Island biogeographic theory and integrated pest management. *In* "Ecological Theory and Integrated Pest Management Practice." (M. Kogan, ed.), pp. 19-35. John Wiley and Sons. New York, NY.

Simberloff, D. (1991). "Review of Theory Relevant to Acquiring Land." Report to Florida Dept. of Natural Resources. Florida State University, Tallahassee, FL. (Cited in Meffe and Carroll, 1994).

Simberloff, D. S., and Abele, L. G. (1976). Island biogeography theory and conservation practice. *Science* 191, 285-286.

Simberloff, D., and Gotelli, N. (1984). Effects of insularisation on plant species richness in the prairie-forest ecotone. *Biol. Cons.* 29, 27-46.

Slosser, J. E., Jacoby, P. W., and Price, J. R. (1985). Management of sand shinnery oak for control of the boll weevil (Coleoptera: Curcullonidae) in the Texas rolling plains. *J. Econ. Entomol.* 78, 383-389.

Smith, J. G. (1976). Influence of crop background on natural enemies of aphids on Brussels sprouts. *Ann. Appl. Biol.* 83, 15-29.

Soulé, M. E. (1991). Theory and strategy. *In* "Landscape Linkages and Biodiversity." (W. E. Hudson, ed.), pp. 91-104. Island Press. Washington, DC.

Speight, H. R., and Lawton, J. H. (1976). The influence of weed cover on the mortality imposed on artificial prey by predatory ground beetles in cereal fields. *Oecologia* 23, 211-233.

Spies, T. A., and Franklin, J. F. (1996). The diversity and maintenance of old growth forests. *In* "Biodiversity in Managed Landscapes." (R. C. Szaro, and D. W. Johnson, eds.), pp. 296-314. Oxford University Press. Oxford, U.K.

Spiller, D. A., and Schoener, T. W. (1990). A terrestrial field experiment showing the impact of eliminating top predators in foliage damage. *Nature* 347, 469-472.

Stinner, R. E., Barfield, C. S., Stimac, J. L., and Dohse, L. (1983). Dispersal movement of insect pests. *Annu. Rev. Entomol.* 28, 319-335.

Stinner, D. H., Paoletti, M. G., and Stinner, B. R. (1989). In search of traditional farm wisdom for more sustainable agriculture: a study of Amish farming and society. *Agric., Ecosys., Environ.* 27, 77-90.

Stinner, R. E., Saks, M., and Dohse, L. (1986). Modeling of agricultural pest displacement. *In* "Insect Flight: Dispersal and Migration" (W. Danthanarayana, ed.), pp. 235-241. Springer-Verlag. New York, NY.

Strong, D. R. (1974). Rapid asymptotic species accumulation in phytophagous insect communities: the pests of cacao. *Science* 185, 1064-1066.

Strong, D. R. (1979). Biogeographical dynamics of insect-host plant communities. *Annu. Rev. Entomol.* 24, 89-119.

Strong, D. R., McCoy, E. D., and Rey, J. R. (1977). Time and the number of herbivore species: pests of sugarcane. *Ecology* 58, 167-175.

Terborgh, J. (1974). Preservation of natural diversity: the problem of extinction prone species. *BioScience* 24, 715-722.

Thacker, J. R. M. (1996). Third international symposium on carabidologists and fragmented habitats, Kauniainen, Finland. *Trends Ecol. Evol.* 11, 103-104.

Thiele, H.-U. (1977). "Carabid Beetles in Their Environment. A Study on Habitat Selection by Adaptions in Physiology and Behavior." Springer-Verlag. Berlin, Germany.

Thresh, J. M. (1981). "Pests, Pathogens and Vegetation: The Role of Weeds and Wild Plants in the Ecology of Crop Pests and Diseases." Pitman Publ., Inc., MA.

Tilman, D. (1994). Competition and biodiversity in spatially structured habitats. *Ecology* 75, 2-16.

Tilman, D., May, R. M., Lehman, C. L., and Nowak, M. A. (1994). Habitat destruction and the extinction debt. *Nature* 371, 65-66.

Tscharntke, T. (1992). Fragmentation of *Phragmites* habitats, minimum viable population size, habitat suitability, and local extinction of moths, midges, flies, aphids, and birds. *Cons. Biol.* 6, 530-536.

USDA (1973). "Monoculture in Agriculture: Extent, Causes and Problems: Report of the Task Force on Spatial Heterogeneity in Agricultural Landscapes and Enterprises." USDA, Washington, DC.

van den Bosch, R., and Telford, A. D. (1964). Environmental modification and biological control. *In* "Biological Control of Insect Pests and Weeds." (P. DeBach, ed.), pp. 459-488. Reinhold, New York, NY.

Vandermeer, J., and Andow, D. A. (1986). Prophylactic and responsive components of an integrated pest management program. *J. Econ. Entomol.* 79, 299-302.

van Emden, H. F. (1965). The role of uncultivated land in the biology of crop pests and beneficial insects. *Sci. Hort.* 17, 121-136.

van Emden, H. F. (1988). The potential for managing indigenous natural enemies of aphids on field crops. *In* "Biological Control of Pests, Pathogens and Weeds: Developments and Prospects." (R. K. S. Wood, and M. J. Way, eds.), pp. 183-201. *Phil. Trans. Royal Soc. Lond.* Series B 318, 1189.

van Emden, H. F., and Williams, G. F. (1974). Insect stability and diversity in agroecosystems. *Annu. Rev. Entomol.* 19, 455-475.

Wellington, W. S. (1983). Biometeorology of dispersal. *Bull. Entomol. Soc. Amer.* 29, 24-29.

Whitham, T. G. (1983). Host manipulation by parasites: within-plant variation as a defense against rapidly evolving pests. *In* "Variable Plants and Herbivores in Natural and Managed Systems." (R. F. Denno, and M. S. McClure, eds.), pp. 15-42. Academic Press. New York, NY.

Wiens, J. A. (1997). Wildlife in patchy environments: metapopulations, mosaics, and management. *In* "Metapopulation Biology: Ecology, Genetics, and Evolution." (I. Hanski, and M. E. Gilpin, eds.), pp. 53-84. Academic Press. San Diego, CA.

Williams, D. W. (1984). Ecology of blackberry-leaf hopper-parasite system and its relevance to California grape agroecosystems. *Hilgardia* 52, 1-33.

Williams, K. S. (1993). Use of terrestrial arthropods to evaluate restored riparian woodlands. *Restor. Ecol.* 1, 107-116.

Willis, E. 0. (1974). Populations and local extinctions of birds on Barro Colorado Island, Panama. *Ecol. Monogr.* 44, 153-169.

Wilson, E. O., and Willis, E. 0. (1975). Applied biogeography. *In* "Ecology and Evolution of Communities." (M. L. Cody, and J. M Diamond, eds.), pp. 522-534. Harvard University Press. Cambridge, MA.

Worthen, W. B., Carswell, M. L., and Kelly, K. A. (1996). Nested subset structure of larval mycophagous fly assemblages: nestedness in a non-island system. *Oecologia* 107, 257-264.

Wu, J., and Loucks, 0. L. (1995). From balance of nature to hierarchical patch dynamics: a paradigm shift in ecology. *Quart. Rev. Biol.* 70, 439-466.

AGROECOSYSTEMS AND CONSERVATION
BIOLOGICAL CONTROL

P. Barbosa

I. INTRODUCTION

Although managed habitats (such as agroecosystems, *sensu* Hill, 1987) can be similar in many ways to unmanaged habitats (Mitchell, 1984), all managed habitats have relatively unique biotic and abiotic forces that shape and direct the interactions that occur in the habitat. In this book a number of ecological and biological principles underlying the tactics of conservation biological control in agroecosystems are proposed and discussed. These discussions are intended to show that conservation biological control is a viable and effective option in the integrated management of insect, mite, and weed pests. However, the successful use of proposed tactics, or tactics that may result from research inspired by the discussions in this book, will be of limited usefulness unless they are developed and implemented within a realistic framework: one that considers the unique traits of agroecosystems. Conservation biological control tactics, and research on new tactics, must recognize both the opportunities afforded in agroecosystems as well as the constraints inherent in such habitats. I suggest that if conservation biological control tactics, or research on new tactics, are to be effective they must be designed with the unique traits of agroecosytems in mind.

The theme of this chapter is that conservation biological control will have the highest likelihood of success if we recognize both the constraints and opportunities afforded by agroecosystems. In this regard, two features are discussed, one representing a constraint and one an opportunity. One feature of agroecosystems, not usually relevant in unmanaged habitats, is the unique temporal and spatial patterns of crop phenology created by farming practices. I suggest that these patterns can have a constraining influence on conservation biological control, but one which can be mitigated by various approaches discussed throughout this book.

Another feature of agroecosystems which may influence how readily conservation biological control is accepted, and thus implemented, is the structure of the fauna (specifically the arthropod community) in agroecosystems. I suggest that in any given agroecosystem, even those that are speciose, a limited number of species are dominant, or pestiferous (Smith and van den Bosch, 1967). The dominance of a few, often related, taxa has important implications for the structure and composition of natural enemy communities and thus for the effectiveness of conservation biological control. A relatively narrow suite of pests reduces the number of natural enemies that need to be targeted for conservation and may facilitate the use of a small number of effective conservation tactics that are also cost-effective. Not having to suggest to a farmer that all natural enemies must be conserved to produce effective control of pests of his/her crop may make conservation biological control a more acceptable option. Thus, in general, understanding how the characteristics of agroecosystems (those discussed here and others) constrain or facilitate the implementation of conservation biological control may lead to the "right" conservation biological control recommendations and thus enhance the likelihood of the success of this control strategy.

II.   THE NATURE OF MANAGED HABITATS AND ITS IMPACT ON CONSERVATION BIOLOGICAL CONTROL

Crop temporal and spatial patterns in agroecosystems throughout the world vary tremendously. There has been little analysis and discussion of whether temporal and spatial patterns that characterize crop phenology in agroecosystems influence the potential success of conservation biological control. Yet it is obvious that spatial and temporal patterns determine the extent and persistence of crop plants and thus the availability of key resources associated with the crop (Karieva, 1990; van Emden, 1990). The spatial and temporal availability of those crop-associated resources (and those in surrounding or adjacent unmanaged habitats, or in managed refuges) also may be critical determinants of whether, when, or how, a natural enemy such as a parasitoid or predator survives and performs in an agroecosystem (see Chapters 4 and 5). Specifically, it may determine whether, when, or how they respond to herbivores or other resources provided by crops (or other plants). Indeed in simulation studies, Corbett and Plant (1993) have noted that the timing of the availability of interplanted (refuge) vegetation relative to the germination of crop plants may determine if the refuge is likely to act as a source of natural enemies or a sink (i.e., taking natural enemies away from crops).

Agroecosystems are managed habitats with concentrations of perennial crops, annual crops, or both. The crop plant's life cycle, to a great extent, dictates the nature of the habitat (i.e., its structure and texture, longevity, and the composition and complexity of its fauna and flora). The latter similarly shapes and determines the intensity and complexity of the interactions that unfold in a given agroecosystem. Temporal and spatial variation in perennial systems may be manifested in a more subtle fashion (as a function of crop phenology or imposed agronomic practices). However, in perennial agroecosystems, although the variation in time and space may be considered less dramatic than that in annual agroecosystems it often is not. In both annual and perennial agroecosystems crop phenology may cause asynchrony between resource availability and the natural enemy stage requiring that resource. Further it may cause the loss or an insufficiency of a resource. Effective conservation of natural enemies must ameliorate and/or compensate for the elimination, reduction, or disruption of needed resources and conditions that result from patterns of crop phenology in agroecosystems.

In annual agroecosystems, the availability of the crop varies in time and space depending on agronomic as well as biological constraints (Fig. 1). In addition, pests may be mobile and move from crop to crop (see Chapter 12). At one end of a gradient, an agroecosystem may consist of a single crop cultivated throughout most or all of the growing season (e.g., potato or corn; or in traditional agriculture a polyculture grown over the entire season). At the other end of the gradient, an agroecosystem may be characterized by a sequence of plantings and harvests of the same crop or by a series of plantings of different crops (see Table 1 in Chapter 12). A third point in this hypothetical gradient is represented by agroecosystems in which a given crop may occur discontinuously at two different times during a season (e.g., *Brassica* crops or legumes such as peas). In these or other types of annual and perennial agroecosystems crop/habitat spatial and temporal patterns may determine the likelihood of success of implemented conservation biological control tactics.

Implementation of conservation tactics in the first crop of a sequential cropping system (Fig. 1A or 1C) without knowing what key resources are, or are not, provided in the other crop phases is likely to be less than fully effective (see example of *Ooencyrtus nezarae* Ishii in Chapter 12). Similarly, in a discontinuous cropping system (Fig. 1B) implementation of conservation biological control during one or both crop phases is likely to have little impact without a thorough plan for the conservation of natural enemies during the interval between the first and second crop plantings. Natural enemies, particularly monophagous species, must survive when the crop and hosts are nonexistent. Alternatively, conservation biological control may involve reliance on or manipulations of unmanaged habitats (or managed refuges) in the landscape of the agroecosystem to compensate for the discontinuous crop pattern (see Colorado potato beetle example in Chapter 7, Section II).

Spatial and temporal crop patterns may determine what resources are available or lacking to natural enemies and thus what conditions or resources must be compensated for by implemented conservation biological control tactics. Of course, the resources required may vary depending on whether the natural enemy is polyphagous or monophagous, a carnivorous natural enemy, or one that feeds directly on plants as well as on prey. Similarly, the same conservation biological control tactic is likely to have a different impact on natural enemies of pests on crops exhibiting different spatial and temporal patterns. What may effectively conserve natural enemies of pests on crops grown following the pattern in Fig. 1A, may not accomplish the same for natural enemies of pests on crops grown following the pattern in Fig.1B. What crop is or is not available to a newly emerged (or formed) adult natural enemy and the nature of its initial interaction in its "new" environment may determine whether the natural enemy remains in the habitat and is subject to efforts to conserve its population, or whether it disperses.

Although not usually considered a phase of host selection for most natural enemies, dispersal is probably an early behavior of newly emerged/teneral adults, undertaken prior to host/prey finding, and causing inter- and intrahabitat movement. This may be particularly true for natural enemies that have a preoviposition maturation

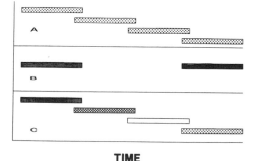

TIME

Figure 1. Hypothetical sequence of crop plants. Bars with different patterns represent different crops or crop cultivars.

Table 1. Lepidoptera Guild of Three Crops Grown in Maryland, U.S.A.

LEPIDOPTERA GUILD OF CABBAGE	LEPIDOPTERA GUILD OF TOMATO	LEPIDOPTERA GUILD OF SOYBEAN
TRICHOPLUSIA NI CABBAGE LOOPER (NOCTUIDAE)	*HELICOVERPA ZEA* TOMATO FRUITWORM (NOCTUIDAE)	*HELICOVERPA ZEA* TOMATO FRUITWORM (NOCTUIDAE)
SPODOPTEA FRUGIPERDA FALL ARMYWORM (NOCTUIDAE)	*PERIDOMA SAUCIA* VARIEGATED CUTWORM (NOCTUIDAE)	*SPODOPTERA FRUGIPERDA* FALL ARMYWORM (NOCTUIDAE)
SPODOPTEA ORNITHOGALLI YELLOW-STRIPED CABBAGEWORM (NOCTUIDAE)	*SPODOPTERA FRUGIPERDA* FALL ARMYWORM (NOCTUIDAE)	*SPODOPTEA ORNITHOGALLI* YELLOW-STRIPED ARMYWORM (NOCTUIDAE)
PERIDOMA SAUCIA VARIEGATED CUTWORM (NOCTUIDAE)	*SPODOPTERA EXIGUA* BEET ARMYWORM (NOCTUIDAE)	*PLATHYPENA SCABRA* GREEN CLOVERWORM (NOCTUIDAE)
PLUTELLA XYLOSTELLA DIAMONDBACK MOTH (PLUTELLIDAE)	*MANDUCA QUINQUEMACULATA* TOMATO HORNWORM SPHINGIDAE)	*ESTIGMENE ACREA* SALT MARSH CATERPILLAR (ARCTIIDAE)
EVERGESTIS RIMOSALIS CROSS-STRIPED CABBAGEWORM (PYRALIDAE)	*MANDUCA SEXTA* TOBACCO HORNWORM (SPHINGIDAE)	*EPARGYREUS CLARUS* SILVER SPOTTED SKIPPER (HESPERIIDE)
ARTOGEIA RAPAE IMPORTED CABBAGEWORM (PIERIDAE)		*LOXOSTEGE SIMILALIS* GARDEN WEBWORM (PYRALIDAE)
		LOXOSTEGE COMMIXTALIS ALFALFA WEBWORM (PYRALIDAE)

Note. Based on personal communication with extension entomologists, Drs. J. L. Hellman, and G. Dively, Department of Entomology, University of Maryland, College Park period (Thorpe and Caudle, 1938; Nishida, 1956). However, dispersal may also occur because of crop phenology. It is likely that a post-overwintering (or second generation) natural enemy may have to disperse to seek needed resources that are no longer available because of crop phenology. Newly emerged natural enemies may have to disperse to find a new host habitat because upon emergence the original crop may not be available, may not be available in the proper state, or may have been replaced by a different crop which fails to supply needed resources. The latter might be a crop that no longer supports an adequate number of hosts/ prey, or suitable host/prey individuals, or a crop that is either too old or young to provide important nonhost/prey resources.

The hosts/prey on crops in existence at the time of emergence may be significantly less suitable than those on the original crop. This may be the case for parasitoids that attack univoltine herbivore hosts on a long season crop or predators that prefer smaller, young prey available mostly early in the season (Fig. 1A). Another example of changes in crop suitability would be that of parasitoids and predators faced with a sequence of crops over the season (Fig. 1A and C) where the host/prey complex, competitors or predators (as well as the crop) changes dramatically. The presence of a different suite of organisms in each crop phase may initiate dispersal if proactive conservation efforts are not initiated.

For parasitoids capable of learning to respond to plant odors, early exposure to and experience with such odors will retain them in a habitat (Kester and Barbosa, 1991). However, as noted above, upon emergence natural enemies may find that the host population in the area from which they originated (or for which they had foraged) is no longer available because individuals in the population have completed their development (Fig. 1B and C). Or, newly emerging natural enemies may find themselves in a host population in which only unacceptable host stages remain in the habitat. For some parasitoids and perhaps other natural enemies learned responses occur early in adult life or not at all. This allows individuals which do not find themselves in favorable habitats to search for hosts in other habitats if hosts are absent when they emerge. The parasitoid thus retains the flexibility of response that facilitates persistence in variable agroecosystems. If such flexibility is counterproductive to control of pests, conservation biological control must aim to compensate for the conditions or situations that result in these responses.

In these and other similar situations the timing of conservation biological control tactics may be critical and may be determined by our understanding of the role of spatial and temporal crop patterns on natural enemy responses. Many agroecosystems do not stand alone in time and space. Nor is the effectiveness of natural enemies in such agroecosystems decoupled in time and space from changes in their crop/habitats. This linkage must be recognized in the development of conservation biological control tactics.

III. THE NATURE OF HERBIVORE COMMUNITIES AND ITS IMPACT ON CONSERVATION BIOLOGICAL CONTROL

Although some managed (as well as unmanaged) habitats are relatively species poor, other agroecosystems can be relatively speciose habitats. For example, Pimentel and Wheeler (1973) reported 591 species in alfalfa agroecosystems of New York state and Heong *et al.* (1991) reported a total of 212 species in irrigated rice grown in five sites in the Philippines. This type of whole community sampling of agroecosystems is rare and thus comparable data for most agroecosystems are not available. Whether speciose or not, there appears to be a tendency for certain functional feeding types to dominate in agroecosystems. Of all the feeding niches in agroecosystems, leaf feeding is among the most common. Indeed, defoliation is perhaps the dominant form of crop damage (Hill, 1987). For example 40% of lepidopteran pests of temperate and tropical agroecosystems are defoliators (Barbosa, 1993).

Lepidoptera are primary defoliators in agroecosystems, although pests in other orders such as the Coleoptera are important (Hill 1983, 1987). The similarity in functional feeding mode, to some degree, exposes these species to similar types of natural enemies. Further, I suggest that what also reduces the number of natural enemies requiring conservation is that only a limited number of taxa (i.e., species in an order or a family) may be important, dominant species in many agroecosystems. These key pests are likely to be attacked by the same natural enemies. If so, conservation biological control would be required for only a limited number of natural enemies. The conservation of a few natural enemies rather than the conservation of all or most natural enemies in an agroecosystem would make the development and acceptance of conservation biological control tactics by farmers easier and more cost-effective.

A preliminary examination of the composition of the Lepidoptera herbivore guilds in agroecosystems was undertaken to evaluate the taxonomic diversity among potential hosts in various agroecosystems (Barbosa, unpublished data). The herbivore guilds of three major cropping systems, which have been cultivated in Maryland, U.S.A. for many years, were considered. The guilds examined included lepidopteran herbivores of tomato, soybean, or cabbage. The lepidopteran species that occur on these crops and which are found on a consistent basis year after year are listed in Table 1. What was immediately apparent from our analysis was that in all crops the herbivore guild was dominated by species in the family Noctuidae. Thus, it was clear that, at least for these crop systems, the most available hosts of larval parasitoids were likely to be related species, i.e., noctuiid larvae.

For each crop system there are considerable periods of overlap of the larval periods of Lepidoptera pests (Fig. 2). Thus, it is likely that whatever conservation biological control tactics were used against these pests (e.g., refugia, weed strips, selected crop cultivars) to conserve natural enemies, the same tactics would likely be effective against all major pests (see Section IV).

The dominance of a limited number of species in agroecosystems can be observed even in those agroecosystems in which Lepidoptera are not the key pest group (Smith and van den Bosch, 1967). The few studies in which the entire arthropod community has been sampled and tabulated support the preliminary analysis noted above. Pimentel and Wheeler (1973) noted that within each major order represented in the community only a few species (in a limited number of families) were dominant primary herbivores (defined as species that completed their development on alfalfa). Among the Lepidoptera, 52.4% of the primary herbivores were species of Noctuidae and Tortricidae. These represented 7 and 5 species in the Noctuidae and Tortricidae, respectively. If four species in the Geometridae were added, species in these three families represented 76% of all primary herbivores: a relatively small proportion of a community of 591 species. Similarly, in the Hemiptera, 58.8% of primary herbivores were species in the Miridae and Cicadellidae. The only other order which included a primary herbivore was the Diptera, which included 1 agromyzid species

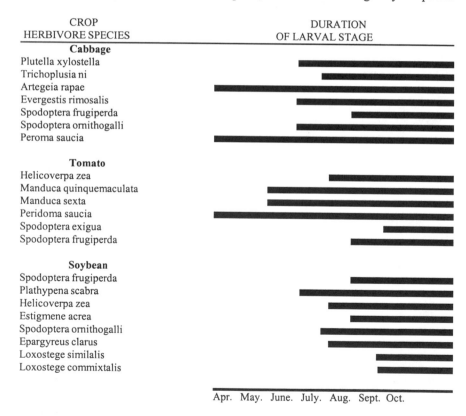

Figure 2. Comparison of larval periods of some herbivores found on cabbage, tomato, and soybean in Maryland, U.S.A.

considered a primary herbivore. Thus, even in a highly diverse agroecosystem such as this, major herbivores are limited in number; and thus so too are the number of their natural enemies (relative to the large natural enemy species complex typically found in alfalfa). That is, only a small number of key natural enemies need to be conserved.

In rice agroecosystems a similar pattern is observed. Overall, for all sites sampled 51.1 % of all arthropods sampled were phytophages (Heong *et al.* 1991). Of these Homoptera, Diptera, and Collembola represented 51.4, 32.3 and 12.4%, respectively, of all arthropods collected. The species in the latter two orders are most likely detritivores. Cicadellidae (comprised mostly of two species *Nephotettix virescens* (Distant) and *N. nigropictus* (Motshulsky)) represented 61.4% of all Homoptera species collected in all sites. The other 38.6% were represented by two species of Delphacidae *Nilaparvata lugens* (Stål) and *Sogatella furcifera* (Horvath). Among the Diptera, the Ephydridae and Chironomidae represented 37.4 and 36.5%, respectively, of all Diptera species collected in all five sites. The dominance of a relatively few major species in these agroecosystems suggest that the suite of natural enemies that may have to be targeted by conservation biological control may be similarly limited.

Even in perennial agroecosystems the pattern seems consistent. Brown and Adler (1989) surveyed managed and abandoned apple orchards from April through September in Virginia, West Virginia, and eastern New York. They found that although 12 species of herbivores were dominant (a designation based on the Berger and Parker index used by Brown and Adler (1989)) in abandoned orchards only six were dominant in managed orchards. In addition, on average the most dominant species constituted only 20 % of the individuals in abandoned orchards and 50% of the individuals in managed (nonorganic) orchards. Thus, in summary, not only were dominant species more frequently dominant in managed orchards but fewer species attained dominance in managed orchards. Determining whether this is a widespread phenomenon will require further research on the community structure of various agroecosystems. This research may be crucial for conservation biological control but data such as those discussed here provide a justification for continued development of new tactics for conservation biological control.

IV.   THE NATURE OF THE NATURAL ENEMY COMMUNITY AND ITS IMPACT ON CONSERVATION BIOLOGICAL CONTROL

Few, if any, studies have been published which have been expressly designed to evaluate the degree to which parasitoids share herbivore hosts in agroecosystems. However, a few studies on the potential for biological control in agroecosystems suggest that despite the priority given in biological control to host specific parasitoids, in some situations co-occurring pest species have parasitoids in common. West and Miller (1989) noted that 7 of 15 noctuiid species in alfalfa and 4 of 8 in peppermint were attacked by *Meteorus communes* (Cresson). The few other studies that exist, report similar results (Martin *et al.,* 1981; Latheef and Irwin, 1983; AliNiazee, 1985; Chamberlin and Kok, 1986; Kok and McAvoy, 1989). Because of the way the data were presented in these studies, however, it is difficult to calculate the extent of parasitoid sharing of hosts.

The differences between agroecosystems and unmanaged ecosystems may have a significant influence on the occurrence and degree of parasitoid sharing among herbivores. For example, Miller (1980) found more parasitoid species attacking *Spodoptera praefica* (Grote) in disrupted habitats (comparable to some agroecosystems) than in nondisrupted sites. He also found that in a native host-parasitoid association, occurring in a disrupted agroecosystem, polyphagous parasitoids characteristically dominated (Miller, 1977; Miller and Ehler, 1978). Ehler and van den Bosch (1974) demonstrated the effectiveness of polyphagous predators and suggested that polyphagous natural enemies may be more suitable as biological control agents in annual crops.

To assess the consequences of the dominance of a limited and related number of species on the likelihood of sharing parasitoids, I tabulated all the hymenopterous parasitoids that are recorded to attack the Lepidoptera species listed in Table 1 (Krombein *et al.,* 1979). The focus was placed on parasitoids since of all major types of natural enemies they are most often presumed to be host specific. All parasitoid species recorded from areas other than Maryland were excluded from the tabulation. This assessment indicated that a relatively low proportion of the parasitoid guild of any given species were host species-specific (although there were some exceptions) (Fig. 3A-D). A large proportion of the parasitoid guild of any given (target) pest tends to utilize a wide variety of co-occurring herbivores, particularly those in the same family as the target pest (Fig. 3A-D). This may be, in part, because of the dominance of noctuiids in these agroecosystems. The parasitoids "specialize" on or share host species in the same family; these host species represent the major pests in the agroecosystem. I suggest that if indeed only a few related pest species dominate in agroecosystems, and these pests are attacked by the same natural enemies then conservation biological control is a practical approach. Fewer tactics will be needed for cost-effective control of pests than if all natural enemies required conservation.

The use of related host species by natural enemies can certainly be viewed not only as adaptive (i.e., increasing fitness) but efficient in the context of biological control. Just finding one or a few host/prey individuals may not be sufficient for maximal natural enemy fitness (i.e., the ability to produce the maximum number of female progeny reaching reproductive age). This is particularly true for natural enemies such as most larval parasitoids of hosts in agroecosystems which have high reproductive potential (Force, 1972; Price, 1973; Miller, 1977). Orientation by natural enemies to cues associated with the crop plant fed on by host/prey herbivores (see Chapters 4 and 5) rather than relying solely on cues associated with hosts/ prey may enable natural enemies to attack a potentially wide variety of species with the same

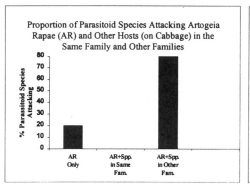

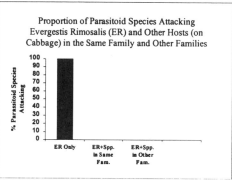

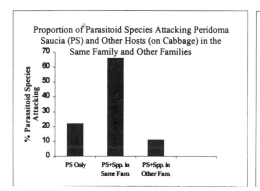

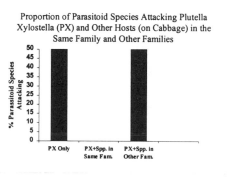

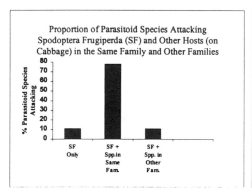

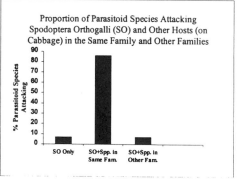

Figure 3A. Relative degree of host specialization among parasitoids of herbivores found on cabbage, tomato and soybean in Maryland, U.S.A.

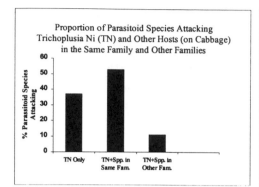

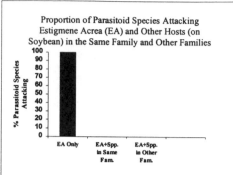

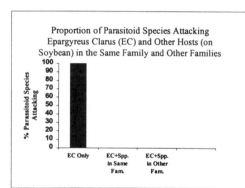

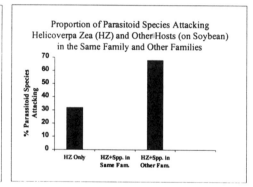

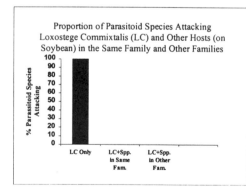

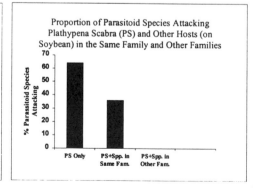

Figure 3B-Continued

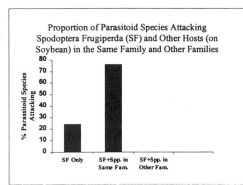

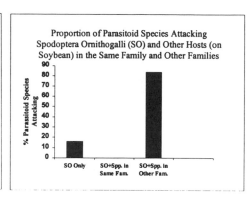

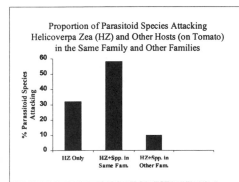

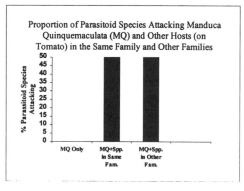

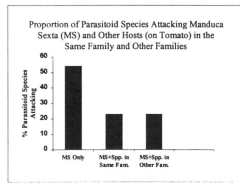

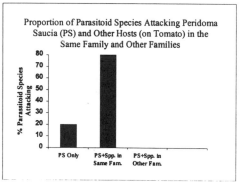

Figure 3C-Continued

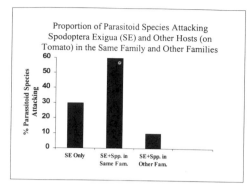

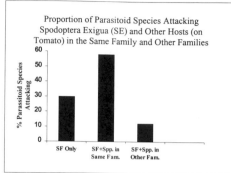

Figure 3D-Continued

feeding niche (and/or physical and behavioral traits). That is, the use of plant cues (niche specificity, if you will) may have evolved to maximize the location of the most abundant hosts (regardless of their taxa) or to enhance searching in areas (i.e., plants) where host herbivores are most likely to be found. Having a few related dominant host species on the same plant minimizes the "information" needed to find hosts/prey and thus, may maximize the likelihood of "success". For some natural enemies such as these, the reliability-detectability conundrum (Vet *et al.,* 1991; Vet and Dicke, 1992) may be less of a dilemma than for other species if related hosts/prey are found using plant cues that persist (see Steinberg *et al.,* 1993), if these signals are replaced as they dissipate, are carried over a sufficient distance, and are reenforced by oviposition.

If the assumptions and speculations presented here are confirmed and found to be appropriate generalizations, they have important implications for conservation biological control. That is, any given tactic can affect and conserve several natural enemies. As was noted in the introduction, agroecosystems can exhibit many unique traits that characterize and influence the interactions among crops, pest herbivores, and natural enemies. A few have been suggested that may influence the success or failure of conservation biological control. Similar phenomena that may influence the success of conservation biological control may exist in agroecosystems (see Lowrance *et al.,* 1984) and must be identified and researched if the implementation of conservation biological control is to succeed.

REFERENCES

AliNiazee, M. T. (1985). Opiine parasitoids (Hymenoptera: Braconidae) of *Rhagoletis pomonella* and *R. zephyria* (Diptera: Tephritidae) in the Willamette Valley, Oregon. *Can. Entomol.* 117, 163-166.

Barbosa, P. (1993). Lepidoptera foraging on plants in agroecosystems: Constraints and consequences. *In* "Ecological and Evolutionary Constraints of Caterpillars." (N. Stamp, and T. Casey, eds.), pp. 523-566. Chapman Hall. New York, NY.

Brown, M. W., and Adler, C. R. L. (1989). Community structure of phytophagous arthropods on apple. *Environ. Entomol.* 18, 600-607.

Chamberlin, J. R., and Kok, L. T. (1986). Cabbage lepidopterous pests and their parasites in southwestern Virginia. *J. Econ. Entomol.* 79, 629-632.

Corbett, A., and Plant, R. E. (1993). Role of movement in the response of natural enemies to agroecosystem diversification: a theoretical evaluation. *Environ. Entomol.* 22, 519-531.

Ehler, L. E., and van den Bosch, R. (1974). An analysis of the natural biological control of *Trichoplusia ni* on cotton in California. *Can. Entomol.* 106,10671073.

Force, D. C. (1972). r- and K- strategists in endemic host-parasitoid communities. *Bull. Entomol. Soc. Amer.* 18,135-137.

Heong, K. L., Aquino, G. B., and Barrion, A. T. (1991). Arthropod community structures of rice ecosystems in the Philippines. *Bull. Entomol. Res.* 81, 407-416.

Hill, D. S. (1983). "Agricultural Insect Pests of the Tropics and Their Control." Cambridge University Press. Cambridge, U.K. (2nd Edition).

Hill, D. S. (1987). "Agricultural Insect Pests of Temperate Regions and Their Control." Cambridge University Press. Cambridge, U.K.

Karieva, P. (1990). The spatial dimension in pest enemy interactions *In* "Critical Issues in Biological Control" (M. Mackauer, L. E. Ehler, and J. Roland, eds.), pp. 213-227. Intercept. Andover, Hants, U.K.

Kester, K. M., and Barbosa, P. (1991). Post-emergence learning in the insect parasitoid, *Cotesia congregata* (Say). *J. Insect Beh.* 4, 727-742.

Kok, L.-T., and McAvoy, T. J. (1989). Fall broccoli pests and their parasites in Virginia. *J. Entomol. Sci.* 24, 258-265.

Krombein, K. V., Hurd, P. D., Jr., Smith, D. R., and Burks, B. D. (1979). "Catalog of Hymenoptera in America North of Mexico." Smithsonian Institution Press. Washington, DC.

Latheef, M. A., and Irwin, R. D. (1983). Seasonal abundance and parasitism of lepidopterous larvae on *Brassica* greens in Virginia. *J. Georgia Entomol. Soc.* 18, 164-168.

Laurence, R., Stinner, B. R., and House, G. J. (1984). "Agricultural Ecosystems. Unifying Concepts." John Wiley and Sons. New York, NY.

Martin, P. B., Lingren, P. D., Greene, G. L., and Grissell, E. E. (1981). The parasitoid complex of three noctuids in a northern Florida cropping system: seasonal occurrence, parasitization, alternate hosts, and influence of host habitat. *Entomophaga* 26,401-419.

Miller, J. C. (1977). Ecological relationships among parasites of *Spodoptera praefica. Environ. Entomol.* 6, 581-585.

Miller, J. C. (1980). Niche relationships among parasitic insects occurring in a temporary habitat. *Ecology* 61, 270-275.

Miller, J. C., and Ehler, L. E. (1978). Parasitization of *Spodoptera praefica* larvae in hay alfalfa. *Environ. Entomol.* 7, 744-747.

Mitchell, R. (1984). The ecological basis for comparative primary production. *In* "Agricultural Ecosystems. Unifying Concepts." (R. Laurence, B. R. Stinner, and G. J. House, eds.), pp. 13-53. John Wiley and Sons. New York, NY.

Nishida, T. (1956). An experimental study of the ovipositional behavior of *Opius fletcheri* Silvestri (Hymenoptera: Braconidae) a parasite of the melon fly. *Proc. Hawaii. Entomol. Soc.* 16, 126-134.

Pimentel, D., and Wheeler, A. G., Jr. (1973). Species and diversity of arthropods in the alfalfa community. *Environ. Entomol.* 2, 659-668.

Price, P. W. (1973). Parasitoid strategies and community organization. *Environ. Entomol.* 2, 623-626.

Smith, R. F., and van den Bosch, R. (1967). Integrated control. *In* "Pest Control: Biological, Physical, and Selected Chemical Methods." (W. W. Kilgore, and R. L. Doutt, eds.), pp. 295-340. Academic Press. New York, NY.

Steinberg, S., Dicke, M., and Vet, L. E. M. (1993). Relative importance of infochemicals from first and second trophic level in long-range host location by the larval parasitoid *Cotesia glomerata. J. Chem. Ecol.* 19, 47-59.

Thorpe, W. H., and Candle, H. B. (1938). A study of the olfactory responses of insect parasites to food plant of their host. *Parasitology* 30, 523-528.

van Emden, H. F. (1990). Plant diversity and natural enemy efficiency in agroecosystems. *In* "Critical Issues in Biological Control." (M. Mackauer, L. E. Ehler, and J. Roland, eds.), pp. 63-80. Intercept. Andover, Hants, U.K.

Vet, L. E. M., and Dicke, M. (1992). Ecology of infochemical use by natural enemies in a tritrophic context. *Annu. Rev. Entomol.* 37, 141-172.

Vet, L. E. M., Wäckers, F. L., and Dicke, M. (1991). How to hunt for hiding hosts: the reliability-detectability problem in foraging parasitoids. *Neth. J. Zool.* 41, 202-213.

West, K. J., and Miller, J. C. (1989). Patterns of host exploitation by *Meteorus communis. Environ. Entomol.* 18, 537-540.

CHAPTER
4

THE INFLUENCE OF PLANTS ON INSECT PARASITOIDS: IMPLICATIONS FOR CONSERVATION BIOLOGICAL CONTROL

Pedro Barbosa and Betty Benrey

I. INTRODUCTION

For most arthropod species and their parasitoids, the most significant component of their habitat is the plants on which they live, eat, and depend. While about 22% of all known species are green plants, about 26% are phytophagous insect species (Strong *et al.,* 1984). Phytophagy (i.e., plant feeding) is even more dominant among pest arthropods. The vast majority of pest species are phytophagous species, and leaf feeding is the most common form of damage observed in agroecosystems (sensu Hill, 1987). For example, in temperate and tropical agroecosystems about 40% of lepidopteran pests are defoliators, and many other species feed on crop plant tissues other than leaves (see Barbosa, 1993).

For parasitoids, in particular, it is the physical and chemical characteristics of the food plants of herbivorous hosts as well as the distribution of the plants in time and space that mediate parasitoid survival, parasitoid-host interactions, and thus the potential effectiveness of parasitoids as biological control agents. The role of plants in the ecology and biology of insect parasitoids is perhaps most clearly manifested in their effects on parasitoid host selection. Critical information provided by plants provides relatively clear cues about actual or potential host location and suitability. These cues may arise directly from plants, or may be incorporated in host tissues, secretions, or excrement. In addition, they may arise as a direct result of the feeding by herbivores on plants. Plants and the habitat conditions they create also provide essential requisites for survival. Food for adult parasitoids, favorable microclimates, alternative hosts, and perhaps even protection from their own natural enemies may be provided by plants in any given habitat. In some species, responses to plants may even be involved in mate finding. Males of some species such as *Diadromus pulchellus Wesmeal, Macrocentrus grandii Goidanich, Microplitis demolitor* (Wilkinson), and *Campoletis sonorensis* (Cameron) (Lecomte and Pouzat, 1985; Ding *et al.,* 1989; McAuslane *et al.,* 1990; Ramachandran and Norris, 1991) perceive and respond to plant volatiles (but see Elzen *et al.,* 1986; Eller *et al.,* 1988; Whitman and Eller, 1990; Udayagiri and Jones, 1992). Although the role of plant chemicals may vary among different parasitoid species, they directly or indirectly determine the survival, fecundity, longevity, host selection behavior, and other aspects of parasitoid-host interactions. These plant-mediated interactions, in turn, have an impact on the persistence of parasitoid populations, i.e., whether they are conserved and are effective as biological controls.

Our intent in this chapter is to provide an overview (rather than a comprehensive review) of some of the ways in which plants directly and indirectly influence parasitoids; and how they mediate host-parasitoid interactions and interactions between parasitoids and their biotic and abiotic environment. It is unlikely that plant factors affecting parasitoids act independent of each other; however, we present them as such for the sake of simplicity. Not only do we aim to answer the question, In what way do plants affect parasitoids? but we also address the question, What are the constraints and advantages of using plant-mediated interactions as a basis for tactics useful in conservation biological control with parasitoids?

II.   INFLUENCE OF PLANT PATCH STRUCTURE AND DIVERSITY (INTER- AND INTRAPATCH TRAITS)

For natural enemies of herbivores the task of finding their host's habitat entails finding the right vegetational association and then the right patch of plants (one with preferred hosts or host stages) within the association. Plants may influence the orientation of parasitoids to certain habitats and along with cues from other sources help orient parasitoids to their hosts. A dramatic illustration of that influence is the attraction of parasitoids to plants (or host habitats) even in the absence of any host. For example, *Cardiochiles nigriceps* Viereck and *Opius flecheri* Silvestri are attracted to or search host-free plants, *Campoletis sonorensis* responds to volatiles emanating from specific plant tissues, and *Macrocentrus grandii is* attracted to volatiles from undamaged plants (suggesting that host associated cues are not always needed) (Nishida, 1956; Vinson, 1975; Elzen *et al.,* 1983, 1984; Whitman and Eller, 1990; Udayagiri and Jones, 1992).

The search for the "right" habitat is imperative because like all other animals, insect parasitoids have requirements for resources, other than hosts. However, these other resources may or may not be found in the same habitat in which hosts are found. Even when needed conditions or resources occur in the habitats of their hosts, they may not occur in all such habitats. For example, optimal microclimatic conditions for a given parasitoid, nectar sources, or pollen may exist in some host habitats (crop systems) but not others. One assumes that the habitats in which parasitoids find hosts also provide other needed requisites at optimum levels but there is little empirical or experimental data which clearly show this to be true, even for unmanaged ecosystems. The objective of conservation biological control is to ensure that the occurrence of as many essential parasitoid resources and hosts coincide in time and space. One way to accomplish this may be by maximizing plant species diversity and structural complexity of vegetation. This assumes that the greater the (biotic and abiotic) variety in a habitat the greater the likelihood that parasitoid requisites will be met.

The capacity of a host's habitat and the plant species therein to meet the needs of a parasitoid is particularly problematic for polyphagous parasitoids whose hosts live in distinctly different habitats. Such parasitoids may either have broad biotic and abiotic tolerances and adaptations or may persist globally as a collection of widely varying subpopulations each with their own particular needs. This dichotomy, if common, may have important implications for the development of conservation biological control tactics. For example, species such as *Praon occidentale* (Baker) and *Lysiphlebus testaceipes* (Cresson) have a broad range of habitats and aphid hosts (Carroll and Hoyt, 1986) and thus their "needs" may differ for each host association. Carroll and Hoyt (1986) reported that *Praon unicum* Smith, like the two species noted above, utilized different aphids on different plants in different habitats. Approaches to the conservation of this parasitoid may differ in each habitat. Further, what occurs in one habitat may affect populations in other habitats. Levels of attack of a host occurring late in the growing season might depend on the density of the parasitoid emerging from another host in the same or another habitat and on the ability of parasitoids to disperse within and between habitats. Clearly, this abundance of host/habitat associations complicates the task of developing tactics that will conserve these parasitoid species.

Plant species diversity and the specific plant species that make up that diversity may determine the survival and abundance of parasitoids. However, whether one should conserve parasitoids by enhancing vegetational structure and species diversity depends on whether such changes have positive effects on pests. In particular, whether they disproportionately favor pest species relative to their effects on parasitoids. Similarly, understanding of the impact of plant/habitat diversity on hyperparasitism or predation of parasitoids must be taken into account in the development of conservation biological control. For example, the rate of hyperparasitism of *P. unicum* was only 12% in *Myzus persicae* on peach whereas on weeds, in the same host, hyperparasitism was 33% (Carroll and Hoyt, 1986). Therefore, the planting of peaches in or near apple orchards was their recommendation for the enhancement of parasitoid densities and the biological control of the apple aphid *Aphis pomi* De Geer in orchards.

A.   Influences of the Size, Number, and Shape of Plant Patches

The sophistication of the responses of parasitoids to chemical and nonchemical cues associated with the habitat/plant of their hosts is so well developed and fine tuned that it is reasonable to expect that many parasitoids are habitat specialist rather than host specialist. The microhabitat specialization exhibited by parasitoids (such as *Asobara tabida* (Nees), *A. rufescens* (Foerster), *Leptopilina clavipes* (Hartig), etc.) of saprophagous and fungivorous drosophilids supports the latter contention (Vet *et al.*, 1984; Dicke *et al.*, 1984; Vet, 1983, 1985; and other studies cited in this chapter). However, the ability of parasitoids to find host plant patches varies. The availability of habitats and the ease with which they can be found (a function of the size, shape, and number of plant patches) can influence the impact of parasitoid species on host populations and how well habitat manipulations are likely to conserve parasitoid populations.

Whether the herbivore's host plant grows isolated from other plants or clumped with other types of plants appears to have a significant influence on parasitoids (Harrington and Barbosa, 1978). Monteith (1960) found that levels of parasitism of the larch sawfly *Pristophora erichsonii* were greater (up to 86%) in stands of pure larch compared to levels in larvae on larch growing in close association with other trees, in which case parasitism was reduced to about 12%. The plant canopy creates conditions which can influence light intensity, color, etc., and thus vision. There has been almost no research conducted on this aspect of host selection by parasitoids. Most available data are anecdotal or subject to multiple interpretations. Nevertheless, changes in plant canopies may play a role in host finding by parasitoids (Sato and Ohsaki, 1987), particularly among visually orienting parasitoids such as tachinids. For example, when the pierid *Pieris napi japonica* occurred on crucifers that were overshadowed by other vegetation, it suffered low levels of parasitism by *Apanteles glomeratus* (sic). If the overshadowing vegetation was cleared subsequent rates of parasitism were higher (Sato and Ohsaki, 1987). The mechanism for the difference was not determined but the influence of canopy shade on visual acuity is a possibility.

Of course other factors may influence parasitoids as canopies change in time and space or as a result of the inherent differences in the canopies of different plant species. In addition, for all plant species as plant patch size changes so does the canopy and its influence on parasitoids. Parasitism rates can vary with the height of a plant (Monteith, 1960). The size and shape of plants in patches have been shown to influence the foraging behavior of parasitic wasps (Arthur, 1962). Further, differences in the structure and density of the forest canopy (resulting from differential susceptibility to defoliation by an herbivore) can have a significant impact on the density of parasitoids and thus their effectiveness. For example, *Brachymeria intermedia* (Nees) exhibits a preference for gypsy moth pupae located on the edges of woodland fields and open or sunny locations (i.e., open or sparse canopies) often created by defoliation by its host (Barbosa *et al.,* 1978).

1. Response to chemical plant patch signals

A variety of plant-associated allelochemicals have been identified which mediate the location of a parasitoid's host habitat. These compounds are primarily aldehydes, alcohols, sulfur-based compounds, esters, and terpenes. They illicit chemoanemotactic responses in a variety of hymenopteran parasitoids (Rutledge, 1997). Rutledge (1997) notes that aldehydes, alcohols, and sulfur-based compounds have been identified only as cues in host-habitat location, and that many of the alcohols, aldehydes, and esters involved in host habitat location are common six-carbon "green leaf volatiles." Although damaged plants may emit higher levels of green leaf volatiles (i.e., six-carbon alcohols, aldehydes, and derivative esters), undamaged plants emit green leaf volatiles (GLVs) to which insect parasitoids respond (Whitman and Eller, 1990). The response of parasitoids is often enhanced if volatiles originate from damaged foliage. This suggests that GLVs help searching parasitoids find patches of vegetation whereas higher levels of GLVs produced in damaged plant tissues draw parasitoids to those areas of the patch with hosts. Responses are not, however, necessarily generalized and dose dependent (Baehrecke *et al.,* 1989). For example, the GLVs which are attractive to the braconid *Microplites croceipes* (Cresson) are not the same GLVs that are attractive to ichneumonid *Netelia heroica* Townes (Whitman and Eller, 1990). The action of a compound may be functionally quite specific. The host plant of *Acrolepiopsis assectella* Zeller emits volatiles that aids its parasitoid *Diadromus pulchellus* Wesmael in finding its host's habitat (Lecomte and Thibout, 1984). Locomotor activity of the parasitoid is stimulated by various sulphur compounds (disulfides and sulfonothioic acid-5-esters) but they do not attract the parasitoid to the source of volatile (Auger *et al.,* 1987; Thibout *et al.,* 1987).

Parasitoids do not simply respond to any vegetation producing GLVs. Parasitoids may exhibit clear (and inherent) preferences among the host plants of their herbivore hosts. For example, *Microplites croceipes* exhibits a preference for cotton *(Gossypium hirsutum* L.) and velvet leaf *(Abutilon theophrasti* Medicus) over groundcherry *(Physalis angulata* L.) (Navasero and Elzen, *1989). Campoletis sonorensis* is attracted to glanded (high terpene) cotton over glandless (low terpene) cotton when given a choice (Elzen *et al.,* 1986). *C. sonorensis* is attracted to volatiles of tobacco, cotton, sorghum, and bluebonnet, less attracted to wild carrot, and is not attracted to volatiles of various other plants tested (Elzen *et al.,* 1983). *Macrocentrus grandii* exhibits flight responses to volatiles of undamaged corn. This response does not require previous experience with (or exposure to) volatiles in association with oviposition experiences (Udayagiri and Jones, 1992). Clearly, parasitoid responses to chemical cues from plant patches are complex and variable, but nevertheless important in host finding.

2.   Response to physical attributes of plant patches

Chemical cues appear to be, by far, the most common signals providing useful information to parasitoids in searching host habitats for hosts. However, this conclusion may be solely a result of the fact that there has been so little research conducted on the role of physical factors. As with most biological phenomena no one factor alone is likely to mediate the complex responses of parasitoids. Indeed, there already is some clear evidence that both vision and smell work together to provide parasitoids the maximum information (Wardle and Borden, 1989; Wäckers 1994; Wäckers and Lewis, 1994).

Weseloh (1986) found that a variety of hymenopteran parasitoid species were attracted to yellow traps. A response to yellow is interpreted as a response to foliage since yellow reflects at wavelengths above 500nm, the region where leaves reflect maximally. Yellow panels are said to act as a "supernormal" foliage type stimulus (Prokopy, 1972). Interestingly, the hosts of the parasitoids attracted to yellow were leaf-feeding or leaf-dwelling herbivores (Weseloh, 1986). The ability of several parasitoid species to distinguish color and learn colors (Wardle, 1990; Wäckers and Lewis, 1994) have been demonstrated in the laboratory. However, it is reasonable to assume that responses of this type occur and can be important in host patch finding by parasitoids in the field.

Plants in an agroecosystems exist as cultivars that have dramatically distinct physical characteristics, while having relatively insignificant differences in their chemical characteristics. Among these are differences in the shape, size, and structural complexity (i.e., architectural complexity) which influence host habitat selection. At a larger scale, patches of different plants or different cultivars of a given crop may provide cues that attract parasitoids and perhaps cues to which only specific parasitoids respond. Even herbivore feeding damage may change the physical profile of a plant, forming patterns to which parasitoids learn to orient. There is some evidence to suggest that parasitoids may have the ability to detect and differentiate plant damage (Sugimoto *et al.,* 1988; Faeth 1990). However, there are too few studies on this phenomenon to develop any generalizations.

3. Landscape influences

The landscape in most environments is characterized by a series of habitats; each of which has certain traits, flora, and fauna which create conditions that make it more or less favorable to any given species in comparison to another adjacent habitat. This landscape diversity provides a mosaic of favorable and unfavorable habitats in which hosts and their natural enemies live and through which they traverse. The degree to which the number of habitats that are favorable to both herbivore and natural enemy can determine the degree to which natural enemies regulate host populations, the area across which control is exerted, and the duration of that control. Similarly, the coincidence of favorable habitats may determine which parasitoid species, of a complex of species, will regulate host populations in any given habitat. For example, although overall parasitism of the gypsy moth *Lymantria dispar* in a series of adjacent mesic and xeric forest habitats was about 12 to 18% the level of parasitism and the species responsible for the parasitism differed from habitat to habitat (Skinner *et al.,* 1993). Parasitism rates of over 40% by *Parasetigena silvestris* (Robineau-Desvoidy) and *Phobocampe disparis* (Viereck) were recorded in mesic habitats. In contrast, parasitism by *Cotesia melanoscelus* (Ratzeburg) and *Blepharipa pratensis* (Meigen) were common in xeric habitats and parasitism ranged from 1 to 9% and from 1 to 4%, respectively. A fifth species, *Compsilura concinnata* (Meigen) was common in all types of habitats and levels of parasitism reached 40% (Skinner *et al.,* 1993). The latter study did not discern whether habitat differences affected parasitoids directly or indirectly (e.g., by supporting host populations of different densities) but the impact of landscape differences was evident. Similar landscape effects on parasitoids have been reported by others (Pschorn-Walcher, 1980; Chapter 6).

Fragmentation of habitats in and around agroecosystems and unmanaged ecosystems is a significant impediment to biological control. Isolation of plant patches reduces both parasitoid diversity and the level of parasitism (Kruess and Tscharntke, 1994). Other studies support the hypothesis that parasitoids respond to habitat fragmentation. Parasitism by species such as *Phacogenes hariolus* (Cress.), *Itoplectis conquisitor (*Say*), Brachymeria intermedia, and Eriborus terebrans* (Gravenhorst) is higher on plants (whether herbaceous or trees) along the edges of their host's habitats compared to those in the interior of stands (Kulman and Hodson, 1961; Barbosa and Frongillo, 1979; Landis and Hass, 1992). Although habitat fragmentation can have an influence, parasitoids respond in different ways. For example, *Glypta fumiferanae* parasitism is higher in the interior of stands rather than along stand edges (Kulman and Hodson, 1961). However, Roland and Taylor (1995) found that although two parasitoids of the forest tent caterpillar *Malacosoma distria* Hübner are affected by forest fragmentation they are differentially affected. Parasitism by *Patelloa pachypyra* (Ald. and Web.) is lower in isolated forest fragments compared to more uniform patterns of parasitism exhibited by *Sarcophaga aldrichi* Parker across all sites.

B. Size and Shape of Plants in Patches

Even the size of nonhost plants can influence parasitoids. Coll and Bottrell (1996) showed that maize height in maize-bean dicultures was the primary factor influencing the movement of the Mexican bean beetle parasitoid *Pediobius faveolatus* (Crawford). In general, the size and shape of plants are important if only because many parasitoids are niche specific (i.e., they parasitize hosts occupying a particular niche regardless of what species occupies the niche) and thus respond strongly to shape and size or orient to particular plant parts (Arthur, 1962). On a relatively small scale, parasitoids may show preferences for different plant parts. The aphid parasitoid *Aphidius rhopalosiphi* spends more time searching for hosts on the leaves of wheat than on the ear, even though the latter is the preferred oviposition site of its host (Gardner and Dixon, 1985). A parasitoid species such as *Exeristes comstockii* attacks 19 host species in 7 genera and 4 families, but all are shoot or cone borers of coniferous trees (Townes, 1960). Modification of the shape of herbivore's host plant can alter the vulnerability of herbivores to parasitoids.

Variation in the structural complexity among plant genotypes can influence parasitism rates. *Bracon mellitor* Say, a parasitoid of the boll weevil *Anthonomus grandis* Boheman, is more efficient on cotton genotypes with narrow twisted bracts than with normal bracts (McGovern and Cross, 1976). Likewise, populations of the cabbage butterfly *Pieris rapae* (L.) suffer greater parasitism on flat, open-leafed *Brassica oleracea* L. than on heading or curly-leafed varieties in which caterpillars feed between folded leaves where they escape attack (Pimentel, 1961). Stary (1970) reported that when aphid density becomes large enough to curl leaves, the aphids become vulnerable to parasitoids to which they are normally immune.

Even when leaf shape changes are not of significance to parasitoids the sheer number and size of leaves may be. Increases in leaf surface area are inversely related to searching efficiency of parasitoids (Need and Burbutis, 1979). Finally, the architectural complexity of a herbivore's host plant may significantly influence parasitoid effectiveness. Andow and Prokrym (1990) reported that the rate of parasitism by *Trichogramma nubilale* was 2.9 times higher on simple surfaces compared to those of parasitoids on structurally complex surfaces.

C. Plant Taxonomic Diversity

1. Abundance and type of host plants in a patch

Habitats vary widely in plant species diversity. The abundance and type of plants in a patch that are hosts of the herbivores attacked by parasitoids and of nonhost plants that provide important resources are determinants of the impact of plant patch species diversity. However, many if not most studies on patterns of parasitism in a habitat rarely provide information relating patterns and levels of parasitism to patch species diversity. Nor do they indicate whether plants providing essential resources occur in the same habitat where parasitoids and herbivores interact.

Regardless of the nature of the diversity in the patch, species diversity can be a determinant of parasitoid survival and effectiveness (Chapters 8 and 9). Indeed, a central theme of research in agroecology has been whether mixtures of crops (polycultures) increase the abundance of natural enemies and thus facilitate the control of pests compared to monocultures (see discussions in Chapter 9). Even in polycultures or habitats with variable plant species composition the level of parasitism will vary. The plant on which hosts occur can determine both the likelihood of a parasitoid finding its hosts and the probability of successful parasitism (Pair *et al.,* 1982; Mueller, 1983; Lewis and Gross, 1989; Felland, 1990). Although the issue of the impact of species diversity on parasitoids is far from straightforward a significant number of studies do support the contention that plant species mixtures (whether they are host plants of parasitoid hosts or not) do enhance the abundance and variety of parasitoids (Andow, 1986; Sheehan, 1986; Russell, 1989).

a. *Direct effects*   The importance of plant diversity in or adjacent to agroecosystems is important because of the direct effects of plants on parasitoids. It is well known that nectar and pollen have a significant influence on parasitoids (Bombosch, 1966; Syme, 1975; Schuster and Calderon, 1986; Hagley and Barber, 1992). Brassicaceae

weeds, such as *Berteroa incana L.* DC., *Barbarea vulgaris R. Br., Lepidium campestre*
(L.) R. Br., and *Brassica kaber* D.C. Wheeler that grow in agroecosystems with
Brassica crops were shown to be good nectar sources for *Diadegma insulare* (Cresson),
increasing the longevity of females 5 to 10 fold and fecundity 100-fold in comparison
to nectars of the flower of other species (ldris and Grafius, 1995). It would not be
unreasonable to assume that the pollen and nectar needs could be met by crops in
an agroecosystem. However, in most situations plant species diversity in or near an
agroecosystem may be essential because different pollens affect fecundity and longevity
differently (Leius, 1963). In addition, access to floral and extrafloral nectar by
parasitoids may vary from plant species to plant species (Baker *et al.,* 1978) because
of seasonal variation in nectar production (Butler *et al.,* 1972) or the constraints on
access due to flower morphology (ldris and Grafius, 1997; see Chapter 9). Access
to a variety of nectar and pollen sources may be essential for the conservation of
a variety of parasitoids.

b. *Indirect effects*   The abundance and distribution of a parasitoid's host population
is a direct consequence of the abundance and distribution of its herbivore's host plant.
Similarly, for polyphagous herbivores the number of host plant species in a patch
may influence its abundance. Needless to say, the abundance and type of host plants
in an area can have a significant indirect influence on parasitoids since they can
determine host availability and thus the potential for parasitoid population increase.
Although self evident, the number of studies documenting this linkage and the
consequences of disruptions of this tritrophic level linkage are less common than
one would expect.

D.   Physical Plant Cues and Chemical Signals within Patches

1.   Responses to physical aspects of plants

Host selection by parasitoids has been shown to involve the use of chemical
cues (from the host and host's habitat) as well as physical cues (e.g., visual cues such
as movement, color, leaf structure, etc.) of the patch canopy. Differences in the structure
and shape of plants in a patch define the structure of the patch canopy. Differential
response by parasitoids to any aspect of canopy structure may result in differentially
parasitization.

However, few studies have investigated the ability of parasitoids to respond
to differences in physical traits of plant or within the canopy. At present we can only
speculate, based on evidence from investigations of learning of microhabitat traits

by parasitoids of nonphytophagous hosts (Rosenheim, 1987), or from laboratory studies of parasitoids of phytophagous hosts (Wardle and Borden, 1990). Wardle and Borden (1990) demonstrated the ability of *Exeristes roborator* (F.), a polyphagous ectoparasitoid, to learn the form of an artificial microhabitat. Parasitoids learned to distinguish cylinders from spheres, forms chosen because they resemble plant structures such as fruits, shoots, and stems, within which their hosts normally reside. Studies of the responses of parasitoids to plant morphology suggests that they may also respond to changes in the plant canopy. However, this remains an area where much more research is needed before generalizations can be proposed.

2. Cues originating from herbivore-plant interactions

Although a great deal of host finding is based on the response of natural enemies to host associated cues, much of host finding is made possible by cues associated with herbivore-modified plants. The data collected to date make it difficult to unambiguously differentiate between chemical cues that influence plant patch finding, finding single plant species or plant individuals, or finding hosts (see Section II, A,1). Nevertheless, the fact remains that plant- and damage-associated cues are essential in host selection. These phenomena are discussed in Sections II,A,1 and Ill,B.

III. INFLUENCE OF SINGLE PLANTS ON WITHIN-PLANT PARASITOID RESPONSES AND SURVIVAL

A. Sources of Food

Individual plants may provide adult parasitoids with essential food in the form of nectar (floral or extrafloral) and pollen (Hagen, 1986; Jervis *et al.,* 1993; Stapel *et al.,* 1997; Lewis *et al.*, in press). Food can also be provided indirectly when herbivores that feed on the plant produce honeydew or other sugar secretions used by parasitoids (Powell, 1986; Whitman, 1994; Jervis and Kidd, 1996). For many parasitoid species, the availability of food can increase fecundity, longevity, survival, and levels of parasitism (Leius, 1963; Shahjahan and Streams, 1973; Syme, 1975; Foster and Ruesink, 1984; Wäckers and Swaans, 1993; ldris and Grafius, 1995,1997, Olson and Nechols, 1995) as well as influence the foraging behavior of searching parasitoids (Wäckers and Swaans, 1993; Wäckers, 1994). The hunger state of searching parasitoids may determine whether they spend time searching for hosts or for food (Wäckers and Swaans, 1993; Lewis and Takasu, 1990). Wäckers (1994) compared the response of starved and satiated parasitoids to the odors produced by flowers and host- infested leaves. Food deprived individuals chose flower odors, while sugar-fed individuals preferred host-associated odors. Thus, whether or not food is available for adult parasitoids in the areas where their hosts are located may determine, in part, tenure time and levels of parasitism and thus their efficiency as biological control agents. Some recent studies on the effectiveness of parasitoids as biological control agents have stressed the importance of providing food to parasitoids in agricultural fields (Power, 1986; Jervis et al., 1993; Wäckers and Swaans, 1993). In some cases the use of food supplements has been implemented with some success (Altieri and Whitcomb, 1979; Powell, 1986; Bugg et al., 1989).

B.   Chemical Cues and Barriers to Searching Parasitoids

Studies conducted with lima beans, corn, cabbage, and cotton, have demonstrated that plants are actively involved in the production and release of host locating cues used by parasitoids (Dicke and Sabelis, 1988, 1989; Takabayashi et al., 1991; Turlings et al., 1990, 1991, 1995; Cortesero et al., 1993; Dicke, 1994; Agelopoulos and Keller, 1994; Mattiaci et al., 1994). When plants are attacked by insect herbivores, they emit odorous signals that attract natural enemies (e.g., Dicke and Sabelis, 1988; Turlings et al., 1990). The release of such induced compounds appears to be triggered by a factor present in the herbivore's saliva (Turlings et al., 1993b; Mattiaci et al., 1995; Paré and Tumlinson, 1997a). Each plant species will emit its own specific blend of herbivore-induced volatiles (Turlings et al., 1993a). There is evidence that specialist and generalist parasitoids may respond differently to plant volatiles (Röse et al., 1997, in press) and that the same plant species will emit different odors when it is attacked by different herbivores (Dicke, 1995) and by different stages of the same herbivore (Takabayashi et al., 1995). Moreover, these induced compounds are not only emitted at the damaged site but also systemically by the entire plant (Turlings and Tumlinson, 1991, 1992; Dicke et al., 1993; Cortesero et al., 1997; Röse et al., 1996). Thus, a plant under herbivore attack is able to emit large amounts of a specific chemical signal that lures in parasitoids (Alborn et al., 1996; Röse et al., 1996; Alborn et al., 1997; Paré and Tumlinson, 1997b).

The implications of these results for conserving and enhancing parasitoid effectiveness are important. That is, the effectiveness of parasitoids could be enhanced by using varieties of plants with desired characteristics; e.g., plants that release greater amounts of volatile compounds that attract parasitoids. An example of the importance of these compounds is found in cotton plants. The parasitoid *Campoletis sonorensis* is attracted to the odors emitted by the leaves and flowers of cotton plants (Elzen *et al.,* 1983, 1984). The volatiles involved in this attraction are terpenoids that are stored in specialized glands. Glandless cultivars developed through plant breeding do not produce these volatiles and are less attractive to the parasitoid (Elzen *et al.,* 1986). Loughrin *et al.* (1995) found that leaves of wild varieties of cotton produced greater amounts of terpenoids than commercial cultivars. Effective control might be achieved more readily by using plant varieties highly attractive to parasitoids. Thus, the activity of natural enemies can be enhanced and pest populations can be suppressed more effectively.

C.   Chemical Cues and Barriers to Developing Parasitoids

Typically, immature stages of parasitoids have intimate nutritional, biochemical, and physiological interactions with their hosts. Plant chemistry, in the form of nutrients and allelochemicals, can affect the survival, development, size, fecundity, longevity, and sex ratio of parasitoids (Vinson and lwantsch, 1980; Vinson and Barbosa, 1987; Godfray, 1994). Tomatine, an allelochemical present in tomato plants *(Lycopersicum esculentum* Mill.) fed upon by the noctuid *Heliothis (=Helicoverpa) zea* (Boddie), has detrimental effects on the time of development, survival, and adult size of the ichneumonid, *Hyposoter exiguae* (Viereck) (Campbell and Duffey, 1979, 1981; Duffey *et al.,* 1986). Similarly, nicotine (the prime alkaloid in tobacco *Nicotiana tobacum* L.) in the diet of the tobacco hornworm *Manduca sexta* (L.) decreases the survivorship of the parasitoid *Cotesia congregata* (Say) (Thurston and Fox, 1972; Barbosa and Saunders, 1985; Barbosa *et al.,* 1982, 1986; Barbosa, 1988; Kester and Barbosa, 1991).

Thus, plant chemistry can affect immature parasitoids directly when the toxic substances, sequestered by the host, have adverse affects on the developing parasitoid. Other examples have been reviewed in Barbosa and Saunders (1985), Boethel and Eikenbary (1986), and Godfray (1994).

D.   Indirect Effects of Plant Quality

Genetic and environmental variation in the nutritional quality of individual plants may affect parasitoids indirectly. Herbivores may vary in their suitability as hosts for parasitoids when feeding on different plants or plant parts. Variation in plant quality can interfere with an herbivore's immune response; e.g., by affecting its ability to encapsulate parasitoid eggs (Cheng, 1970; Rhoades, 1983; Benrey and Denno, 1997). The effectiveness of the encapsulation reaction depends on the physiological condition of the host, which can be weakened under conditions of poor plant nutrition or by the presence of toxins (Muldrew, 1953; Salt, 1956, 1964; van den Bosch, 1964; Vinson and Barbosa, 1987). When the aphid *Neomyzus circumflexus* (Buck.) was reared on poor diets higher parasitization was found because of its inability to encapsulate the parasitoid's eggs (El-Shazly, 1972). Similarly, encapsulation rates on the eggs of the parasitoid *Cotesia glomerata* (L.) by its host *Pieris rapae* were found to be higher on the host plant on which larval growth was high (Benrey and Denno, 1997).

Variation in the nutritional quality of plants or plant parts can affect parasitoids indirectly, when it leads to an increase in an herbivore's development time. In general, parasitoids attack only certain stages of their hosts. If these stages are prolonged, hosts will be available to parasitoids for longer periods of time (Feeny, 1976; Clancy and Price, 1987; Damman, 1987). Within-plant variation in development rate of larvae of the cabbage butterfly *Pieris rapae* resulted in higher levels of parasitism on slowly developing larvae (Benrey and Denno, 1997).

The effects of resistant plant cultivars on parasitoids have been discussed in several studies (see reviews by Bergman and Tingey, 1979; Boethel and Eikenbary, 1986; Hare, 1992; Bottrell *et al.* 1998; Cortesero and Lewis, in press). Plant traits that confer resistance against herbivores are known to have positive and negative effects on natural enemies (Obrycki and Tauber, 1984; Elzen *et al.*, 1986). Studies have shown that by feeding on certain plant cultivars, herbivores may be at a higher risk of being attacked by parasitoids and that parasitoid performance may vary among hosts on different plant cultivars (Hare and Luck, 1991; ldris and Grafius, 1997; Benrey *et al.*, submitted). Hare and Luck (1991) found that *Aphytis melinus* DeBach which emerged from California red scale *Aonidiella aurantii* (Maskell) reared on lemon produced more female progeny with a higher egg complement than parasitoids that emerged from hosts reared on grapefruit, orange, and mandarin. Similarly, studies with crucifers (ldris and Grafius, 1996; Benrey *et al.*, submitted) and beans (Benrey *et al.*, submitted) show that parasitoid performance varies among hosts reared on wild and cultivated plant species. Percentage parasitism by *Diadegma insulare* was higher and its development time shorter in diamondback moths feeding on cultivated crucifers than on wild species (ldris and Grafius, 1996). For *Stenocorse bruchivora* (Crawford), a parasitoid of the bean weevil *Zabrotes subfasciatus* (Boheman), cultivated plant species also provided the most suitable hosts. Parasitoid development time was shorter and survival was higher on hosts feeding on cultivated seeds (Benrey *et al.*, submitted). In both the studies with the diamondback moth and the bean weevil the herbivores preferred to oviposit on the cultivated plants, perhaps also as a result of their higher plant quality for the herbivore. Thus, the choice made by the herbivore will indirectly affect the parasitoid's performance via plant mediated host quality.

E. Physical Features of Plants

1. Morphology and plant structures

Plant morphology and plant structures can affect parasitoids in several ways. Factors such as surface area, foliar pubescence, glandular trichomes, waxy leaf surfaces, and leaf toughness can impede or facilitate parasitoid search (Woets and van Lenteren, 1976; Vinson, 1976; Obrycki and Tauber, 1984; Obrycki, 1986). A negative influence of plant pubescence on levels of parasitism has been shown in a number of crop plants such as tobacco (Elsey and Chaplin, 1978), potatoes (Obrycki *et al.,* 1983), cotton (Treacy *et al.,* 1986), and peanuts and soybeans (McAuslane *et al.,* 1995). In cotton for example, there is an inverse relationship between trichome density and parasitism by *Trichogramma pretiosum (*Treacy *et al.,* 1986).

Pubescent varieties of cucumbers cause reduced rates of parasitization by the whitefly parasitoid *Encarsia formosa* Gahan compared to nonpubescent varieties (van Lenteren *et al.,* 1977). Hulspas-Jordaan and van Lenteren (1978) showed that the walking speed of the same parasitoid was inversely related to the hairiness of the leaves. Hairy leaves also retained honeydew which resulted in increased parasitoid mortality and an increase in the time that the parasitoids spent grooming. Similarly, on tobacco plants, glandular trichomes reduce rates of parasitism of tobacco hornworm *Manduca sexta* eggs (Rabb and Bradley 1968) compared to that of eggs on a variety that lacked these type of trichomes (Elsey and Chaplin, 1978). In contrast, Casagrande and Haynes (1976) found no adverse effect of wheat leaf pubescence on three parasitoids of the cereal leaf beetle *Oulema melanopus* (L.).

2. Visual Cues

Few studies have investigated the importance of plant-associated visual cues in the process of host location by parasitoids. Nevertheless, some studies have shown that parasitoids respond to color and other visual cues (reviewed by Vinson, 1976; Wäckers and Lewis, 1994*). Microplitis croceipes* (Cresson), a parasitoid of the noctuid *Helicoverpa zea,* is capable of learning visual stimuli; concentrating its search on plant structures where it is more likely to find hosts (Wäckers and Lewis, 1994). Way and Murdie (1965) found that Brussels sprout cultivars with light green glossy leaves were more attractive to parasitoids than the darker green, waxy-leaf cultivars.

Wäckers and Lewis (1994) and Wäckers (1994) showed that parasitoids can show innate preferences for specific visual cues and they may be able to learn to associate these cues with hosts or food. Food-deprived parasitoids preferred to search and displayed intensified searching behavior on yellow targets compared to sugar-fed individuals, who concentrated their search on green leaf tissue (Wäckers, 1994). The attraction to the yellow color is apparently associated by the wasp with the presence of nectar, since yellow is the most common flower color (Wäckers, 1994). In contrast, other parasitoids do not appear to respond to color. For example, the number of visits by *Diadegma insulare* (a parasitoid of the diamondback moth) to yellow and white flowers of several crucifer species did not differ (Idris and Grafius, 1997).

F.   Morphology of Plants and Microclimate

Although microclimatic conditions such as humidity, light intensity, and temperature are known to have direct effects on the physiology of insects (Willmer, 1986), very little is known on the specific ways in which these conditions affect parasitoid efficiency. Differences in plant structure can affect the microclimate, and microclimates may vary between plants, affecting, in turn, the foraging behavior of parasitoids.

Differences in plants and plant parts can influence the "microclimate" immediately along the surfaces of plants. Large leaves, for example, tend to have warmer temperatures than smaller leaves (Willmer, 1986). There may also be variations in microclimatic conditions over a single leaf. For example, upper leaf surfaces tend to be warmer than lower leaf surfaces (Willmer, 1982). These differences may influence the herbivore's choice for oviposition sites and its dispersion patterns and as a consequence the probability of being found by a parasitoid. For example, on the same plant, young instars of *Pieris brassicae* (L.) tend to feed on the underside of the leaf (perhaps as a strategy to avoid desiccation), whereas late instars feed on the surface of the leaf (Willmer, 1980). The parasitoid of *P. brassicae, C. glomerata,* parasitizes only young instars (Laing and Levin, 1982). By concentrating their search on the underside of the leaf, female parasitoids could maximize the rate of encounter with suitable hosts. Thus, larval dispersion patterns that result from microclimatic conditions in the plant or the direct effects of microclimate may influence rates of parasitism.

Flowers by virtue of their shapes, like leaves, can also have very specialized microclimates. This may affect parasitoids directly or may facilitate or impede their search for hosts depending on whether both food sources and hosts co-occur at the same place, or time.

IV. CONCLUSIONS: CONSTRAINTS AND OPPORTUNITIES

It should be clear that many, if not all, of the factors discussed in this chapter may operate synergistically or antagonistically. Indeed, the traditional experimental approach has been to study single factors and then to study combinations of factors, often assuming an additive effect that may not be justified in these types of interactions. The interactions that we have described may be so complex that the most parsimonious way to study the influence of plants on parasitoids is to accept that there is a high degree of correlation among variables and to use experimental designs appropriate for multivariate statistical analyses. Such an approach allows researchers to ask what set of correlated variables best explain plant mediated host-parasitoid interactions rather than ascertaining one-on-one cause and effect relationships that are less relevant to these kinds of inquiries.

Tactics developed based on these interactions may enhance the survival and effectiveness of many parasitoids or a particular species. However, direct and indirect effects may not be equally useful as bases of conservation biological control tactics. We suggest that interactions that are subject to manipulation must enhance fitness and/or effectiveness directly and in a timely fashion. That is, we suggest that although the outcome of certain interactions (i.e., certain indirect effects) may be favorable to the survival/performance of parasitoids, they may not necessarily enhance their effectiveness (i.e., ability to control pests). Other (direct) effects are both pivotal to the fitness of parasitoids and have measurable impacts on levels of control of target pests. Nevertheless, the manipulation of plants in agroecosystems to favor parasitoids is essential if conservation biological control is to be successful.

REFERENCES

Agelopoulos, N. A., and Keller, M.A.. (1994). Plant-natural enemy association in the tritrophic system *Cotesia rubecula-Pieris* rapae-Brassicaceae (Cruciferae). 1: Sources of infochemicals. *J. Chem. Ecol.* 20, 1725-1734.

Alborn, H.T., Röse, U. S. R., and McAuslane, H. J. (1996). Systemic induction of feeding deterrents in cotton plants by feeding of *Spodoptera* spp. larvae. *J. Chem. Ecol.* 22, 919-932.

Alborn, H.T., Turlings, T. C. J., Jones, T. H., Stenhagen, G., Loughrin, J. H., and Tumlinson, J. H. (1997). An elicitor of plant volatiles identified from beet armyworm oral secretion. *Science* 276, 945-949.

Altieri, M. A., and Whitcomb, W. H. (1979). The potential use of weeds in the manipulation of beneficial insects. *Hort. Sci.* 14, 12-18.

Andow, D. A. (1986). Plant diversification and insect population control in agroecosystems. *In* "Some Aspects of Integrated Pest Management." (D. Pimentel, ed.), pp. 277-368. Department of Entomology, Cornell University. Ithaca, NY.

Andow, D. S., and Prokrym, D. R. (1990). Plant structural complexity and host-finding by a parasitoid. *Oecologia* 82, 162-165.

Arthur, A. P. (1962). Influence of host tree on abundance of *Itoplectis conquisitor,* a polyphagous parasite of the European pine shoot moth *Rhyacionia buoliana. Can. Entomol.* 94, 337-347.

Auger, J., Lecomte, C., and Thibout, E. (1987). A case of strict chemical dependence: Allium - the leek-moth - its entomophage. *In* "Insects - Plants." (V. Labeyrie, G. Fabres, and D. Lachaise, eds.), pp. 366-367. Dr. W. Junk Publishers. Dordrecht. The Netherlands.

Baehrecke, E. H., Williams, H. J., and Vinson, S. B. (1989). Electroantennogram responses of *Campoletis sonorensis* (Hymenoptera: Ichneumonidae) to chemicals in cotton *(Gossypium hirsutum* L.). *J. Chem. Ecol.* 15, 37-45.

Baker, H. G., Ogler, P. A., and Baker, I. (1978). A comparison of the amino acid compliments of floral and extrafloral nectars. *Bot. Gaz.* 139, 322-332.

Barbosa, P. (1988). Natural enemies and herbivore-plant interactions: Influence of plant allelochemicals and host specificity. *In* 'Novel Aspects of Insect-Plant Interactions." (P. Barbosa, and D. Letourneau, eds.), pp. 201-210. John Wiley and Sons. New York, NY.

Barbosa, P. (1993). Lepidoptera foraging on plants in agroecosystems: constraints and consequences. *In* "Ecological and Evolutionary Constraints of Caterpillars." (N. Stamp, and Casey, T., eds.), pp. 523-566. Chapman and Hall. New York, NY.

Barbosa, P., and Frongillo, E. A. (1979). Host parasitoid interactions affecting reproduction and oviposition by *Brachymeria intermedia. Entomophaga* 24, 139-143.

Barbosa, P., and Saunders, J. A. (1985). Plant allelochemicals: Linkages between herbivores and their natural enemies. *In* 'Chemically Mediated Interactions Between Plants and Other Organisms." (G. A. Cooper-Driver, T. Swain, and E. E. Conn, eds.), pp. 107-137. Recent Advances in Phytochemistry. Vol. 19 Plenum Press. New York, NY.

Barbosa, P., Frongillo, E. A., and Cranshaw, W. (1978). Orientation of field populations of *Brachymeria intermedia* to host and host-habitat cues. *Entomophaga* 23, 63-67.

Barbosa, P., Saunders, J. A., Kemper, J., Trumbule, R., Olechno, J., and Martinat, P. (1986). Plant allelochemicals and insect parasitoids: effects of nicotine on *Cotesia congregata* and *Hyposoter annulipes. J. Chem. Ecol.* 12, 1319-1328.

Barbosa, P., Saunders, J. A., and Waldvogel, M. (1982). Plant mediated variation in herbivore suitability and parasitoid fitness. *In* "Insect-Plant Relationships." H. Visser, and

A. K. Minks, (eds.), pp. 63-71. Proc. 5th Intern. Symp. Plant-Insect Relationships. Wageningen, Pudoc. Wageningen. The Netherlands.

Benrey, B., and Denno, R. F. (1997). The slow-growth-high-mortality hypothesis: a test using the cabbage butterfly. *Ecology.* 78, 897-999.

Benrey, B., Callejas, A., Rios, L., and Denno, R. F. The effects of plant domestication on the interaction between phytophagous insects and parasitoids. *Biol. Cont.* (in press).

Bergman, J. M., and Tingey W. M. (1979). Aspects of interaction between plant genotypes and biological control. *Bull. Entomol. Soc. Amer.* 25, 275-279.

Boethel, D. J., and Eikenbary, R. D. (1986). "Interactions of Plant Resistance and Parasitoids and Predators of Insects." Ellis Horwood, Ltd. New York, NY.

Bombosch, S. (1966). Distribution of enemies in different habitats during the plant growing season. *In* 'Ecology of Aphidophagous Insects." (I. Hodek, ed.), pp. 171-175. Academia, Prague, Czechoslovakia.

Bottrell, D. G., Barbosa, P., and Gould, F. (1998). Manipulating natural enemies by plant variety selection and modification: a realistic strategy? Annu. Rev. Entomol. (in press).

Bugg, G. D., Ellis, R. T., and Carlson, R. T. (1989). Ichneumonidae (Hymenoptera) using extrafloral nectar of faba bean *(Vicia faba* L., Fabaceae) in Massachusetts. *Biol. Agric. Hort.* 6, 107-114.

Butler, G. D., Loper, G. M., McGregor, S. E., Webster, J. L., and Margolis, H. (1972). Amounts and kinds of sugars in the nectars of cotton *(Gossypium* spp.) and the time of their secretion. *Agron. J.* 64, 364-368.

Campbell, B. C., and Duffey, S. S. (1979). Tomatine and parasitic wasps: potential incompatibility of plant antibiosis with biological control. *Science* 205, 700-702.

Campbell, B. C., and Duffey, S. S. (1981). Alleviation of a-tomatine-induced toxicity to the parasitoid *Hyposoter exiguae,* by phytosterols in the diet of the host, *Heliothis zea. J. Chem. Ecol.* 7, 927-946.

Carroll, D. P., and Hoyt, S. C. (1986). Hosts and habitats of parasitoids (Hymenoptera: Aphididae) implicated in biological control of apple aphid (Homoptera). *Environ. Entomol.* 15, 1171-1178.

Casagrande R. A., and Haynes, D. L. (1976). The impact of pubescent wheat on the population dynamics of the cereal leaf beetle. *Environ. Entomol.* 5, 153-159.

Cheng, L. (1970). Timing of attack of *Lypha dubia* Fall (Diptera: Tachinidae) on the winter moth, *Operophtera brumata* (L.) (Lepidoptera: Geometridae) as a factor affecting parasite success. *J. Anim. Ecol.* 39, 313-320.

Clancy, K. M., and Price, P. W. (1987). Rapid herbivore growth enhances enemy attack sublethal plant defenses remain a paradox. *Ecology* 68, 736-738.

Coll, M., and Bottrell, D. G. (1996). Movement of an insect parasitoid in simple and diverse plant assemblages. *Ecol. Entomol.* 21, 141-149.

Cortesero, A. M., and Lewis, W. J. (1997). Understanding and manipulating plant attributes to enhance biological control. *Biol. Cont.* (in press).

Cortesero, A. M., De Moraes, C. M., Stapel, J. O., Tumlinson, J. H., and Lewis, W. J. (1997). Comparisons and contrasts in host foraging strategies of two larval parasitoids with different degrees of host specificity. *J. Chem. Ecol.* 23, 1589-1606.

Cortesero, A.M., Monge, J. P., and Huignard, J. (1993). Response of the parasitoid *Eupelmus vuilleti* to the odours of the phytophagous host and its host plant in an olfactometer. *Entomol. Exp. Appl.* 69, 109-116.

Damman, H. (1987). Leaf quality and enemy avoidance by larvae of a pyralid moth. *Ecology* 68, 87-97.

Dicke, M. (1994). Local and systemic production of volatile herbivore-induced terpenoids: their role in plant-carnivore mutualism. *J. Plant Physiol.* 143, 465-472.

Dicke, M. (1995). Why do plants 'talk'? *Chemoecol.* 6, 159-165.

Dicke, M., and Sabelis, M. W. (1988). How plants obtain predatory mites as bodyguards. *Neth. J. Zool.* 38, 148-165.

Dicke, M. and Sabelis, M. W. (1989). Does it pay plants to advertize for bodyguards? Towards a cost-benefit analysis of induced synomone production. *In* "Causes and Consequences of Variation in Growth Rate and Productivity of Higher Plants." (H. Lambers, M. L. Cambridge, H. Konings, and T. L. Pons, eds.), pp. 341-358. SPB Academic Publ. NL-The Hague.

Dicke, M., van Baarien, P., Wessels, R., and Dilkman, H. (1993). Herbivory induces systemic production of plant volatile that attract predators of the herbivores: extraction of endogenous elicitor. *J. Chem. Ecol.* 19, 581-599.

Dicke, M., van Lenteren, J. C., Boskamp, G. J. F., and van Dongen-van Leeuwen, E. (1984). Chemical stimuli in host-habitat location by *Leptopilina heteroma* (Thomson) (Hymenoptera: Eucoilidae), a parasite of *Drosophila*. *J. Chem. Ecol.* 10, 695-712.

Ding, D., Swedenborg, P. D., and Jones, R. L. (1989). Chemical stimuli in host seeking behavior of *Macrocentrus grandii* Goidanich (Hymenoptera: Braconidae). *Ann. Entomol. Soc. Amer.* 82, 232-236.

Duffey, S. S., Bloem, K. A., and Campbell, B. C. (1986). Consequences of sequestration of plant natural products in plant-insect-parasitoid interactions. *In* "Interactions of Plant Resistance and Parasitoids and Predators of Insects." (D. J. Boethel, and R. D. Eikenbary, eds.), p. 31-60. John Wiley and Sons. New York, NY.

Eller, F. J., Tumlinson, J. H., and Lewis, W. J. (1988). Beneficial arthropod behavior mediated by airborne semiochemicals: source of volatile mediating the host-location flight behavior of *Microplitis croceipes* (Cresson) (Hymenoptera: Braconidae), a parasitoid of *Heliothis zea* (Boddie) (Lepidoptera: Noctuidae). *Environ. Entomol.* 17, 745-753.

El-Shazly, N. Z. (1972). Der einfluss aussere faktoren auf die hamocytare abwehrreaktion von *Neomyzus circumflexus* (Buck.) (Homoptera: Aphididae). *Z. angew. Entomol.* 70, 414-436.

Elsey, K. D., and Chaplin, J - F. (1978). Resistance of tobacco introduction 1112 to the tobacco budworm and green peach aphid. *J. Econ. Entomol.* 71, 723-725.

Elzen, G. W., Williams, H. J., and Vinson, S. B. (1983). Response by the parasitoid *Campoletis sonorensis* (Hymenoptera: Ichneumonidae) to synomones in plants: implications for host habitat location. *Environ. Entomol.* 12, 1873-1877.

Elzen, G. W., Williams, H. J., and Vinson, S. B. (1984). Isolation and identification of cotton synomones mediating searching behavior by parasitoid *Campoletis sonorensis*. *J. Chem. Ecol.* 10, 1251-1264.

Elzen, G. W., Williams, H. J., and Vinson, S. B. (1986). Wind tunnel flight response by hymenopterous parasitoid *Campoletis sonorensis* to cotton cultivars and lines. *Entomol. Exp. Appl.* 42, 285-289.

Faeth, S. H. (1990). Structural damage to oak leaves alters natural enemy attack on a leafminer. *Entomol. Exp. Appl.* 57, 57-63.

Feeny, P. (1976). Plant apparency and chemical defense. *In* "Biochemical Interaction between Plants and Insects." (J. W. Wallace, and R. L. Mansell, eds.), pp. 1-40. Plenum Press. New York, NY.

Felland, C. M. (1990). Habitat-specific parasitism of the stalk borer (Lepidoptera: Noctuiidae) in northern Ohio. *Environ. Entomol.* 19, 162-166.

Foster, M. A., and Ruesink, W.G. (1984). Influence of flowering weeds associated with reduced tillage in corn on a black cutworm (Lepidoptera: Noctuidae) parasitoid, *Meteorus rubens* (Nees von Esenbeck). *Environ. Entomol.* 13, 664-668.

Gardner, S. M., and Dixon, A. F. G. (1985). Plant structure and the foraging success *of Aphidius rhopalosiphi* (Hymenoptera: Aphididae). *Ecol. Entomol.* 10, 171-179.

Godfray, H. C. J. (1994). "Parasitoids: Behavior and Evolutionary Ecology." Princeton University Press. Princeton, NJ.

Hagen, K. S. (1986). Ecosystem analysis: Plant cultivars (HPR), entomophagous species and food supplements. *In* "Interactions of Plant Resistance and Parasitoids and Predators of Insects." (D. Boethel, and R. D. Eikenbary, eds.), pp. 151-198. Ellis Horwood Ltd. New York, NY.

Hagley, E. A. C., and Barber, D. R. (1992). Effect of food sources on the longevity and fecundity of *Pholetesor ornigis* (Weed) (Hymenoptera: Braconidae). *Can. Entomol.* 124, 341-346.

Hare, J. D. (1992). Effects of plant variation on herbivore-natural enemy interactions. *In* "'Plant Resistance to Herbivores and Pathogens: Ecology, Evolution and Genetics." (R. S. Fritz, and E. L. Simms, eds.), pp. 278-300. The University of Chicago Press. Chicago, IL.

Hare, J. D., and Luck, R. F. (1991). Indirect effects of citrus cultivars on life history parameters of a parasitic wasp. *Ecology* 72, 1576-1585.

Harrington, E. A., and Barbosa, P. (1978). Host-habitat influences on oviposition by *Parasetigena silvestris,* a larval parasitoid of the gypsy moth. *Environ. Entomol.* 7, 466-468.

Hill, D. S. (1987). "Agricultural Insects Pests of Temperate Regions and Their Control." Cambridge University Press. Cambridge, U.K.

Huispas-Jordaan, P. M., and van Lenteren, J. C. (1978). The relationship between host-plant leaf structure and parasitization efficiency of the parasitic wasp *Encarsia formosa* Gahan (Hymenoptera: Aphelinidae). *Med. Fac. Landbouww. Rijksuniv. Gent.* 43, 431-440.

Idris, A. B., and Grafius, E. (1995). Wildflowers as nectar sources for *Diadegma insulare* (Hymenoptera: Ichneumonidae), a parasitoid of diamondback moth, *Plutella xylostella* L. (Lepidoptera: Plutellidae). *J. Econ. Entomol.* 24, 1726-1735.

Idris, A. B., and Grafius, E. (1996). Effects of wild and cultivated host plants on oviposition, survival, and development of diamondback moth (Lepidoptera: Plutellidae) and its parasitoid *Diadegma insulare* (Hymenoptera: Ichneumonidae). *Environ. Entomol.* 25, 825-833.

Idris, A. B., and Grafius, E. (1997). Nectar-collecting behavior of *Diadegma insulare* (Hymenoptera: Ichneumonidae), a parasitoid of diamondback moth (Lepidoptera: Plutellidae). *Environ. Entomol.* 26, 114-120.

Jervis, M. A., and Kidd, N. A. C. (1996). Phytophagy. *In* "Insect Natural Enemies Practical Approaches in Their Study and Evaluation." (M. Jervis, and N. Kidd, eds.), pp 375-394. Chapman and Hall. London, U.K.

Jervis, M.A., Kidd, N. A. C., Fitton, M. G., Huddleston, T., and Dawah, H. A. (1993). Flower-visiting by hymenopteran parasitoids. *J. Nat. Hist.* 27, 67-105.

Kester, K. M., and Barbosa, P. (1991). Behavioral and ecological constraints imposed by plants on insect parasitoids: implications for biological control. *Biol. Cont.* 1, 94-106.

Kruess, A., and Tscharntke, T. (1994). Habitat fragmentation, species loss, and biological control. *Science* 264, 1581-1584.

Kulman, H. M., and Hodson, A. C. (1961). Parasites of the jack-pine budworm, *Choristoneura pinus,* with special reference to parasitism at particular stand locations. *J. Econ. Entomol.* 54, 221-224.

Laing, J. E., and Levin, D. B. (1982). A review of the biology and a bibliography of *Apanteles glomeratus* (L.) (Hymenoptera: Braconidae). *Biocont. News Inform.* 3, 7-23.

Landis, D. A., and Haas, M. J. (1992). Influence of landscape structure on abundance and within-field distribution of European corn borer (Lepidoptera: Pyralidae) larval parasitoids in Michigan. *Environ. Entomol.* 21, 409-416.

Lecomte, C., and Pouzat, J. (1985). Réponses électroantennographiques de deux parasitoïdes ichneumonides, *Diadromus pulchellus* et *D. collaris,* aux odeurs de végétaux, du phytophage-hôte *Acrolepiopsis assectella* et du partenaire sexuel. *Entomol. Exp. Appl.* 39, 295-306.

Lecomte, C., and Thibout, E. (1984). Etude olfactométrique de l'action de diverses substances allélochimiques végétales dans la recherche de l'hôte par *Diadromus pulchellus* (Hymenoptera, Ichneumonidae). *Entomol. Exp. Appl.* 35, 295-303.

Leius, K. (1963). Effects of pollen on fecundity and longevity of adult *Scambus buolianae* (Htg.) (Hymenoptera: Ichneumonidae). *Can. Entomol.* 95, 444-446.

Lewis, W. J., and Gross, H. R. (1989). Comparative studies on field performance of *Heliothis* larval parasitoids *Microplitis croceipes and Cardiochiles nigriceps at* varying densities and under selected host plant conditions. *Fla. Entomol.* 72, 6-14.

Lewis, W. J., and Takasu, K. (1990). Use of learned odours by a parasitic wasp in accordance with host food and needs. *Nature* 348, 635-636.

Lewis, W. J., Stapel, J. O., Cortesero, A. M., and Takasu, K. Understanding how parasitoids balance food and host needs: importance for successful biological control. *Biol. Cont.* (in press).

Loughrin, J. H., Manuklan, A., Heath, R. R., and Tumlinson, J. H. (1995). Volatiles emitted by different cotton varieties damaged by feeding beet armyworm larvae. *J. Chem. Ecol.* 21, 1217-1227.

Mattiacci, L., Dicke, M., and Posthumus, M. A. (1994). Induction of parasitoid attracting synomone in Brussels sprout plants by feeding of *Pieris brassicae* larvae: role of mechanical damage and herbivore elicitor. *J. Chem. Ecol.* 20, 2229-2247.

Mattiacci, L., Dicke, M., and Posthumus, M. A. (1995). β-Glucosidase: an elicitor of herbivore-induced plant odor that attracts host-searching parasitic wasps. *Proc. Natl. Acad. Sci.* 92, 2036-2040.

McAuslane, H. J., Vinson, S. B., and Williams, H J. (1990). Influence of host plant on mate location by parasitoid *Campoletis sonorensis* (Hymenoptera: Ichneumonidae). *Environ. Entomol.* 19, 26-31.

McAuslane, H. J., Johnson, F.A., Colvin, D. L., and Sojack, B. (1995). Influence of foliar pubescence on abundance and parasitism of *Bemisia argentifolii* (Homoptera: Aleyrodidae) on soybean and peanut. *Environ. Entomol.* 24, 1135-1143.

McGovern, W. L., and Cross, W. H. (1976). Effects of two cotton varieties on levels of boll weevil parasitism (Col.: Curculionidae). *Entomophaga* 21, 123-125.

Monteith, L. G. (1960). Influence of plants other than food plants on their host and host-finding by tachinid parasites. *Can. Entomol.* 92, 641-652.

Mueller, T. F. (1983). The effect of plants on the host relations of a specialist parasitoid *of Heliothis larvae. Entomol. Exp Appl.* 34, 78-84.

Muldrew, J. A. (1953). The natural immunity of the larch sawfly *(Pristiphora erichsonii* [Htg.1]) to the introduced parasite *Mesoleius tenthredinis* Morley, in Manitoba and Saskatchewan. *Can. J. Zool.* 31, 313-332.

Navasero, R. C., and Elzen, G. W. (1989). Responses of *Microplitis croceipes* to host and nonhost plants of *Heliothis virescens* in a wind tunnel. *Entomol. Exp. Appl.* 53, 57-63.

Need, J. T., and Burbutis, P. P. (1979). Searching efficiency of *Trichogramma nubilale. Environ. Entomol.* 8, 224-227.

Nishida, T. (1956). An experimental study of the ovipositional behavior of *Opius fletcheri* Silvestre (Hymenoptera: Braconidae) a parasite of the melon fly. *Proc. Hawaiian Entomol. Soc.* 16, 126-134.

Obrycki, J. J. (1986). The influence of foliar pubescence on entomophagous species. *In* "Interactions of Plant Resistance and Parasitoids and Predators of Insects." (D. J. Boethel, and R. D. Eikenbary, eds.), pp. 61-83. Ellis Horwood Ltd. New York, NY.

Obrycki, J. J., and Tauber, M. J. (1984). Natural enemy activity on glandular pubescent potato plants in greenhouse: an unreliable predictor of effects in the field. *Environ. Entomol.* 13, 679-683.

Obrycki, J. J., Tauber, M. J., and Tingey, W. M. (1983). Predator and parasitoid interaction with aphid-resistant potatoes to reduce aphid densities: a two-year field study. *J. Econ. Entomol.* 76, 456-462.

Olson, D. L., and Nechols, J. R. (1995). Effects of squash leaf trichome exudates and honey on adult feeding, survival, and fecundity of the squash bug (Heteroptera: Coreidae) egg parasitoid *Gryon pennsylvanicum* (Hymenoptera: Scelionidae). *Environ. Entomol.* 24, 454-458.

Pair, S. D., Laster, M. L., and Martin, D. F. (1982). Parasitoids of Heliothis spp. (Lepidoptera: Noctuidae) larvae in Mississippi associated with sesame interplantings in cotton, 1971 -1974: implications of host-habitat interactions. *Environ. Entomol.* 11, 509-512.

Paré, P. W., and Tumlinson, J. H. (1997a). Induced synthesis of plant volatiles. *Nature* 385, 30-31.

Paré, P. W., and Tumlinson, J. H. (1997b). *De novo* biosynthesis of volatile induced by insect herbivory in cotton plants. *Plant Physiol.* 141, 1161-1167.

Pimentel, D. (1961). An evaluation of insect resistance in broccoli, brussels sprouts, cabbage, collards, and kale. *J. Econ. Entomol. 54,* 156-158.

Powell, W. (1986). Enhancing parasitoid activity in crops. *In* "Insect Parasitoids." (J. Waage, and D. Greathead, eds.), pp. 319-340. Academic Press. London, U.K.

Prokopy, R. J. (1972). Responses of apple maggot flies to rectangles of different colors and shades. *Environ. Entomol.* 1, 720-726.

Pschorn-Walcher, H. (1980). Populations fluktuationen und parasitierung der birken-erienminiermotte *(Coleophora serratella* L.) In abhèngigkelt von der habitat-diversitat. *Z. Ang. Entomol.* 89, 63-81.

Rabb, R. L., and Bradley, J. R. (1968). The influence of host plants on parasitism of eggs of the tobacco hornworm. *J. Econ. Entomol.* 61, 1249-1252.

Ramachandran, R., and Norris, D. M. (1991). Volatiles mediating plant-herbivore-natural enemy interactions: electroantennogram responses of soybean looper, *Pseudoplusia includens,* and a parasitoid, *Microplitis demolitor,* to green leaf volatiles. *J. Chem. Ecol.* 17, 1665-1690.

Rhoades, D. F. (1983). Herbivore population dynamics and plant chemistry. *In* "Variable Plants and Herbivores in Natural and Managed Systems." (R. F. Denno, and M. S. McClure, eds.), pp. 155-204. Academic Press. New York, NY.

Roland, J., and Taylor, P. D. (1995). Herbivore-natural enemy interactions in fragmented and continuous forests. *In* "Population Dynamics. New Approaches and Synthesis." (N. Cappucino, and P. W. Price, eds.), pp. 195-208. Academic Press. San Diego, CA.

Röse, U. S. R., Lewis, W. J., and Tumlinson, J. H. Specificity of systemically released cotton volatiles as attractants for specialist and generalist parasitic wasps. J. *Chem. Ecol.* (in press).

Röse, U. S. R., Manukian, A., Heath, R. R, and Tumlinson, J. H. (1996). Volatile semiochemicals released from undamaged cotton leaves. *Plant Physiol.* 111, 487-495.

Röse U. S. R., Alborn, H. T., Makranczy, G., Lewis, W. J., and Tumlinson, J. H. (1997). Host recognition by the specialist endoparasitoid *Microplitis croceipes* (Hymenoptera: Braconidae): Role of host and plant-related volatiles. *J. Insect Beh.* 10, 313-330.

Rosenheim, J. A. (1987). Host location and exploitation by the cleptoparasitic wasp *Argochrysis armilla:* the role of learning (Hymenoptera: Chrysididae). *Beh. Ecol. Sociobiol.* 21, 401-406.

Russell, E. P. (1989). Enemies hypothesis: a review of the effect of vegetational diversity on predatory insects and parasitoids. *Environ. Entomol.* 18, 590-599.

Rutledge, C. E. (1997). A survey of identified kairomones and synomones used by insect parasitoids to locate and accept their hosts. *Chemoecol.* (in press).

Salt, G. (1956). Experimental studies in insect parasitism. IX. The reactions of a stick insect to an alien parasite. *Proc. Roy. Soc. Lond. B* 146, 93-108.

Salt, G. (1964). The ichneumonid parasite *Nemeritis canescens* (Gravenhorst) in relation to the wax moth *Galleria mellonella (L.) Trans. Roy. Entomol. Soc.* 116, 1-14.

Sato, Y., and Ohsaki, N. (1987). Host-habitat location by *Apanteles glomeratus* and effect of food-plant exposure on host-parasitism. *Ecol. Entomol.* 12, 291-297.

Schuster, M. F., and Calderon, M. (1986). Interactions of host plant resistant genotypes and beneficial insects in cotton ecosystems. *In* "Interactions of Plant Resistance and Parasitoids and Predators of Insects." (D. J. Boethel, and R. D. Eikenbary, eds.), pp. 84-97. Halsted Press. New York, NY.

Shahjahan, M., and Streams, F. A. (1973). Plant effects on host finding by *Leiophron pseudopallipes,* a parasitoid of the tarnished plant bug. *Environ. Entomol.* 2, 921-925.

Sheehan, W. (1986). Response by specialist and generalist natural enemies to agroecosystem diversification: a selective review. *Environ. Entomol.* 15, 456-461.

Skinner, M., Parker, B. L., Wallner, W. E., Odell, T. M., Howard, D., and Aleong, J. (1993). Parasitoids in low-level populations of *Lymantria dispar (Lep.:* Lymantriidae) in different forest physiographic zones. *Entomophaga* 38, 15-29.

Stapel, J. O., Cortesero, A. M., De Moraes, C. M., Tumlinson, J. H., and Lewis, W. J. (1997). Effects of extrafloral nectar, honeydew, and sucrose on searching behavior and efficiency of *Microplitis croceipes* (Hymenoptera: Braconidae) in cotton. *Environ. Entomol.* 26, 617-623.

Stary, P. (1970). "Biology of Aphid Parasites with Respect to Integrated Control." Junk. The Hague, The Netherlands.

Strong, D. R., Lawton, J. H., and Southwood, T. R. E. (1984). "Insects on Plants. Community Patterns and Mechanisms." Harvard University Press. Cambridge, MA.

Sugimoto, T., Shimono, Y., Hata, Y., Nakai, A., and Yahara, M. (1988). Foraging for patchily distributed leaf miners by the parasitoid *Dapsilarthra rufiventris* (Hymenoptera: Braconidae). III. Visual and acoustic cues to a close range patch location. *Appl. Entomol. Zool.* 23, 113-121.

Syme, P. D. (1975). The effects of flowers on the longevity and fecundity of two native parasites of the European pine shoot moth in Ontario. *Environ. Entomol.* 4, 337-346.

Takabayashi, J., Dicke, M., and Posthumus, M. A. (1991). Variation in composition of predator-attracting allelochemicals emitted by herbivore-infested plants: relative influences of plant and herbivore. *Chemoecol.* 2, 1-6.

Takabayashi, J., Takahashi, S., Dicke, M., and Posthumus, M. A. (1995). Developmental stage of herbivore *Pseudaletia separata* affects production of herbivore-induced synomone by corn plants. *J. Chem. Ecol.* 3, 273-287.

Thibout, E., Lecomte, C., and Auger, J. (1987). *Diadromus pulchellus:* search for a host and specificity. *In* "Les Insectes Parasitoides." (M. Bouletreau, and G. Bonnot, eds.), pp. 7-10. INRA, Les Colloques de l'INRA. No. 48. Institut National De La Recherche Agronomique. Paris, France.

Thurston, R., and Fox, P. M. (1972). Inhibition by nicotine of emergence of *Apanteles congregatus* from its host, the tobacco hornworm. *Ann. Entomol. Soc. Amer.* 65, 547-550.

Townes, H. (1960). Host selection patterns in some nearctic ichneumonids. *Proc. Intern. Congr. Entomol.* 2, 738-741.

Treacy, M. F., Benedict, J. H., Segers, J. C., Morrison, R. K., and Lopez, J. D. (1986). Role of cotton trichome density in bollworm (Lepidoptera: Noctuidae) egg parasitism. *Environ. Entomol.* 15, 365-368.

Turlings, T. C. J., and Tumlinson, J. H. (1991). Do parasitoids use herbivore-induced plant chemical defenses to locate hosts? *Fla. Entomol.* 74, 42-50.

Turlings C. J., and Tumlinson, J. H. (1992). Systemic chemical signaling by herbivore-injured corn. *Proc. Natl. Acad. Sci.* 89, 8399-8402.

Turlings, T. C. J., Tumlinson, J. H., and Lewis, W. J. (1990). Exploitation of herbivore-induced plant odors by host-seeking wasps. *Science 250,* 1251-1253.

Turlings, T. C. J., McCall, P. J., Alborn, H. T., and Tumlinson, J. H. (1993a). An elicitor in caterpillar oral secretions that induces corn seedlings to emit chemical signals attractive to parasitic wasps. *J. Chem. Ecol.* 19, 411-425.

Turlings, T. C. J., Tumlinson, J. H., Eller, F. J., and Lewis, W. J. (1991). Larval-damaged plants: source of volatile synomones that guide the parasitoid *Cotesia marginiventris* to the micro-habitat of its hosts. *Entomol. Exp. Appl.* 58, 75-82.

Turlings, T. C. J., Wäckers, F., Vet, L. E. M., Lewis, W. J., and Tumlinson, J. H. (1993b). Learning of host-finding cues by hymenopterous parasitoids. *In* "Insect Learning: Ecological and Evolutionary Perspectives." (D. R. Papaj, and A. Lewis, eds.), pp. 51-78. Chapman and Hall. New York, NY.

Turlings, T. C. J., Loughrin, J. H., Röse, U., McCall, P. J., Lewis, W. J., and Tumlinson, J. H. (1995). How caterpillar-damaged plants protect themselves by attracting parasitic wasps. *Proc. Natl. Acad. Sci.* 9, 4169-4174.

Udayagiri, S., and Jones, R. L. (1992). Flight behavior of *Macrocentrus grandii* Goidanich (Hymenoptera: Braconidae), a specialist parasitoid of European corn borer (Lepidoptera: Pyralidae): factors influencing response to corn volatiles. *Environ. Entomol.* 21, 1448-1456.

van den Bosch, R., (1964). Encapsulation of the eggs of *Bathyplectes curculionis* (Thomson) (Hymenoptera: Ichneumonidae) in larvae of *Hypera brunneipennis* (Boheman) and *Hypera postica* (Gyllenhal) (Coleoptera: Curculionidae). *J. Insect Pathol.* 6, 343-367.

van Lenteren, J. C. van, Woets, J., van der Poel, N., van Boxtel, W., van de Merendonk, S., van der Kamp, R., Nell, H., and Sevenster-van der Lelie, L. A. (1977). Biological control of the greenhouse whitefly *Trialeurodes vaporariorum* (Westwood) (Homoptera: Aleyrodidae) by *Encarsia formosa* Gahan (Hymenoptera: Aphelinidae) in Holland, an example of successful applied ecological research. *Med. Fac. Landbouww. Rijksuniv. Gent.* 42, 1333-1342.

Vet, L. E. M. (1983). Host-habitat location through olfactory cues by *Leptopilina clavipes* (Hartig) (Hym.: Eucoilidae), a parasitoid of fungivorous *Drosophila:* the influence of conditioning. *Neth. J. Zool.* 33, 225-248.

Vet, L. E. M. (1985). Olfactory microhabitat location in some eucoilid and alysiine species (Hymenoptera), larval parasitoids of Diptera. *Neth. J. Zool. 35,* 720-730.

Vet L. E. M., Janse, C., van Achterberg, C., and van Alphen, J. M. (1984). Microhabitat location and niche segregation in two sibling species of drosophilid parasitoids: *Asobara tabida* (Nees) and *A. rufescens* (Foerster) (Braconidae: Alysiinae). *Oecologia* 61, 182-188.

Vinson, S. B. (1975). Biochemical coevolution between parasitoids and their hosts. *In* "Evolutionary Strategies of Parasitic Insects and Mites." (P. W. Price, ed.), pp. 14-48. Plenum Press. New York, NY.

Vinson, S. B. (1976). Host selection by insect parasitoids. *Annu. Rev. Entomol.* 21, 109-133.

Vinson, S. B., and Barbosa, P. (1987). Interrelationships of nutritional ecology of parasitoids. *In* 'Nutritional Ecology of Insects, Mites, and Spiders and Related Invertebrates." (F. Slansky, and J. G. Rodriguez, eds.), pp. 673-695. John Wiley and Sons. New York, NY.

Vinson, S. B., and Iwantsch, G. F. (1980). Host suitability for insect parasitoids. *Annu. Rev. Entomol.* 25, 397-419.

Wäckers, F. L. (1994). The effect of food deprivation on the innate visual and olfactory preferences in the parasitoid *Cotesia rubecula. J. Insect Physiol.* 40, 641-649.

Wäckers, F. L., and Lewis, W. J. (1994). Olfactory and visual learning and their combined influence on host site location by *Microplitis croceipes. BioCont.* 4, 105-112.

Wäckers, F. L., and Swaans C. P. M. (1993). Finding floral nectar and honeydew in *Cotesia rubecula:* random or directed? *Proc. Exper. Appl. Entomol.* 4, 67-72.

Wardle, A. R. (1990). Learning of host microhabitat colour by *Exeristes roborator* (F.) (Hymenoptera: Ichneumonidae). *Anim. Behav.* 39, 914-923.

Wardle, A. R., and Borden, J. H. (1990). Learning of host microhabitat form by *Exeristes roborator* (F.) (Hymenoptera: Ichneumonidae). *J. Insect Beh.* 3, 251-263.

Way, M. J., and Murdie, G. (1965). An example of varietal variations in resistance of Brussels sprouts. *Ann. Appl. Biol.* 56, 326-328.

Weseloh, R. M. (1986). Host and microhabitat preferences of forest parasitic Hymenoptera: inferences from captures on colored sticky panels. *Environ. Entomol.* 15, 64-70.

Whitman, D. (1994). Plant bodyguards: Mutualistic interactions between plants and the third trophic level. *In* "Functional Dynamics of Phytophagous Insects." (T. N. Ananthakrishnan, ed.), pp. 133-159. Oxford and IBH Publishing. New Delhi.

Whitman, D. W., and Eller, F. J. (1990). Parasitic wasps orient to green leaf volatiles. *Chemoecol.* 1, 69-76.

Willmer, P. G. (1980). The effects of a fluctuating environment on the water relationships of larval Lepidoptera. *Ecol. Entomol.* 5, 271-292.

Willmer, P. G. (1982). Hygrothermal determinants of insect activities patterns: the Diptera of water lily leaves. *Ecol. Entomol.* 7, 221-231.

Willmer, P. G. (1986). Microclimatic effects on insects at the plant surface. *In* "Insects and the Plant Surface." (B. E. Juniper, and T. R, E. Southwood, eds.), pp. 65-80. Edward Arnold. London, U.K.

Woets, J., and van Lenteren, J. C. (1976). The parasite-host relationship between *Encarsia formosa* (Hymenoptera: Aphilinidae) and *Trialurodes vaporarium* (Homoptera: Aleyroydidae). VI. The influence of the host plant on the greenhouse whitefly and its parasite *Encarsia formosa. Proc. 3rd. Conf. Biol. Control Glasshouses O.I.L.B./S.R.O.P.* 76, 125-137.

CHAPTER
5

INFLUENCE OF PLANTS ON INVERTEBRATE
PREDATORS: IMPLICATIONS TO
CONSERVATION BIOLOGICAL CONTROL

P. Barbosa and S. D. Wratten

I. INTRODUCTION

It is evident that a diverse and abundant complex of invertebrate predators can be highly effective at regulating the density and/or degree of population fluctuation of phytophagous arthropods. Thus, in agroecosystems as well as unmanaged habitats, an effective predator complex can reduce the damage caused by phytophagous insect herbivores. However, the effectiveness of predators is influenced by many biotic factors, not the least of which is plants. The effects of plants are illustrated in examples of interactions between predators and both cultivated and uncultivated plants. However, the most dramatic illustrations of the impact of plants on predators often are observed in response to relatively small changes in different crop cultivars (Treacy *et al.* 1985, Scott *et al.,* 1988; Powell and Lamert, 1993; Rapusas *et al., 1996*).

In comparison to parasitoids and parasites, predaceous invertebrates are highly polyphagous. Thus, any habitat that contains a wide variety and abundance of prey provides optimal conditions for these species. Although exceptions exist, habitats that are structurally, biologically, or temporally diverse provide greater microhabitat diversity and a concomitant variety of potential prey. Thus, plant species diversity in agroecosystems or in refugia provide indirect benefits to predators by enhancing the likelihood that they will find prey, particularly during periods of scarcity. In addition, plant richness may have indirect benefits for biological control when predator populations build up prior to encounters with target pests.

Physical factors, both in terms of plant architecture (at any scale) or the microclimatic plants create, can have significant direct impacts on predaceous species. The physical structure of individual plants (or groups of plants) influence predator dispersion within a habitat (or dispersal between habitats), their ability to escape inter- and intraspecific predation, their functional response, etc. Plants may also have other direct effects on predators. The allelochemicals produced by plants for a variety of functions may also serve predators as cues to aid in finding prey or may have detrimental consequences if contacted directly or in prey tissues consumed. Recent research has shown that plants may directly affect predators because many species feed on plants during some portion of their lives (Alomar and Wiedermann, 1996).

In this chapter we provide an overview of the many ways in which plants in the habitat of predators can influence their survival, development, behavior, and interactions with prey. That is, we review the ways in which plants may mediate the predator-prey interactions that influence predator effectiveness as natural controls of pest species. We suggest that the influences of plants on invertebrate predators are both direct and indirect, and each type of influence implies different approaches to conservation biological control. Indeed, the likelihood of success of conservation biological control tactics may differ depending on whether they are based on direct or indirect interactions. That is, tactics based on direct or indirect interactions among plants, predators, and their prey may not be equally useful in conservation biological control.

The intent of this chapter is to highlight the various ways in which plants can influence invertebrate predators rather than to present a comprehensive review of all studies on this topic. The examples provided suggest the potential for one or more of the following options for the effective conservation of predaceous invertebrates: (1) altering the morphology and chemistry of crop varieties, (2) altering the growth form of crops (i.e., plant architecture, canopy structure, etc.), (3) diversifying the vegetation in and around crop plants, and (4) controlling and/or manipulating the size and distribution of crop patches and other landscape features (see Chapters 7 to 9).

What is known about the influence of plants on invertebrate predators provides insights into the actual and potential constraints and opportunities available for the conservation of predaceous biological control agents. The fact that any one plant trait affects a predator does not necessarily mean that it can be "engineered" into a tactic for conservation biological control. There are many constraints that influence our ability to translate our understanding of plant-predator-prey interactions into conservation biological control tactics. It may be possible to conserve invertebrate predators by altering the morphology and chemistry of crop plants, altering the growth form of crop plants, diversifying crop systems, and/or manipulating or altering landscape features in ways that favor predators. However, these and other tactics that might be implemented will likely succeed only if they provide conditions and/or resources that are equal to, or greater than, those in the agroecosystem or surrounding habitats.

II. PLANT MORPHOLOGY AND CHEMISTRY

A.   Plant Chemical Cues

Although predator responses to prey odors or to chemical odors associated with prey tissues, excretions, or secretions have received a great deal of attention (Nordlund *et al.*, 1977; Carter and Dixon, 1984; Sabelis *et al.1984*, Hågvar and Hofsvang, 1989; van der Meiracker *et al.*, 1990; McEwen *et al.*, 1993), less is known about the responses of predators to plant chemical volatiles. The one major exception to this generalization is the understanding we have recently gained from research on predaceous mites. Although the examples of non-mite predator responses to plant chemicals are not numerous, responses to plants, often in the absence of prey, have been clearly demonstrated (Kersten, 1969; Obata, 1986; Ponsonby and Copland, 1995; Kielty *et al.*, 1996; Rapusas *et al.*, 1996; and other references below). These data suggest that many other predators also may respond to plant odors, but their responses have yet to be investigated.

Some of the data suggesting the importance of plant volatiles include work on common bark beetle predators such as the clerid beetle *Thanasimus dubius (F.)* which exhibit oriented flight responses to compounds found in the tree hosts of their bark beetle prey (i.e., α- and β-pinene) (Mizell *et al.*, 1984), or other bark beetle predators which are stimulated to oviposit (Fitzgerald and Nagel, 1972; Baisier *et al.*, 1988) by tree compounds. Other terpenoids, such as the sesquiterpenoids caryophyllene and β-caryophyllene, have been found to be attractive to green lacewings such as *Chrysopa (sic) carnea* Stephens (Flint *et al.*, 1979). Interestingly, terpenoid compounds also are the plant chemical cues most commonly found to elicit responses of parasitoids and predaceous mites. Both damaged plants and undamaged plants emit so-called green leaf volatiles (i.e., six-carbon alcohols, aldehydes, and derivative esters) and relatively persistent terpenoids (Whitman and Eller, 1990; Steinberg *et al.*, 1993). Although direct evidence of mite predator response to injured plants is available, only a few studies suggest this type of interaction for non-mite predators (Greany and Hagen, 1981).

Predators also may respond to the chemical and physical profile of the entire plant. Some species of syrphid females oviposit almost as many eggs on aphid-free plants as they oviposit on aphid-infested plants. In addition, the preferential oviposition by syrphids such as *Platycheirus peltatus* (Meig.) on Brussels sprout plants compared to bean plants suggest that this predator is responding to a particular plant compound or blend rather than exhibiting a general response to plants (Chandler, 1968).

The response of predatory mites to plant volatiles has been the subject of rigorous and elegant experimentation in the past 10 years. Although the choice of plant upon which mites land is under the control of the wind currents that carry and deposit them, volatile plant kairomones play a role in the decision to stay on the plant or to make themselves subject to wind dispersal (Sabelis and Dicke, 1985). Predatory mites can distinguish between prey-infested and uninfested plants by olfactory responses to volatile kairomones emitted by injured leaves (Dicke, 1986, 1988, 1994; Dicke and Sabelis, 1988; Dicke *et al.,* 1990a). Three terpenoids as well as methyl salicylate have been found to attract the predatory mite *Phytoseilus persimilis* Athias-Henriot (Dicke *et al.,* 1990b). Experience with prey-infested leaves of one plant species leads to preferential selection by predators of the latter species over infested leaves of another species (Dicke *et al.,* 1990c). Thus, these natural enemies can discriminate different plant-herbivore combinations (Sabelis and van de Baan, 1983; Dicke, 1988; Dicke and Groeneveld, 1986; Sabelis and Dicke, 1985). This discrimination may result because chemical blends can differ among different plant-herbivore systems (Dicke *et al.,* 1990b; Takabayashi *et al.,* 1991).

In general, terpenoids are the most commonly produced volatiles by damaged plants to which carnivores (including parasitoids as well as predators) respond. Interestingly, some prey such as the spider mite *Tetranychus urticae* Koch are attracted by a volatile kairomone produced by undamaged plants but disperse in response to volatiles from plants infested by conspecifics (Dicke, 1986). Further, females placed on a leaf (and prevented from dispersing) and exposed to volatiles from infested cotton seedlings exhibited a reduced oviposition rate compared to those exposed to volatiles from uninfested plants (Bruin *et al.,* 1992).

In addition to behavior modification, plant chemicals can have a direct influence on the many predators that feed directly on plants (Alomar and Wiedemann, 1996). Similarly, the effects of plant chemicals may have an indirect effect on predators when plant compounds are taken into the body of prey. Development time of *Hippodamia convergens* Guerin was significantly increased and adult weight and survival were reduced when this beetle fed on *Schizaphis graminum* (Rondani) (aphids) reared on resistant sorghum cultivars, compared to when fed aphids reared on susceptible cultivars (Rice and Wilde, 1989). Although aphids on resistant cultivars may differ from aphids on susceptible cultivars in a variety of ways (e.g., size) it is likely the basis of the resistance to aphids (chemically based antibiosis) is also the basis for the negative effects observed among predators. Other similar examples

(Martos *et al.*, 1992) and related phenomena (Rothschild *et al.,* 1973; Pasteels, 1978; Moraes and McMurtry, 1987) have been reported (also see Chapter 9).

B. Influence of Plant Morphology

The morphology of plants also has the potential of influencing predators directly (see below) or indirectly by determining the type and abundance of herbivores on a plant (Banerjee, 1987). The differential responses of predators to even apparently small changes in plants are dramatically illustrated when they occur on different cultivars. The plant surfaces upon which predators search for prey can have favorable and detrimental effects on their ability to find and capture prey. Morphological plant traits such as the presence, type, and density of trichomes (Elsey, 1974; Belcher and Thurston, 1982; Obrycki and Tauber, 1984; Obrycki, 1986; Treacy *et al.,* 1987; Kauffman and Kennedy, 1989) and the presence or absence of epidermal wax (Way and Murdie, 1965; Shah, 1982; Carter *et al.*, 1984; Kareiva and Sahakian, 1990; Espelie *et al.,* 1991) can be important determinants of predator effectiveness. Oviposition of syrphids differs when presented with waxy or glossy varieties of Brussels sprout, regardless of the presence of aphid prey (Chandler, 1968). Similarly, predators such as *Hippodamia convergens* Guérin Méneville, *Orius insidiosus* (Say) and *Chrysoperla carnea* (Stephens) more effectively reduce populations of prey on cabbage cultivars with glossy surfaces compared to those on a "standard" normal wax variety (Eigenbrode *et al.,* 1995, 1996).

Predators may also assess plant morphology to enhance the likelihood of finding prey or to determine the parts of the plant that will support high densities of prey. First and second instars of the predator *Anthocoris confusus* (Reuter) spend about 34% of their time on the veins on the underside of sycamore leaves. This is where 78% of sycamore aphid are found (Dixon and Russel, 1972). *A. confusus* females also are able to distinguish between young and old plants and preferentially oviposit on the younger ones; which usually support the highest aphid prey densities (Evans, 1976). Similarly, predators such as coccinellids follow veins of leaves where prey are most likely to be found (Banks, 1957; Dixon, 1959, 1970; Wratten, 1973).

Plants may also develop structures that are utilized by, or enhance the survival and performance of, predators. Domatia, for example, appear to be important shelters for predatory mites and their existence on plants may result in higher mite densities (Walter and O'Dowd, 1992; Karban *et al.*, 1995; Walter, 1996; Agrawal and Karban, 1997; also see Chapters 9 and 17 (Section IV,B)).

The influence of plant morphology varies, as one might expect, with species, growth stage, plant type, and other factors. For example, predator age may determine the impact of plant morphology. First instar *Chrysopa rufilabris* destroy eight and four times more eggs on the smooth leaf of a glabrous variety than on hirsute (moderately hairy) and pilose (very hairy) varieties of cotton, respectively. In contrast, the eggs destroyed by second instars on the smooth variety represented a 2½ and a two fold increase over that found on hirsute and pilose varieties, respectively (Treacy *et al.,* 1985).

Although discussions such as this compartmentalize the responses of predators, clearly, predators may and probably do simultaneously respond to several types of plant cues. Thus, species such as the anthocorid discussed above may also respond to plant chemicals. Indeed, other anthocorid species do (Drukker *et al.,* 1995). *Anthocoris nemorum* L. responds to chemical signals from leaves of sting nettle, goat willow, and tomato (Dwumfour, 1992). The overall effectiveness of any plant trait in enhancing the survival and performance of predators may also depend on the influence of the same trait(s) on pest species. Thus, the usefulness of plant traits in conservation biological control is likely to be dependent on whether the same traits have a detrimental or positive effect on pest species (Treacy *et al.,* 1985; Kartohardjono and Heinrichs, 1984).

III. PLANT ARCHITECTURE AND CANOPY STRUCTURE

Current evidence suggests that for many predators the search for prey is a process which rarely involves chemical odor cues and relies on so-called "random" search. For those predators that truly search randomly, plant features may facilitate or hinder search. For example, the density of the plant canopy throughout which prey are scattered, the structure of leaves over which predators must forage, and the seasonal changes in leaf and plant size all may produce significantly different levels of prey availability or accessability. O'Neil and Stimac (1988) noted that predators of the velvetbean caterpillar *Anticarsia gemmatalis* maintain a constant per capita rate of attack even though soybean leaf area increases through time (i.e., the size of a predator's searching universe increases). They concluded that the predators must be compensating for leaf area changes by searching more area. Not all predators may have this ability and thus their effectiveness may decrease as the plant canopy increases.

The architectural complexity of the plant/habitat may determine predator species diversity. Hatley and MacMahon (1980) found that the more structurally complex the habitat (experimentally created by clipping shrub foliage to reduce foliage density or tying foliage together to increase foliage density) the greater the spider species densities and diversity. Interference created by the physical traits of plants may be sufficient to alter the movements of herbivores (Risch, 1980) and natural enemies (Coll and Bottrell, 1996). Even in the absence of plant or canopy structure which impedes movement, differences in the structure of plants (i.e., their architecture) and thus in plant patch canopy structure can have a major impact on predators. Grevstad and Klepetka (1992) found that the rates of aphid prey consumption by coccinellids (such as *H. convergens, H. variegata* (Goeaze), and C. *septempunctata* L.) differed on crucifer cultivars that differed in leaf structure, leaf surface texture, and stem and petiole architecture. These plant differences had significant influences on movement and thus on foraging time, foraging site selection, rate of prey encounter, and ratio of aphids killed to aphids contacted. Effectiveness of predators may be reduced simply because of the inability of predators to hang on to the plant (Juniper and Southwood, 1986; Karieva and Sahakian, 1990).

IV. PLANT SPECIES DIVERSITY IN AND AROUND AGROECOSYSTEMS

A.   Consequences of Plant Species Diversity

A number of studies have shown that for some predators, the number and type of plant species in their habitat can influence abundance and predation rate (Andow and Risch, 1985; Nentwig, 1988; Bugg and Ellis, 1990). Whereas in some cases the type of plant species appears unimportant (Bugg and Dutcher, 1989; Bugg *et al.,* 1990), in other circumstances the particular plant species present determines if species diversity has an effect or not (Robinson *et al.,* 1972). However, the reason for an observed influence is often less than obvious, or there may be confounding factors which make drawing unambiguous conclusions impossible. The density of plants in polycultures often confound the influence of plant species diversity. For example, foraging rate per individual *Coleomegilla maculata* (De Geer) is significantly reduced by increasing density, but not diversity of plants in its habitat. Thus, intercropping may be observed to decrease predation rate and abundance of this and other predators, since polycultures are often more dense than their respective monocultures (Risch *et al.,* 1982; Gold *et al.,* 1989).

Indeed, plant density may be important even in comparisons between different crop monocultures. For example, Culin and Yeargan (1983) speculated that the conditions created by the density of the vegetation in alfalfa fields compared to soybeans was responsible for differences in the number of individuals and species richness of ground surface spiders. In contrast, specific plant species traits may be important. Certain predators such as *Geocoris* spp., coccinellids, and ants may be more or less active in certain crop agroecosystems (Nordlund *et al.*, 1984) but whether this is a direct effect (e.g., due to the presence of extrafloral nectaries) or an indirect effect (e.g., due to the presence of preferred, susceptible, or abundant prey) is less than clear.

Plant diversity may be a critical feature of a predator's habitat because of the phytophagous habits of many predators and their requirement for nutrients supplied by plants. Pollen is a particularly important plant resource for a variety of predators including coccinellids and mites (Huffaker *et al.*, 1970; Hagen, 1976; Isenhour and Yeargan, 1981; Chapter 17 (Section IV, A, 2)). In refugia or unmanipulated habitats adjacent to agroecosystems the variety of plant species often influence the degree to which predators will have abundant and long-term sources of pollen, nectar, and other plant resources. Some species of coccinellids can complete their life cycles with pollen as their sole food (Smith, 1960). Indeed, high predator densities are observed during flowering periods of crops (Huffaker *et al.*, 1970; Hagen, 1976; Isenhour, 1977; Groden *et al.*, 1990; Coll and Bottrell, 1992). Plant nectar also is an important nutrient for predators. Reductions in predator density have been observed in nectariless varieties of crops such as cotton (Schuster *et al.*, 1976; Adjei-Maafo and Wilson, 1983; Scott *et al.*, 1988). The importance of plants as sources of nutrients vary among predator species. For example, *Orius* species appear to be more dependent on these nutrients than other predators such as *Nabis or Geocoris* species (Kiman and Yeargan, 1985*)*. *Orius insidiosus* (Say) responds to volatiles of corn silk, where the eggs of its prey are often laid (Reid and Lampman, 1989).

The importance of these nutrients may also vary with the circumstances in which predators find themselves. They may, for example, be extremely important during periods of prey scarcity. The need for pollen (of a certain type or at a particular point in time) may favor the colonization of monocultures by predators over that of polycultures or refugia even when the latter has a high species diversity (Wetzler and Risch, 1984). Andow and Risch (1985) noted that a higher density of evenly spaced food rewards (both aphids and pollen) in a corn monoculture resulted in decreased predator emigration and greater abundance.

Thus, the potential impact of plant diversification on predators may depend on the particular effects of the plants and how predators respond to them. Nevertheless, it is clear that species-rich refugia can enhance the availability of natural enemies. In New Zealand, shelterbelts can hold more than 1000 spider individuals per m^2 whereas in adjacent fields typical densities are less than 100/m^2 (A. McLachlan and S. Wratten, pers. commun.). Why such a phenomenon occurs and the circumstances under which it leads to control of target pests still requires further study.

B.   Consequences of Differences in Plant Quality

Although predators, particularly young predator stages, may feed on plants merely to take in water, there is also the potential of imbibing plant allelochemicals which may have an impact on their survival, behavior, development, and fecundity. In some cases plant allelochemicals have little or no effect on predators. For example, the incorporation of rutin (a widespread allelochemical known to detrimentally affect some herbivores) into diet had no significant effect on *Geocoris punctipes* (Cohen and Urias, 1988) and a minor effect on *Podisus maculiventris* (Say) (Stamp *et al.,* 1991). Predators such as *G. punctipes* may be adapted to the chemistry of some of the host plants of their prey. On the other hand, predators such as *G. punctipes* may be detrimentally affected by feeding on certain cultivars of a crop (e.g., soybean) compared to other cultivars (Rogers and Sullivan, 1986). The latter results are consistent with the observation that *G. punctipes* feeds and is abundant on certain plants more so than other plant species (Naranjo and Stimac, 1985). Thus, conservation of biological control agents by enhancing plant species diversity may be a valuable tactic but its success may depend on the blend of species present and their effects on key predators as well on target pest species.

Plant allelochemicals can cause predators to reject prey (Smiley *et al.,* 1985; Bowers and Larin, 1989). Similarly, plant allelochemicals are sometimes sequestered in prey tissues taken up by predators when they eat prey (Rothschild *et al.,* 1973) and may cause detrimental changes in the behavior and survival, development, and fecundity of predators (Pasteels, 1978). In contrast, predators may benefit by feeding on prey which have been reared on specific host plants. Predatory *Phytoseilus persimilis* Athias-Henriot fed adult female *Tetranychus urticae* Koch reared on lima bean *(Phaseolus vulgaris* L.) were significantly heavier than those reared on nightshade *(Solanum douglasii* Dunal) (Moraes and McMurtry, 1987). In some interactions specific changes in the nutrition of prey have been found to effect predators. Consumption of aphids reared on a diet with reduced sugar concentration or without iron resulted in diminished weight and fecundity in the predator *Aphidoletes aphidimyza* (Rond.) (Kuo, H.-L., 1982).

Although the size and distribution of plant patches may be important to predators, some would suggest that the quality of the plants in the patch may have an overwhelming influence on the interaction between predator and prey. Studies such as that of Haggstrom and Larsson (1995) have asked whether larvae feeding on low quality plants are preyed upon to a greater extent than larvae feeding on high quality plants. They concluded that larvae feeding on unsuitable food plants suffer from more predation than conspecifics on suitable plants because those on unsuitable plants are exposed to predation for a longer period of time. However, the daily predation rate was higher on low quality plants, which may have been a consequence of lower levels of survival and smaller size of prey on the unsuitable host plant. In either case, it is obvious that the indirect effects of plants can be as important and consequential as their direct effects.

The importance of the structure of patches on predator-prey interactions is dependent on the particular plant(s), herbivores, and predators interacting in the patch. However, it is clear that with the appropriate mix of species, the existence of plant refugia, in or near agroecosystems, can have a significant impact on the abundance (Nentwig, 1988; Bugg and Ellis, 1990) and effectiveness of predators (see Chapters 8 and 9). Similarly, the size of patches (or degree of habitat patchiness), often referred to as fragmentation. has a significant effect on predators. However, whether the influence of increasing patchiness, for example, results in a greater or lesser effectiveness of predators depends on the predator and host species involved and the response of the predator and/or the prey to habitat fragmentation (see Karieva, 1987; Chapters 2, 8, and 9).

V. CONCLUSIONS

The responses of both predators and parasitoids to plants are sophisticated and complex and the manipulation of (crop and noncrop) plants, or plant patches, without a detailed understanding of predator-prey-plant interactions can produce counterproductive or unexpected results. We have a more detailed understanding of the responses of parasitoids to plants and host traits (such as moth scale kairomones) than we have for many predators (Beevers et al., 1981). Yet, it has become clear that pest control as a result of the manipulation of parasitoids or widespread application of plant or host compounds is not simple or straightforward (Lewis et al., 1975a,b; Lewis et al., 1979; Gardner and van Lenteren, 1986).

A further constraint on the use of conservation tactics for predators is that tactics (such as the provision of refuges) may provide conditions that are so favorable for natural enemies that they fail to colonize crops and impose the desired mortality of pest species. The phenomenon of refuges acting as natural enemy sinks rather than as sources (implying movement to crops) has been noted in several studies (Perrin, 1975; Bugg et al., 1987; Kemp and Barrett, 1989). In his study of the influence of common knotweed on predators such as *Geocoris* spp. (Bugg *et al.,* 1987) concluded that even though this weed provides resources like nectar in a field situation, it also provides a sufficient abundance of alternate prey to preclude movement of predators from weed species to crop species. Therein lies the dilemma.

A related constraint is that a tactic for the conservation of predators must provide stimuli that are sufficiently strong to effectively compete with similar stimuli occurring in surrounding habitats or other agroecosystems. Further, it must be equal to or more effective than competing habitats at the "appropriate" time. However, its influence can not occur too frequently or with such intensity that it results in habituation or "resistance" to the resources provided or to manipulated cues. A balance must be developed between the conservation biological control tactic and extant stimuli or resources, such that natural enemies tend to move to areas in which they are needed, when they are needed.

Having said all that, it is also quite clear that without understanding how plants affect the predators of phytophagous pest species we are unlikely to develop effective tactics for conservation biological control. Chapters 8, 9, and 15 to 17 provide excellent examples of such tactics.

REFERENCES

Adjei-Maafo, I. K., and Wilson, L. T. (1983). Factors affecting the relative abundance of arthropods on nectaried and nectariless cotton. *Environ. Entomol.* 12, 349-352.

Agrawal, A. A., and Karban, R. (1997). Domatia mediate plant-arthropod mutualism. *Nautre* 387, 562-563.

Alomar, O., and Wiedenmann, R. N. (1996). "Zoophytophagous Heteroptera: Implications for Life History and Integrated Pest Management." Proc. Thomas Say Publ. Entomol. Entomol. Soc. Amer. Lanham, MD.

Andow, D. A., and Risch, S. J. (1985). Predation in diversified agroecosystems: relations between a coccinellid predator *Coleomegilla maculata* and its food. *J. Appl. Ecol.* 22, 357-372.

Baisier, M., Grégoire, J.-C., Delinte, K., and Bonnard, 0. (1988). The role of spruce monoterpene derivatives as oviposition stimuli for *Rhizophagous grandis, a* predator of the bark beetle, *Dendroctonus micans. In* "Mechanisms of Woody Plant Defenses Against Insects. Search for Pattern." (W. J. Mattson, J. Levieux, and C. Benard-Dagan, eds.), pp. 359-368. Springer Verlag, New York, NY.

Banerjee, B. (1987). Can leaf aspect affect herbivory? A case study with tea. *Ecology* 68, 839-843.

Banks, C. J. (1957). The behavior of individual coccinellid larvae on plants. *Brit. J. Anim. Beh.* 5, 12-24.

Beevers, M., Lewis, W. J., Gross, H. R., and Nordlund, D. A. (1981). Kairomones and their use for management of entomophagous insects. X. Laboratory studies on manipulation of host finding behavior of *Trichogramma pretiosum* with a kairomone extracted from *Heliothis zea* (Boddie) moth scales. *J. Chem. Ecol.* 7, 635-648.

Belcher, D. W., and Thurston, R. (1982). Inhibition of movement of larvae of the convergent lady beetle by leaf trichomes of tobacco. *Environ. Entomol.* 11, 91-94.

Bowers, M. D., and Larin, Z. (1989). Acquired chemical defense in the lycaenid butterfly, *Eumaeus atala. J. Chem. Ecol.* 15, 133-146.

Bruin, J., Dicke, M., and Sabelis, M. W. (1992). Plants are better protected against spider mites after exposure to volatiles from infested conspecifics. *Experientia* 48, 525-529.

Bugg, R. L., and Dutcher, J. D. (1989). Warm-season cover crops for pecan orchards: horticultural and entomological implications. *Biol. Agric. Hort.* 6, 123-148.

Bugg, R. L., and Ellis, R. T. (1990). Insects associated with cover crops in Massachusetts. *Biol. Agric. Hort.* 7, 47-68.

Bugg, R. L., Ehler, L. E., and Wilson, L. T. (1987). Effect of common knotweed *(Polygonum aviculare)* on abundance and efficiency of insect predators of crop pests. *Hilgardia* 55, 1-53.

Bugg, R. L., Phatak, S. C., and Dutcher, J. D. (1990). Insects associated with cool-season cover crops in southern Georgia: implications for pest control in truck-farm and pecan agroecosystems. *Biol. Agric. Hort.* 7, 17-45.

Carter, M. C., and Dixon, A. F. G. (1984). Honeydew: an arrestant stimulus for coccinellids. *Ecol. Entomol.* 9, 383-387.

Carter, M. C., Sutherland, D., Dixon, A. F. G. (1984). Plant structure and the searching efficiency of coccinellid larvae. *Oecologia* 63, 394-397.

Chandler, A. E. F. (1968). Some host-plant factors affecting oviposition by aphidophagous Syrphidae (Diptera). *Ann. Appl. Biol.* 61, 415-423.

Cohen, A. C., and Urias, N. M. (1988). Food utilization and egestion rates of the predator *Geocoris punctipes* (Hemiptera: Heteroptera) fed artificial diets with rutin. *J. Entomol. Sci.* 23, 174-179.

Coll, M., and Bottrell, D. G. (1992). Mortality of European corn borer larvae by natural enemies in different corn microhabitats. *Biol. Cont.* 2, 95-103.

Coll, M., and Bottrell, D. G. (1996). Movement of an insect parasitoid in simple and diverse plant assemblages. *Ecol. Entomol.* 21, 141-149.

Culin, J. D., and Yeargan, K. V. (1983). Comparative study of spider communities in alfalfa and soybean ecosystems: ground-surface spiders. *Ann. Entomol. Soc. Amer.* 76, 832-838.

Dicke, M. (1986). Volatile spider-mite pheromone and host-plant kairomone, involved in spaced-out gregarious in the spider mite *Tetranychus urticae. Physiol. Entomol.* 11, 251-262.

Dicke, M. (1988). Prey preference of the phytoseiid mite *Typholodromus pyri:* 1. Response to volatile kairomones. *Exp. Appl. Acarol. 4,* 1-13.

Dicke, M. (1994). Local and systemic production of volatile herbivore-induced terpenoids: their role in plant-carnivore mutualism. *J. Plant Physiol.* 143, 465-472.

Dicke, M., and Groeneveld, A. (1986). Hierarchical structure in kairomone preference of the predatory mite *Amblyseius potentillae:* dietary component indispensable for diapause induction affects prey location behavior. *Ecol.* Entomol. 11, 131-138.

Dicke, M., and Sabelis, M. W. (1988). How plants obtain predatory mites as bodyguards. *Neth. J. Zool.* 38, 148-165.

Dicke, M., Sabelis, M. W., Takabayashi, J., Bruin, J., and Posthumus, M. A. (1990a). Plant strategies of manipulating predator-prey interactions through allelochemicals: prospects for application in pest control. *J. Chem. Ecol.* 16, 3091-3118.

Dicke, M., van Beek, T. A., Posthumus, M. A., BenDom, N., van Bokhoven, H., and De Groot, A. E. (1990b). Isolation and identification of volatile kairomone that affects acarine predator-prey interactions. Involvement of host plant in its production. *J. Chem. Ecol.* 16, 381-396.

Dicke, M., van der Maas, K. J., Takabayashi, J., and Vet, L. E. M. (1990c). Learning affects responses to volatile allelochemicals by predatory mites. *Proc. Exp. Appl. Entomol.* 1, 31-36

Dixon, A. F. G. (1959). An experimental study of the searching behavior of the predatory beetle *Adalia decempunctata (L.). J. Anim. Ecol.* 28, 259-281.

Dixon, A. F. G. (1970). Factors limiting the effectiveness of the coccinellid beetle, *Adalia bipunctata* (L.), as a predator of the sycamore aphid, *Drepanosiphum platanoides* (Schr.). *J. Anim. Ecol.* 39, 739-751.

Dixon, A. F. G., and Russel, R. J. (1972). The effectiveness of *Anthocorus nemorum* and *A. confusus* as predators of the sycamore aphid, *Drepanosiphum platanoides.* II. Searching behaviour and the incidence of predation in the field. *Entomol. Exp. Appl.* 15, 35-50.

Drukker, B., Scutareanu, P., and Sabelis, M. W. (1995). Do anthocorid predators respond to synomones from psylla-infested pear trees under field conditions? *Entomol. Exp. Appl.* 77, 193-203.

Dwumfour, E. F. (1992). Volatile substances evoking orientation in the predatory flowerbug *Antocoris nemorum* (Heteroptera: Anthocoridae). *Bull. Entomol. Res.* 82, 465-469.

Eigenbrode, S. D., Moodie, S., and Castagnola, T. (1995). Predators mediate host plant resistance to a phytophagous pest in cabbage with glossy leaf wax. *Entomol. Exp. Appl.* 77, 335-342.

Eigenbrode, S. D., Castagnola, T., Roux, M.-B., and Steljes, L. (1996). Mobility of three generalist predators is greater on cabbage with glossy leaf wax than on cabbage with a wax bloom. *Entomol. Exp. Appl.* 81, 335-343.

Elsey, K. D. (1974). Influence of plant host on searching speed of two predators. *Entomophaga* 19, 3-6.

Espelie, K. E., Bernays, E. A., and Brown, J. J. (1991). Plant and insect cuticular lipids serve as behavioral cues for insects. *Arch. Insect Biochem. Physiol.* 17, 223-233.

Evans, H. F. (1976). The effect of prey density and host plant characteristics on oviposition and fertility in *Anthocorus confusus* (Reuter). Ecol. Entomol. 1, 157-161.

Fitzgerald, T. D., and Nagel, W. P. (1972). Oviposition and larval bark-surface orientation of *Medetera aldrichii* (Diptera: Dolichopodidae): response to a prey-liberated plant terpene. *Ann. Entomol Soc. Amer.* 65, 328-330.

Flint, H. M., Salter, S. S., and Walters, S. (1979). Caryophellene: an attractant for the green lacewing. *Environ. Entomol.* 8, 1123-1125.

Gardner, S. M., and van Lenteren, J. C. (1986). Characterization of the arrestment responses of *Trichogramma evanescens. Oecologia* 68, 265-270.

Gold, C. S., Altieri, M. A. and Bellotti, A. C. (1989). The effects of intercropping and mixed varieties of predators and parasitoids of cassava whiteflies (Hemiptera: Aleyrodidae) in Columbia. *Bull. Entomol. Res.* 79, 115-121.

Greany, P. D., and Hagen, K. S. (1981). Prey selection. *In* "Semiochemicals: Their Role in Pest Control." (D. A. Nordlund, R. L. Jones, and W. J. Lewis, eds.), pp. 121-135. Wiley Interscience. New York, NY.

Grevstad, F. S., and Klepetka, B. W. (1992). The influence of plant architecture on the foraging efficiencies of a suite of ladybird beetles feeding on aphids. *Oecologia* 92, 399-404.

Groden, E., Drummond, F. A., Casagrande, R. A., and Haynes, D. L. (1990). *Coleomegilla maculata* (Coleoptera: Coccinellidae): its predation upon the Colorado potato beetle (Coleoptera: Chrysomelidae) and its incidence in potatoes and surrounding crops. *J. Econ. Entomol.* 83, 1306-1315.

Hagen, D. S. (1976). Role of nutrition in insect management. *Proc. Tall Timbers Conf. Ecol. Anim. Contr. by Habitat Manag.* 6, 261-262.

Haggström and Larsson, S. (1995). Slow larval growth on a suboptimal willow results in high predation mortality in the leaf beetle *Galerucella lineola. Oecologia* 104, 308-315.

Håvar, E. B., and Hofsvang, T. (1989). Effect of honeydew and hosts on plant colonization by the aphid parasitoid *Ephedrus cerasicola. Entomophaga* 34, 495-501.

Hatley, C. L., and MacMahon, J. A. (1980). Spider community organization: seasonal variation and the role of vegetation architecture. *Environ. Entomol.* 9, 632-639.

Huffaker, C. B., van de Vrie, M., and McMurtry, J. A. (1970). Tetranychid populations and their possible control by predators: an evaluation. *Hilgardia* 40, 391-458.

Isenhour, D. J. (1977). Seasonal fluctuations of *Orius insidiosus* and Thysanoptera in adjacent soybeans and corn in Missouri. M.S. Thesis. University of Missouri.

Isenhour, D. J., and Yeargan, K. V. (1981). Effect of crop phenology on *Orius insidiosus* populations on strip-cropped soybean and corn. *J. Ga. Entomol. Soc.* 16, 310-322.

Juniper, B., and Southwood, T. R. E. (1986). "Insects and the Plant Surface." Edward Arnold Ltd. London, U.K.

Karban, R., English-Loeb, G., Walker, M. A., and Thaler, J. (1995). Abundance of phytoseiid mites on *Vitis* species: effects of leaf hairs, domatia, prey abundance, and plant phenology. *Exp. Appl. Acarol.*, 19, 189-197.

Karieva, P. M. (1987). Habitat fragmentation and the stability of predator-prey interactions. *Nature* 326, 388-390.

Karieva, P. M., and Sahakian, R. (1990). Tritrophic effects of a simple architectural mutation in pea plants. *Nature* 345, 433-434.

Kartohardjono, A., and Heinrichs, E. A. (1984). Populations of the brown planthopper, *Nilaparvata lugens* (Stål) (Homoptera: Delphacidae), and its predators on rice varieties with different levels of resistance. *Environ. Entomol.* 13, 359-365.

Kauffman, W. C., and Kennedy, G. G. (1989). Relationship between trichome density in tomato and parasitism of *Heliothis* spp. (Lepidoptera: Noctuidae) eggs by *Trichogramma* spp. (Hymenoptera: Trichogrammatidae). *Environ. Entomol.* 18, 698-704.

Kemp, J. C., and Barrett, G. W. (1989). Spatial patterning: impact of uncultivated corridors on arthropod populations within soybean agroecosystems. *Ecology* 70, 114-128.

Kersten, U. (1969). Zur morphologie und biologie von *Anatis ocellata* (L.) (Coleoptera: Coccinellidae). *Z. Angew. Entomol.* 63, 412-445.

Kielty, J. P., Allen-Williams, L. J., Underwood, N., and Eastwood, E. A. (1996). Behavioral responses of three species of ground beetle (Coleoptera: Carabidae) to olfactory cues associated with prey and habitat. *J. Insect Beh.* 9, 237-250.

Kiman, Z. B., and Yeargan, K. V. (1985). Development and reproduction of the predator *Orius insidiosus* (Hemiptera: Anthocoridae) reared on diets of selected plant material and arthropod prey. *Ann. Entomol Soc. Amer.* 78, 464-467.

Kuo, H.-L. (1982). Auswirkungen qualitativ unterschiedlicher ernährung von pfirsichlausen *(Myzus persicae)* auf ihren reaberischen feind *(Aphidoletes aphidimyza)*. *Entomol Exp. Appl.* 31, 211-224. (In German: English summary).

Lewis, W. J., Beevers, M., Nordlund, D. A., Gross, H. R., and Hagen, K. S. (1979). Kairomones and their use for management of entomophagous insects. IX. Investigation of various kairomone-treatment patterns for *Trichogramma* spp. *J. Chem. Ecol.* 5, 673-680.

Lewis, W. J., Jones, R. L., Nordlund, D. A., Gross, H. R. (1975a). Kairomones and their use for management of entomophagous insects. II. Mechanisms causing increase in rate of parasitization by *Trichogramma* spp. *J. Chem. Ecol.* 1, 349-360.

Lewis, W. J., Jones, R. L., Nordlund, D. A., Sparks, A. N. (1975b). Kairomones and their use for management of entomophagous insects. I. Evaluation for increasing rates of parasitization by *Trichogramma* spp. in the field. *J. Chem. Ecol.* 1, 343-347.

Martos, A., Givovich, A., and Niemeyr, M. (1992). Effect of dimboa, an aphid resistance factor in wheat, on the aphid predator *Eriopis connexta* Germar (Coleoptera: Coccinellidae). *J. Chem. Ecol.* 18, 469-479.

McEwen, P. K., Clow, S., Jervis, M. A., and Kidd, N. A. C. (1993). Alteration in searching behaviour of adult female green lacewings *Chrysoperla carnea* (Neur.: Chrysopidae) following contact with honeydew of the black scale *Saissetia oleae* (Hom.: Coccidae) and solutions containing acid hydrolysed L-tryptophan. *Entomophaga* 38, 347-354.

Mizell, R. F., III, Frazier, J. L., and Nebeker, T. E. (1984). Response of the clerid predator *Thanasimus dubius* (F.) to bark beetle pheromones and tree volatiles in a wind tunnel. *J. Chem. Ecol.* 10, 177-187.

Moraes, G. J. de, and McMurtry, J. A. (1987). Physiological effect of the host plant on the suitability of *Tetranychus urticae* as prey for *Phytoseilus persimilus* (Acari: Tetranychidae, Phytoseiidae). *Entomophaga* 32, 35-38.

Naranjo, S. E., and Stimac, J. L. (1985). Development, survival, and reproduction *of Geocoris punctipes* (Hemiptera: Lygaeidae): Effects of plant feeding on soybean and associated weeds. *Environ. Entomol.* 14, 523-530.

Nentwig, W. (1988). Augmentation of beneficial arthropods by strip management. 1. Succession of predaceous arthropods and long-term change in the ratio of phytophagous and predaceous arthropods in a meadow. *Oecologia* 76, 597-606.

Nordlund, D. A., Chalfant, R. B., and Lewis, W. J. (1984). Arthropod populations, yield and damage in monocultures and polycultures of corn, beans and tomatoes. *Agric., Ecosys., Environ.* 11, 353-367.

Nordlund, D. A., Lewis, W. J., Jones, R. L., Gross, H. R., Jr., and Hagen, K. S. (1977). Kairomones and their use for management of entomophagous insects. VI. An examination of the kairomones for the predator *Chrysopa carnea* Stephens at the oviposition sites of *Heliothis zea* (Boddie). *J. Chem. Ecol.* 3, 507-511.

Obata, S. (1986). Mechanisms of prey finding in the aphidophagous ladybeetle, *Harmonia axyridis* (Coleoptera: Coccinellidae). *Entomophaga* 31, 303-311.

Obrycki, J. J. (1986). The influence of foliar pubescence on entomophagous species. *In* "Interactions of Plant Resistance and Parasitoids and Predators of Insects." (D. J. Boethel, and R. D. Eikenbary, eds.), pp. 61-83. Ellis Horwood Ltd. Chichester, U.K.

Obrycki, J. J., and Tauber, M. J. (1984). Natural enemy activity on glandular pubescent potato plants in the greenhouse: an unreliable predictor of effects in the field. *Environ. Entomol.* 13, 679-683.

O'Neil, R. J., and Stimac, J. L. (1988). Measurement and analysis of arthropod predation on velvetbean caterpillar, *Anticarsia gemmatalis* (Lepidoptera: Noctuidae), in soybeans. *Environ. Entomol.* 17, 821-826.

Pasteels, J. M. (1978). Apterous and brachypterous coccinellids at the end of the food chain, *Cionura erecta* (Asclepiadaceae) - *Aphis nerii. Entomol. Exp. Appl.* 24, 379-384.

Perrin, R. M. (1975). The role of the perennial stinging nettle, *Urtica dioca,* as a reservoir of beneficial natural enemies. *Ann. Appl. Biol.* 81, 289-297.

Ponsonby, D. J., and Copland, M. J. W. (1995). Olfactory response by the scale insect predator *Chilocoris nigritus* (F) (Coleoptera: Coccinellidae). *Biocont. Sci. Technol.* 5, 83-93.

Powell, J. E., and Lambert, L. (1993). Soybean genotype effects on big-eyed bug feeding on development of *Microplitis croceipes* and leaf consumption by its *Heliothis* spp. hosts. *J. Agric. Entomol.* 1, 169-175.

Rapusas, H. R., Bottrell, D. G., and Coll, M. (1996). Intraspecific variation in chemical attraction of rice to insect predators. *Biol. Cont.* 6, 394-400.

Reid, C. D., and Lampman, R. L. (1989). Olfactory responses of *Orius insidiosus* (Hemiptera: Anthocoridae) to volatile of corn silks. *J. Chem. Ecol.* 15, 1109-1115.

Rice, M. E., and Wilde, G. E. (1989). Antibiosis effect of sorghum on the convergent lady beetle (Coleoptera: Coccinellidae), a third-trophic level predator of the greenbug (Homoptera: Aphididae). *J. Econ. Entomol.* 82, 570-573.

Risch, S. J. (1980). The population dynamics of several herbivorous beetles in a tropical agroecosystem: the effect of intercropping corn, beans, and squash in Costa Rica. *J. Appl. Ecol.* 17, 593-612.

Risch, S. J., Wrubel, R., and Andow, D. (1982). Foraging by a predaceous beetle, *Coleomegilla maculata* (Coleoptera: Coccinellidae), in a polyculture: effects of plant density and diversity. *Environ. Entomol.* 11, 949-950.

Robinson, R. R., Young, J. H., and Morrison, R. D. (1972). Strip-cropping effects on abundance of predatory and harmful cotton insects in Oklahoma. *Environ. Entomol.* 1, 145-149.

Rogers, D. J., and Sullivan, M. J. (1986). Nymphal performance of *Geocoris punctipes* (Hemiptera: Lygaeidae) on pest-resistant soybeans. *Environ. Entomol.* 15, 1032-1036.

Rothschild, M., von Euw, J., and Reichstein, T. (1973). Cardiac glycosides in a scale insect *(Aspidiotus),* a ladybird *(Coccinella)* and a lacewing *(Chrysopa). J. Entomol.* 48, 89-90.

Sabelis, M. W., and Dicke, M. (1985). Long-range dispersal and searching behavior. *In* "Spider Mites. Their Biology, Natural Enemies and Control. World Crop Pests." (W. Helle, and M. W. Sabelis, eds.), Vol. 1B, pp. 141-160. Elsevier. Amsterdam, The Netherlands.

Sabelis, M. W., and van de Baan, H. E. (1983). Location of distant spider mite colonies by phytoseiid predators: demonstration of specific kairomones emitted by *Tetranychus urticae and Panonychus ulmi. Entomol. Exp. Appl.* 33, 303-314.

Sabelis, M. W., Vermaat, J. E., and Groeneveld, A. (1984). Arrestment responses of the predatory mite, *Phytoseiulus persimilis,* to steep odour gradients of a kairomone. *Physiol. Entomol.* 9, 437-446.

Schuster, M. F., Lukefahr, M. J., and Maxwell, F. G. (1976). Impact of nectariless cotton on plant bugs and natural enemies. *J. Econ. Entomol.* 69, 400-402.

Scott, W. P., Snodgrass, G. L., and Smith, J. W. (1988). Tanished plant bug (Hemiptera: Miridae) and predaceous arthropod populations in commercially produced selected nectaried and nectariless cultivars of cotton. *J. Entomol. Sci.* 23, 280-286.

Shah, M. A. (1982). The influence of plant surfaces on the searching behavior of coccinellid larvae. *Entomol. Exp. Appl.* 31, 377-380.

Smiley, J. T., Horn, J. M., and Rank, N. E. (1985). Ecological effects of salicin at three trophic levels: new problems from old adaptations. *Science* 229, 649-651.

Smith, B. C. (1960). A technique for rearing coccinellid beetles on dry foods, and influence of various pollens on the development of *Coleomegilla maculata* Lengi Timb. *Can J. Zool.* 38, 1047-1049.

Stamp, N. E., Erskine, T., and Paradise, C. J. (1991). Effects of rutin-fed caterpillars on an invertebrate predator depend on temperature. *Oecologia* 88, 289-295.

Steinberg, S., Dicke, M., and Vet, L. E. M. (1993). Relative importance of infochemicals from first and second trophic level in long-range host location by the larval parasitoid *Cotesia glomerata. J. Chem. Ecol.* 19, 47-59.

Takabayashi, J., Dicke, M., and Posthumus, M. A. (1991). Variation in composition of predator-attracting allelochemicals emitted by herbivore-infested plants: relative influence of plant and herbivore. *Chemoecol.* 2, 1-6.

Treacy, M. F., Zummo, G. R., and Benedict, J. H. (1985). Interactions of host-plant resistance in cotton with predators and parasites. *Agric., Ecosys., Environ.* 13, 151-157.

Treacy, M. F., Benedict, J. H., Lopez, J. D., and Morrison, R. K. (1987). Functional response of a predator (Neuroptera: Chrysopidae) to bollworm (Lepidoptera: Noctuidae) eggs on smoothleaf, hirsute, and pilose cottons. *J. Econ. Entomol.* 80, 376-379.

van den Meiracker, R. A. F., Hammond, W. N. O., and van Alphen, J. J. M. (1990). The role of kairomones in prey finding by *Diomus* sp. and *Exochomus* sp., two coccinellid predators of the cassava mealybug, *Phenococcus manihoti. Entomol. Exp. Appl.* 56, 209-217.

Walter, D. E. (1996). Living on leaves: mites, tomentia, and leaf domatia. *Annu. Rev. Entomol.* 41, 101-114.

Walter, D. E., and O'Dowd, D. J. (1992). Leaf morphology and predators: effect of leaf domatia on the abundance of predatory mites (Acari: Phytoseiidae). *Environ. Entomol.* 21, 478-484.

Way, M. J., and Murdie, G. (1965). An example of varietal variation in resistance of Brussels sprouts. *Ann. Appl. Biol.* 56, 326-328.

Wetzler, R. L., and Risch, S. J. (1984). Experimental studies of beetle diffusion in simple and complex crop habitats. *J. Anim. Ecol.* 53, 1-19.

Whitman, D. W., and Eller, F. J. (1990). Parasitic wasps orient to green leaf volatiles. *Chemoecol. 1,* 69-76

Wratten, S. D. (1973). The effectiveness of the coccinellid beetle *Adalia bipunctata* as a predator of the lime aphid, *Eucallipterus tiliae* L. *J. Anim. Ecol.* 42, 785-802.

CHAPTER
6

ECOLOGICAL CONSIDERATIONS IN THE
CONSERVATION OF EFFECTIVE PARASITOID
COMMUNITIES IN AGRICULTURAL SYSTEMS

D. A. Landis and F. D. Menalled

I. INTRODUCTION

Many studies examining the conservation of parasitoids for biological control begin by asking the question: given the parasitoid community which currently attacks the pest, which species are most important and how can we modify the system to allow them to be more effective? In this context, conservation biological control programs are grounded in applied population dynamics with their focus on understanding and providing those basic biological needs that maximize stability in single population pest-enemy systems (Murdoch and Briggs, 1996). However, relying solely on these types of studies may give a limited view of the potential for parasitoid conservation. If we begin a biological control program by examining the parasitoid communities which occur in degraded systems such as farmlands, it is possible that appropriate species or combinations of species are no longer present. There is increasing evidence that disturbances in agricultural systems cause just such an effect. Disturbance in agroecosystems can influence insect biodiversity (Burel and Baudry, 1995; McLaughlin and Mineau, 1995) and in the case of parasitoids can reduce species richness (Kruess and Tscharnke, 1994), abundance (Ryszkowski et al., 1993), and effectiveness (Marino and Landis, 1996). Alternatively, populations of appropriate species may be present but due to the structure of farms or the agricultural landscape may be incapable of effectively suppressing pest populations (see Chapter 3). In some cases they may be prone to local extinction and persistence may occur only by the existence of a viable metapopulation composed of a series of interconnected local populations (Murdoch et al., 1985).

There have been a number of general treatments of conservation biological control (van den Bosch and Telford, 1964; Debach and Rosen, 1991; Mahr and Ridgway, 1993; Rabb *et al.*, 1976), although few have focused specifically on parasitoids (Altieri *et al.*, 1993; Powell, 1986). The general approaches to natural enemy conservation are well known and include: reducing direct mortality, providing supplementary resources, controlling secondary enemies, and manipulating host plant attributes (Rabb *et al.*, 1976). In addition, much is known about conserving natural enemies in relation to the effects of weeds and noncrop plants (Altieri and Whitcomb, 1979; van Emden, 1965), vegetational diversity (van Emden, 1990), the influence of spatial structure on dispersal dynamics (Wratten and Thomas, 1990), and the role of agroecosystem diversification (Andow, 1991; Sheehan, 1986). Many examples of parasitoid conservation can be found in the literature on habitat management to enhance biological control (Altieri and Letourneau, 1982; Bugg and Waddington, 1994; Wratten, 1994; Pickett and Bugg, in preparation). In that light, we have chosen to approach the topic of parasitoid conservation for biological control from a slightly different angle.

To elucidate options for the effective parasitoid conservation in agricultural systems we review three aspects of ecological theory we believe to be important to the conservation of parasitoids in agricultural systems: (1) impacts of disturbance on plant and animal communities, (2) importance of metapopulation dynamics in the conservation of parasitoids in highly disturbed systems such as farmlands, and (3) parasitoid community dynamics in agroecosystems. We do this by first contrasting the disturbance regimes of unmanaged ecosystems to those of managed agricultural systems which have replaced them. We maintain that the difficulty in using parasitoids for biological control in annual crops stems primarily from the intensity and uniformity of the disturbance regimes imposed on agricultural landscapes. We further argue that to effectively conserve parasitoids in agricultural systems one must address the root causes for parasitoid failure in these systems, i.e., disturbance and lack of population persistence at various scales. The conservation of viable metapopulations and communities is presented as a means of enhancing biological control. This entails fundamentally understanding and managing disturbance regimes rather than focusing attention on the symptoms they create (see Chapter 7).

One of the stated purposes of this volume is to develop hypotheses which can be used to move the science of conservation biological control forward. We acknowledge from the outset that due to the infinitely complex nature of host-parasitoid systems and the diversity of cropping systems in which they are important the concepts we develop are likely to be rather generalized and exceptions will easily be found.

However, we hope they form a starting point for increased study and discussion of the role of conservation of parasitoids in agricultural landscapes.

II. DISTURBANCE REGIMES IN UNMANAGED AND AGRICULTURAL SYSTEMS

Parasitoids are the most important natural enemies of many crop pests and act as keystone species in some ecosystems (LaSalle, 1993). Many of the factors that are known to limit parasitoid effectiveness in cropping systems can be viewed within the context of disturbance. Ecologists define disturbance as "any relatively discrete event in time that disrupts ecosystem, community, or population structure and changes resources, substrate availability, or the physical environment" (Pickett and White, 1985). Examples of disturbance generated by abiotic factors include fires, windstorms, floods, landslides, and other physical forces. Disturbance can also originate from biotic sources such as insect or disease outbreaks.

The initial outcome of disturbance is the loss of organisms from the community or ecosystem (Reice, 1994). This loss is followed by the gradual recolonization of the disturbed area by individuals which were not removed or which colonize from undisturbed source areas. Because recolonization is drastically influenced by the type of disturbance, characterizing the disturbance regime of an area is critical to understanding the pattern of recolonization and succession. The disturbance regime for a particular site is the combination of disturbance frequency, magnitude, area (and spatial distribution within the area), predictability, and turnover rate. The result of periodic disturbances of plant communities is that an area is transformed into a mosaic of different successional stages (Sousa, 1984).

Few plant species are adapted to live in frequently disturbed environments, thus early-successional habitats have relatively low species diversity. Also, stable late- successional habitats present low species diversity due to competitive exclusion, which eliminates many species. It has been observed that in mid-successional habitats, early- successional plants species coexist with shade tolerant species. Thus, the intermediate disturbance hypothesis proposes that with an intermediate frequency of disturbances, plant species diversity should be at its maximum (Connell, 1978). Plant species diversity is positively correlated with plant structural and chemical diversity. Due to the bottom-up influence of vegetation on mobile organism, these factors are assumed to be positively correlated with the distribution, abundance, and diversity of insects which utilize those habitats (Bazzaz, 1996; Gardner *et al.,* 1995; Sousa, 1984). Therefore, it is ultimately this disturbance regime which is of critical importance in shaping the structure of the parasitoid community and ecological interactions which take place in an ecosystem (Pickett and White, 1985; Price, 1994; Reice, 1994).

Disturbance regimes differ between unmanaged ecosystems and agroecosystems. Whereas even frequently disturbed terrestrial ecosystems may have only one disturbance event every several years (e.g., fire in grasslands), most agricultural ecosystems experience multiple and intense disturbances each growing season. In addition, many of the these events are originated by human activities, are severe, and occur uniformly over large areas. As conversion of land to farming has occurred over the centuries, the outcome has typically been a fragmentation of unmanaged habitats and an associated isolation of unmanaged populations in these areas (Merriam, 1988). At a regional scale, the aggregate effect of these farming practices reduces edaphic, hydric, and physiographic heterogeneity and results in a limited diversity of highly disturbed habitats (crop types) of like successional stage, managed with relatively similar techniques (Table 1).

The implications of these severe, frequent, and extensive disturbances on parasitoid conservation can be examined at three different levels: the within crop level, the farm level, and the landscape level. We acknowledge that these three discrete levels of analysis represent a continuum and overlap may exist among patterns and processes observed at each one of them. Moreover, scale of spatial heterogeneity may differ among parasitoid taxa. For simplicity in our analysis, we will discuss the relationship between disturbance and parasitoid conservation at each one of these spatial levels of analysis.

A.   Crop Scale Disturbance Regimes and Parasitoids

Disturbance regimes at the crop scale are usually quite different from those that existed in the unmanaged habitats they have replaced. A typical annual crop ecosystem undergoes a near constant series of disturbance events. Within a season, agricultural production frequently begins by removing the original vegetation covering the ground and mixing the upper layers of the soil profile. Several secondary tillage events may follow in advance of planting. This may be followed by various nutrient and pesticide applications, cultivation, and harvest; which may be followed by additional tillage or herbicide applications. Each of these production practices is a significant disturbance event to some portion of the microbial, plant, and animal communities in the field.

As a result of the frequent and intense disturbance regimes, agricultural systems are recognized as particularly difficult environments for many parasitoids to live in and function effectively (Townes, 1972). Nowhere is this more true than in annual monocultural cropping systems (Powell, 1986). Several studies have specifically identified crop type as an important factor in the establishment of imported natural enemies. In these studies crop type has been equated with habitat stability, which is directly related to disturbance regime. Hall and Ehler (1979) and Beirne (1975) found that the lowest rates of establishment (of predators and parasitoids combined) have occurred in annual crop habitats. Stilling (1990) found a similar result for parasitoids alone. The rate of successful biological control resulting from importation efforts is also influenced by crop type. Hall and Ehler (1980) found that the lowest rates of success occurred in annual crop habitats. Parasitoid failure in these systems is often a result of the direct and indirect effects of pesticides, tillage, cultivation, lack of adequate food resources (e.g., pollen and nectar), scarcity of alternate hosts, and lack of shelter (Rabb *et al.,* 1976; Powell, 1986; Dutcher, 1993). In the next two sections we analyze the direct and indirect influences of agricultural practices on parasitoid abundance and diversity.

1. Direct effects of disturbance on parasitoids

Pesticide application is perhaps the most obvious example of a within-field disturbance which limits parasitoid effectiveness. Insecticides can directly kill large numbers of parasitoids and have long-lasting effects on the structure of communities (Debach and Rosen, 1991). Managing pesticide impacts is one of the most important conservation measures to preserve viable and effective parasitoid communities (see Chapter 11). Mensah and Madden (1993) reported on a successful program to control the psyllid *Ctenarytaina thysanura* (Ferris and Klyver) on the commercial oil crop *Boronia megastigma* (Nees) by modifying the placement and timing of insecticide applications. Survival of the psyllid parasitoids *Psyllaephagus* spp. (Encyrtidae) and *Cocophagus* sp. (Aphelinidae) was highest when sprays were selectively applied to plant stems rather than foliage. Timing sprays to occur when psyllid nymphs were active (but most of the parasitoids were enclosed in host mummies) further protected the parasitoid community and allowed a reduction in treatment frequency from 10 sprays per year to 3. By modifying the frequency, timing, and spatial distribution of the disturbance regime in this system the natural enemy community (including predators) provided control of the psyllid and resulted in significantly increased yields and profits. Planting operations, tillage, burning of crop residues, and other cultural practices can also have direct impacts on parasitoids. Mohyuddin (1991) discussed the conservation of two egg parasitoids *Parachrysocharis javensis* (Girault) and the encyrtid *Ooencyrtus papilionis* Ashmead, which attack *Pyrillia perpusilla* (Walker) in sugarcane. These parasitoids overwinter in crop residues, where nearly 100% of the host eggs may be parasitized. The normal practice of burning residues after harvest destroys these overwintering sources and delays colonization of fields in the spring. By delaying the burning of material at field edges until the following spring, parasitism was increased to nearly 80% versus less than 5% in fields where residues were completely burned the previous season.

2.   Indirect effects of disturbance on parasitoids

While a disturbance may directly kill parasitoids, as shown above, its indirect effects are often longer-lasting and can have similar negative impacts on parasitoid populations. Although monocultures are highly suitable for colonization by herbivores, they are frequently very poor environments for natural enemies (Price, 1991). In many cases this is due to disruption in the temporal and/or spatial availability of food and shelter. A herbicide applied weeks before a parasitoid enters a field may have little direct impact. However, because many parasitoids require regular access to pollen or nectar sources (Jervis *et al.,* 1993; Zhao *et al.,* 1992), eliminating flowering weeds from crop fields may indirectly render the field uninhabitable.

Insecticides and weed control practices may indirectly influence parasitoid communities by reducing primary and alternate host populations within crops and surrounding habitats. Weed control practices can also indirectly influence parasitoid success by altering host availability. Barczak (1988) documented the parasitoid complex of *Aphis fabae* Scopoli on various crop and noncrop plants. He found the parasitoid community of *A. fabae* on *Chenopodium album* L. included those species most important in attacking *A. fabae* on beets. In addition, the abundance and percent parasitism of *A. fabae* on beets was greatest in weedy fields containing *C. album.* He concluded that *C. album* served as a reservoir of parasitoids which attack *A. fabae* in crop fields.

Within-crop plant structure and diversity results from species composition and a particular set of disturbances such as planting density, cultivation, and herbicide applications. The structural characteristics of a plant stand influence movement, behavior, and attack by parasitoids. Coll and Bottrell (1996) used a release-recapture experiment to compare movement of the eulophid *Pediobius foveolatus* (Crawford) in four types of habitats: beans planted at a high density, beans planted at low density, beans intercropped with short maize, and beans intercropped with tall maize. Both bean density and presence of maize by itself did not affect parasitoid movements. However, maize height was the primary factor influencing parasitoid movement and resulting in lower immigration and emigration rates into and from the bean-tall maize plots. In the long run, wasps accumulated in these mixed-species, structurally complex habitats. The authors suggested that the lower emigration rate observed from bean-tall maize crops was a response to the high shade present within the crop. Due to the greater accumulation of natural enemies in the mixed species habitats, parasitism rates of the Mexican bean beetle *Epilachna varivestis* Mulsant were higher than in monocultures.

B. Farm-Level Disturbance Regimes and Parasitoids

Within a farm, production technologies such as mechanization which determines crop rotation, presence of hedgerows, and field size and shape set up a particular disturbance scenario. Production practices also determine the spatial distribution of crops and refuge habitats for natural enemies (Drinkwater *et al.,* 1995). Within a farm, the interaction of fields, surrounding habitats, and disturbance regime influences the existence of parasitoid species. The presence of a diverse and abundant local community of parasitoids enhances the probability of success of biological control (Fig. 1).

Several studies point to the importance of surrounding habitats in determining the structure of natural enemy communities that exist within particular crops. Landis and Haas (1992) found that parasitism of the European corn borer *Ostrinia nubilalis* (Hübner) by its ichneumonid larval parasitoid *Eriborus terebrans* (Gravenhorst) was significantly higher at the borders of maize fields than in field interiors and that the greatest parasitism was observed at wooded field edges. *E. terebrans* abundance in the spring was greatest at wooded edges even in second year corn fields containing overwintering populations of the parasitoid, indicating an attractive quality of the edges rather than simple dispersal effects (Dyer and Landis, 1997). Other studies showed that without access to plant nectar or aphid honeydew, *E. terebrans* survived less that 1.5 days in the field (Landis and Marino, 1998). Access to sugar was vital to stress tolerance in the lab and greenhouse and improved survival in crop and noncrop habitats in the field (Dyer and Landis, 1996). Overall it was concluded that the lack of nectar and honeydew sources in large maize fields, combined with high temperatures before canopy closure, forced *E. terebrans* to leave fields to seek food and shelter in wooded habitats. This ultimately resulted in lower parasitism of *0. nubilalis* in field interiors (Dyer, 1995).

Olszak (1991, 1994) studied the role of various shrubs in supporting alternate hosts of parasitoids which attack the rosy aphid *Dysaphis plantaginea* (Pass.) and green apple aphid *Aphid pomi* (De Geer). He found that elder *Sambucus nigra* L.

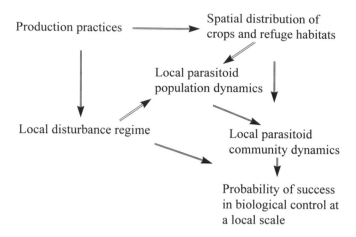

Figure 1. Farm level disturbance and parasitoid dynamics. Diagrammatic representation of the influence of production practices on biological control at the local scale.

and snowball bushes *Viburnum* spp. support a number of aphid species which serve as a reservoir of important apple aphid parasitoids. He suggested that planting these shrubs in the vicinity of apple orchards could improve biological control of aphids.

The above discussion on the influence of surrounding crop habitats on parasitoid communities was focused on the presence or absence of wooded hedgerow because this is a well studied variable. Other hedgerow factors such as successional stage, size, shape, and specific composition together with farm variables such as field size and shape could also be of importance. To our knowledge, no particular study has addressed the influence of these variables in determining parasitoid diversity.

C. Landscape-Level Disturbance Regimes and Parasitoids

Farming procedures fragment unmanaged ecosystems, which in turn may result in changes in insect community structure and function. This has important implications for the conservation of natural enemies which often utilize undisturbed habitats for shelter or food resources (van Emden, 1963, 1965). On a landscape scale, the spatial distribution of disturbance events in agriculture is quite different from that of unmanaged ecosystems. In unmanaged systems, the underlying geographical, topographical, environmental, or community gradients influence the spatial pattern of disturbance (Pickett and White, 1985). Small disturbed areas intermingle with undisturbed remnant patches creating an interacting mosaic. For example, fire in a prairie may leave patches of burned and unburned habitat in relation to soil moisture gradients. If that identical piece of land was farmed, the same gradients may have little or no effect on the spatial pattern of disturbance. The entire field would likely be treated with the same herbicide or rate of fertilizer irrespective of the changing soil gradients; although this situation could change with adoption of site-specific farming practices. As a result, the spatial distribution of disturbance events on agricultural landscapes is related to patterns of past and current land ownership. The within- and between-field variation in disturbance is less than for the same land in its unmanaged state.

Kruess and Tscharntke (1994) showed that in fragmented agricultural landscapes, disturbance and habitat isolation decreased biodiversity. Because parasitoid population growth starts only after the successful establishment of their hosts, parasitoids are very susceptible to changes due to habitat distribution and represent one of the most likely sets of species to be lost when natural habitats are transformed to farmlands. Due to the top-down control that parasitoids may exert on phytophagous insects this decrease in natural enemies increases the risk of pest outbreaks. Thus, disturbance and heterogeneity at the landscape scale may have important impacts on the diversity, abundance, and effectiveness of natural enemies.

Several studies have shown the relationship between landscape structural complexity and natural enemy biodiversity. Gut et al. (1982) found that more diverse and effective predator communities occurred in complex landscapes composed of pear orchards located in areas of mixed crops and woodlands. In simple landscapes with extensive orchard production predator communities were sparse or ineffective, particularly early in the season (see also Liss et al., 1986). A similar result was observed in apple orchards in Hungary by Szentkiralyi and Kozar (1991). They found that numbers of natural enemies were higher and less variable in orchards surrounded by diverse vegetation. In areas of intensive apple production with low vegetational diversity and high disturbance (i.e., 7 to 12 insecticide/acaricide applications/year), species richness of natural enemies was the lowest. Further, they found that increased habitat diversity moderated the negative impacts of within-orchard disturbance. Insecticide treated orchards within diverse vegetational settings had more natural enemy species than those in less diverse more highly disturbed settings.

The interaction between landscape structure and crop edge habitat type has been demonstrated by Corbett and Rosenheim (1996). They studied patterns of vineyard colonization by several myrmarids using rubidium to label the parasitoids. The *Anagrus* complex (*A. epos* Girault, and *A. erythroneurae* Trijopitzin and Chiappini) require host eggs to overwinter. Because both the grape leafhopper *Erythroneura elegantula* Osborn and the variegated leafhopper *E. variabilis* overwinter as adults the parasitoids cannot overwinter in grape vineyards. *Anagrus* colonize vineyards in the spring both from riparian corridors or from prune trees planted in association with vineyards. In these habitats they overwinter in eggs of wild hosts on blackberries or on the prune leafhopper *Edwardsiana prunicola* Edwards. Corbett and Rosenheim (1996) released labeled *Anagrus* in prune refuges and sampled adjacent vineyards to determine the proportion of labeled parasitoids. They found that prune tree refuges directly contributed 1-34% of the early-season *Anagrus* populations in vineyards. However, proximity of prune refuges increased *Anagrus* colonization from other habitats (i.e. riparian corridors) apparently by exerting a windbreak effect. In addition, the number of *Anagrus* contributed by external overwintering habitats (other than prune refuges) was inversely correlated with the distance to these habitats. Vineyards close to riparian corridors had 4-5 times more colonizers than those at greater distances. Thus, prune refuges contribute both directly as overwintering sites and indirectly through windbreak effects to increase *Anagrus* populations in vineyards.

Rsyzkowski *et al.* (1993) studied the biomass of different trophic groups in field crops located in simple or complex landscapes in eastern Europe. The complex landscapes contained small crop fields with varied rotations and considerable crop diversity. The simple landscape consisted of large crop fields with no uncultivated habitats, simplified crop rotations, and low crop diversity. They found that mean parasitoid biomass was always greater in the perennial crop alfalfa compared to the annual crops of wheat, barley, maize, and sugar beet. In Poland, mean parasitoid biomass was 90% greater in annual crops in the complex versus simple landscape. Similarly for alfalfa, parasitoid biomass was 43% greater in complex landscape. In contrast, in Romania there was no change in parasitoid biomass between complex and simple landscapes for annual crops and a 76% decrease for alfalfa in the complex landscape. The parasitoid community of the leaf beetle *Oulema gallecianna* Heyd. was studied over seven years in Bohemia, Czech Republic (Sedivy, 1995). In an area characterized by intensive cereal production, parasitoid species richness was consistently lower than in a more varied landscape with smaller fields and "ecological corridors" (mean richness $\pm$ SE = 4.4 1.32 versus 7.0 $\pm$ 0.53, n = 7). Overall parasitism in the intensive cereal landscape was less than half of that in the varied landscape; influenced heavily by the dominant eulophid species, *Necrenmus leucarthros* (mean percent parasitism $\pm$ SE = 13.7 $\pm$ 3.03 versus 30.3 $\pm$ 2.03, N = 7).

We have analyzed the importance of within-crop, whole farm and landscape agricultural practices in determining the pattern of local habitat modification, landscape heterogeneity, and parasitoid abundance. These findings demonstrate that many of the factors which limit parasitoid abundance and effectiveness in agriculture can be interpreted in light of current ecological theory regarding the impacts of disturbance on plant and animal communities. Therefore, the relationships between parasitoid biology, cropping system, and disturbance should be understood to enhance the possibilities of success of a biological control program. In the next sections, we will review the mechanisms of population regulation at the regional scale and relate them to several ecological and evolutionary aspects of parasitoid community dynamics.

III. PARASITOID METAPOPULATIONS IN AGRICULTURAL SYSTEMS

Traditionally, biological control programs are focused on the establishment of natural enemy populations which are stable at the local level (Murdoch and Briggs 1996). However, as we have seen, the extensive and frequently disturbed plant monocultures of early-successional, annual crops represent severe environments for parasitoid establishment and persistence. High habitat instability and frequent local population extinctions, often seasonally due to crop rotation, pesticides, etc., pose serious limitations to the success of parasitoids as biological control agents.

A possible way to achieve success is by focusing conservation efforts on the parasitoid metapopulation, a series of genetically interconnected local populations. Murdoch et al. (1985) showed that in many cases local nonequilibrium populations may persist thanks to regulation mechanisms occurring at the global scale (see Murdoch et al., 1996 for a counter example). In situations of high local extinction rates, recolonization from adjacent patches is a key factor in persistence of parasitoid-host systems (Kruess and Tscharntke, 1994). Therefore, if parasitoids are to be used to control insect pests in highly disturbed agricultural systems it may be necessary to maintain an effective metapopulation at the landscape level. The persistence of a viable metapopulation is a function of several factors, among them number of patches, patch size, between-patch dispersal rate, and degree of correlation of local extinctions (Hanski, 1989) (Fig. 2).

In the past years there has been significant interest in modeling how habitat fragmentation affects patch occupancy (Hanski, 1994), species persistence (Bierzychudek, 1988; Dytham, 1995), and metapopulation persistence (Murdoch, 1994). Hanski et al. (1996) defined the minimum viable metapopulation (MVM) size as the minimum number of interacting local populations necessary for long-term persistence of a metapopulation. This concept is directly linked to the minimum amount of suitable habitat (MASH). Modeling the nonequilibrium metapopulation dynamics of the butterfly *Melitaea cinixia* L. in Finland, they suggested a minimum of 15-20 well connected patches. While these concepts have obvious implications for conservation biological control (see Chapter 2), the impacts of habitat fragmentation in agricultural landscapes on MVM and MASH for parasitoids is largely unknown (see Chapters 2 and 8).

Consider a migratory pest which arrives in the northern extent of its range and colonizes a newly emerging habitat (i.e., an annual crop). The natural enemies which attack the pest must either be present in the field (e.g., overwintering from the previous year), arrive with the host (e.g., transovarially transmitted diseases), or arrive as immigrants from the local metapopulation (see Chapter 12). Because the latter must frequently occur with parasitoids an important question becomes what types of landscape structure favor biological control by these species?

Marino and Landis (1996) studied the parasitoid community of one migratory species the true armyworm *Pseudaletia unipuncta* Haworth in a complex versus simple landscape in Michigan. Both landscapes were primarily composed of early-successional crop land (60% to 71%), although this early-successional matrix was more highly fragmented in the complex landscape by abundant and highly interconnected mid- to late-successional fencerows and woodlands. In contrast, in the simple landscape, fencerow and woodlot removal seriously reduced the amount and connectivity of

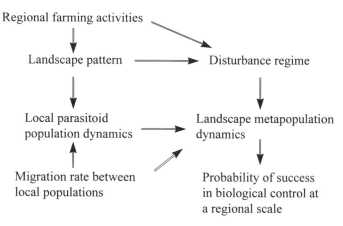

Figure 2. Landscape-level production practices and regional parasitoid metapopulation dynamics. Diagrammatic representation of the factors affecting probability of success of biological control programs at a regional scale.

later-successional habitats, resulting in a less fragmented early-successional matrix. All seven parasitoid species reared from armyworm larvae collected in these landscapes were generalists. In the simple landscape parasitism was low, with two braconids *Meteorus communes* (Cresson) and *Glytapanteles militaris* (Walsh) equally abundant. In the complex landscape parasitism was significantly higher and due almost exclusively to *M. communes*. Marino and Landis (1996) discussed how the interaction of alternate hosts and landscape structure could account for these observations. All of the alternate hosts of *M. communes* in Michigan feed on trees and shrubs while those of *G. militaris* feed primarily on herbaceous hosts, several on field crops. They suggested that in the complex landscape, colonization of crop habitats by *M. communes* from local populations existing in adjacent later-successional habitats was favored, resulting in greater parasitism. In the simple landscape while both species were present, *G. militaris* (with its links to early-successional habitats) was favored but unable to contribute greatly to overall parasitism.

Although metapopulation dynamics were not explicitly addressed by Marino and Landis (1996), it may be that the presence of *M. communes* in crop fields is mediated by processes taking place at the landscape scale. To apply nonequilibrium dynamics concepts in biological control ecologists must incorporate the specific biological needs of parasitoids and rural landscape patterns into metapopulation models. To do so, we must understand short- and long-term community dynamics, degree of environmental stochasticity which conditions extinction probabilities, and dispersal behavior of parasitoids and alternate hosts. In this process, landscape ecology plays a fundamental role by linking these concepts with landscape variables such as fragment composition, fragment size, connectivity between patches, and spatial distribution of elements.

IV. PARASITOID COMMUNITY DYNAMICS IN AGRICULTURAL SYSTEMS

To properly conserve parasitoids in rural landscapes the concepts previously discussed should be complemented with our knowledge of the ecology, evolution, and impact of parasitoid communities. In recent years there has been an explosion of interest in the factors which structure parasitoid communities (Askew and Shaw, 1986; Sheehan, 1991; LaSalle and Gauld, 1993; Hawkins, 1994; Hawkins and Sheehan, 1994). The components of parasitoid community structure which have most frequently been evaluated include species richness, comparisons of species biology (i.e., idiobiont versus koinobiont life histories), degree of host specialization (generalists versus specialists), and the degree of impact on host populations. Biological control practitioners may well ask which types of parasitoid communities are most effective, which should be promoted, and how can we do this?

Using previously published host records, Hawkins (1994) attempted to take a broad look at the ecological factors which may serve to structure parasitoid communities. He examined host feeding niche, food plant type, habitat type (unmanaged or cultivated), latitude, and climate in search for patterns in parasitoid species richness, biology, degree of host mortality, and hyperparasitoid richness. He makes a case for the overriding importance of host feeding niche in determining susceptibility to parasitoid attack and subsequent increase in parasitoid species richness. He also found that parasitoid richness is positively correlated with percent parasitism and host mortality.

In an evolutionary context, Price (1991, 1994) analyzed the interactions between plants, insect herbivores, and parasitoids along a gradient of vegetational succession. He observed that the intensity and mechanisms of these bottom-up influences varies from disturbed sites with low stability, such as herbaceous patches, to mature and stable forests. In this analysis, five main patterns of host/parasitoid interactions were detected. First, parasitoid species richness increases with plant succession. Second, as succession progresses there is a tendency toward more generalized parasitoid species per host species; with specialists on early stages and generalists on later stages of succession. Third, an increase in the number of parasitoid species per host herbivore is positively correlated with the overall mortality inflicted on the host species. Fourth, due to an increase in parasitoid species richness, there is an increased possibility of insect herbivore host population regulation in late-successional habitats. Finally, there is a change in host finding behavior with vegetational succession: genetically fixed innate responses to insect host and plant odors in early stages are replaced by associative learning in stable, late- successional habitats.

In an attempt to combine these levels of understanding, Landis and Marino (unpubl. data) addressed the question of what types of habitats should be included in agricultural landscapes in order to conserve maximum parasitoid community richness? They found that the potential parasitoid communities of lepidopteran pests on the four major crops in the midwestern U.S.A. (corn, soybean, wheat, and alfalfa) were dominated by generalist parasitoids. They then examined the alternate hosts of these generalists in order to determine what types of habitats may be required to conserve parasitoids in agricultural landscapes. It was determined that over 60% of the alternate hosts of these generalist parasitoids feed on late-successional plants such as trees and shrubs. It appears that to conserve species-rich parasitoid communities of the lepidopteran pest complex, agricultural landscapes may need to include late-successional habitats such as woodlots interconnected by hedgerows and fencerows. This should provide the best opportunity for viable metapopulations of these parasitoids to exist at the regional scale and contribute to biological control.

V. CONCLUSIONS

In this chapter we have reviewed several ecological concepts in relation to parasitoid conservation in agricultural systems. Clearly, applying these concepts in conservation efforts will be a challenge due to the complexity and frequently case specific nature of the interactions. However, several general ideas emerge that can guide parasitoid conservation efforts in the future.

1. Many of the proximate factors identified as limiting the effectiveness of parasitoids in agricultural systems (e.g., pesticides, lack of adult food, lack of alternative hosts, etc.) can be viewed as the direct results of the disturbance regimes we impose in these systems. Subsequently, conservation of parasitoids by amelioration of these conditions must ultimately be achieved by managing disturbance not just the symptoms it produces. For example, while food sprays may substitute for lack of pollen and nectar resources in agricultural systems, they do nothing to address the level of overall disturbance. The alternative, establishment of perennial flowering plant habitats, addresses both the proximate need and the ultimate cause of the problem (also see Chapter 7).

2. Disturbance occurs at various spatial scales: within crops, within farms, and at the landscape level. The impact of disturbance at each of these levels can directly limit parasitoid effectiveness. Moreover, disturbance at different levels may interact to further affect parasitoid abundance. To effectively conserve parasitoids for biological control may require that we actively manage the disturbance regimes at several spatial scales. While eliminating a pesticide treatment within a field may permit the establishment or persistence of a parasitoid population, if viable metapopulations do not exist at the landscape-level to provide immigrants, the within-field effort may be ineffective.

3. Due to the current high intensity of the disturbance regimes in agriculture many of the solutions for parasitoid conservation appear to lie in terms of reducing the overall level of disturbance. In this respect, agronomic practices should be examined with an eye toward how they contribute to the disturbance regime in the system. Practices such as cover cropping, intercropping, and reduced tillage may tend to relax

the overall disturbance regime even though they may require some new disturbances in order to manage weeds, e.g., the frequent requirement for a burn-down herbicide in no-tillage systems. Alternatively, some new technologies such as transgenic corn expressing *Bacillus thuringiensis* toxins may appear to reduce disturbance by eliminating pesticide treatments, but may in fact represent a more pervasive disturbance through the potential for cascading multitrophic-level impacts (but see Chapter 10).

4.   Habitat management in agricultural landscapes emerges as an integrating theme in the conservation of parasitoid populations and communities. Parasitoids require several basic resources in their environment: food, water, shelter from adverse conditions, and in many cases alternate hosts and mates. Furthermore, to complete their life cycle they require a threshold host population with individuals at a particular developmental stage or size. To achieve the objectives of biological control, parasitoids must be present at the right time and in sufficient numbers to have an impact on the target hosts density. If parasitoids are to effectively suppress a particular pest these basic resources must be accessible in relatively close temporal and spatial association. Within agricultural landscapes habitats should be managed to provide these resources in the appropriate spatial and temporal scales (also see Chapter 3).

5. Agricultural development typically transforms landscapes from a primeval matrix of mid- to late-successional habitats interspersed with small early-successional areas created by local disturbances, to a matrix of intensely disturbed, early-successional habitats interspersed with small remnant or regenerated patches of later-successional stages. While this situation favors the exploitation of crop plants by pest insects it is extremely unfavorable to many parasitoid species. To effectively conserve parasitoids in these primarily early-successional agricultural landscapes the creation and management of mid- to late-successional habitats may be required. In essence, this is a process of refragmenting highly disturbed landscapes by adding a network of more stable habitats of varying successional stages. These habitats should serve multiple functions as cross- wind trap strips, filter strips, riparian buffer zones, or agroforestry production systems.

        Clearly, the application of these concepts to agricultural systems will require new partnerships in research and extension. To achieve the goal of ecologically based pest management, biological control practitioners may need to access expertise from diverse fields such as agronomy, entomology, land use planning, sociology, population, community and landscape ecology (National Research Council, 1996). A multidisciplinary approach will facilitate the design of sustainable agroecosystems where the conservation of an effective community of natural enemies is coupled with highly productive farmlands.

Acknowledgments

This work was supported in part by USDA SARE grant LWF 62-016-03508. DAL also wishes to acknowledge the support of Wageningen Agricultural University, The Netherlands, where the manuscript was developed while on sabbatical leave in the C. T. deWit Research School for Production Ecology. We thank D. K. Letourneau, P.C. Marino, P. Whitaker, and M. J. Haas for their helpful comments on the manuscript.

REFERENCES

Altieri, M. A., and Letourneau, D. K. (1982). Vegetation management and biological control in agroecosystems. *Crop. Prot.* 1, 405-430.

Altieri, M. A., and Whitcomb, W. H. (1979). The potential use of weeds in manipulation of beneficial insects. *Hort. Sci.* 14, 12-18.

Altieri, M. A., Cure, J. R., and Garcia, M. A. (1993). The role and enhancement of parasitic Hymenoptera biodiversity in agroecosystems. *In* "Hymenoptera and Biodiversity." (J. LaSalle, and I. D. Gauld, eds.), pp. 257-275. C.A.B International. Oxon, U.K.

Andow, D. A. (1991). Vegetational diversity and arthropod population response. *Annu. Rev. Entomol.* 36, 561-586.

Askew, R. R., and Shaw, M. R. (1986). Parasitoid communities: their size, structure and development. *In* "Insect Parasitoids." (J. Waage, and D. Greathead, eds.), pp. 225-264. Academic Press. London, U.K.

Barczak, T. (1988). Parasitoids of the black bean aphid, *Aphis fabae* complex, in the Bydgoszcz district, Poland. *In* "Ecology and Effectiveness of Aphidophaga." (E. Niemczyk, and A. F. G. Dixon, eds.), pp. 105-109. SPB Academic Publishing. The Hague.

Bazzaz, F.A. (1996). "Plants in Changing Environments. Linking Physiological, Population, and Community Ecology." Cambridge University Press. Cambridge, U.K.

Beirne, B. P. (1975). Biological control attempts by introductions against pest insects in the field in Canada. *Can. Entomol.* 107, 225-236.

Bierzychudek, P. (1988). Can patchiness prevent pest outbreaks? *Trends Ecol. Evol.* 3, 62-63.

Bugg, R. L., and Waddington, C. (1994). Using cover crops to manage arthropod pests of orchards: a review. *Agric., Ecosys., Environ.* 50, 11-28.

Burel, F., and Baudry, J. (1995). Species biodiversity in changing agricultural landscapes: a case study in the Pays d' Auge, France. *Agric., Ecosys., Environ.* 55, 193-200.

Coll, M., and Bottrell, D. G. (1996). Movement of an insect parasitoid in simple and diverse plant assemblages. *Ecol. Entomol.* 21, 141-149.

Connell, J. H. (1978). Diversity in tropical rain forest and coral reef. *Science* 199, 1302-1310.

Corbett, A., and Rosenheim, J. A. (1996). Impact of a natural enemy overwintering refuge and its interaction with the surrounding landscape. *Ecol. Entomol.* 21, 155-164.

Debach, P., and Rosen, D. (1991). "Biological Control by Natural Enemies." 2nd ed. Cambridge University Press. Cambridge, U.K.

Drinkwater, L. E., Letourneau, D. K., Workneh, F., Bruggan, A. H. C., and Shennan, C. (1995). Fundamental differences between conventional and organic tomato agroecosystems in California. *Ecol. Appl.* 5, 1098-1112.

Dutcher, J. D. (1993). Recent examples of conservation of arthropod natural enemies in agriculture. *In* "Pest Management: Biologically Based Technologies." (R. D. Lumsden, and J. L. Vaughn, eds.), pp. 101-108. Amer. Chem. Soc. Washington, DC.

Dyer, L. E. (1995). Non-crop habitats and the conservation of *Eriborus terebrans* (Gravenhorst) (Hymenoptera: Ichneumonidae), a parasitoid of the European corn borer, *Ostrinia nubilalis* (Hübner) (Lepidoptera: Pyralidae). Ph.D. Dissertation, Department of Entomology, Michigan State University, E. Lansing, MI.

Dyer, L. E., and Landis, D. A. (1996). Effects of habitat, temperature and sugar availability on longevity of *Eriborus terebrans* (Hymenoptera: Ichneumonidae). *Environ. Entomol.* 25, 1192-1201.

Dyer, L. E., and Landis, D. A. (1997). Influence of non-crop habitats on the distribution of *Eriborus terebrans* (Hymenoptera: Ichneumonidae) in corn fields. *Environ. Entomol.* 26, 924-932.

Dytham , C. (1995). The effect of habitat destruction pattern on species persistence: a cellular model. *Oikos* 74, 340-344.

Gardner, S. M., Cabido, M. R., Valladares, G. R., and Diaz, S. (1995). The influence of habitat structure on arthropod diversity in Argentine semi-arid Chaco forest. *J. Veg. Sci.* 6, 349-356.

Gut, L. J., Jochums, C. E., Westigard, P. H., and Liss, W. J. (1982). Variations in pear psylla *(Psylla pyricola* Foerster) densities in southern Oregon orchards and its implications. *Acta Hort.* 124, 101-111.

Hall, R. W., and Ehler, L. E. (1979). Rate of establishment of natural enemies in classical biological control. *Bull. Entomol. Soc. Amer.* 25, 280-282.

Hall, R. W., and Ehler L. E. (1980). Rate of success in classical biological control of arthropods. *Bull. Entomol. Soc. Amer.* 26, 111-114.

Hanski, I. (1989). Metapopulation dynamics: does it help to have more of the same? *Trends Ecol. Evol.* 4, 113-114.

Hanski, I. A. (1994). Patch occupancy dynamics in fragmented landscapes. *Trends Ecol. Evol.* 9, 131-135.

Hanski, I., Moilanen, A., and Gyllenberg, M. (1996). Minimum viable metapopulation size. *Amer. Nat.* 147, 527-541.

Hawkins, B. A. (1994). "Pattern and Process in Host-Parasitoid Interactions. Cambridge University Press. Cambridge, U.K.

Hawkins, B. A., and Sheehan, W. (1994). "Parasitoid Community Ecology." Oxford University Press. Cambridge, U.K.

Hawkins, B. A., Thomas M. B., and Hochberg, M. E. (1993). Refuge theory and biological control. *Science* 262, 1429-1432.

Jervis, M. A., Kidd, N. A. C., Fitton, M. G., Huggleston T., and Dawah, H. A. (1993). Flower-visiting by hymenopteran parasitoids. *J. Nat. Hist.* 27, 67105.

Kruess, A., and Tscharntke, T. (1994). Habitat fragmentation, species loss, and biological control. *Science* 264, 1581-1586.

Landis, D. A., and Haas, M. J. (1992). Influence of landscape structure on abundance and within-field distribution of *Ostrinia nubilalis* Hübner (Lepidoptera: Pyralidae) larval parasitoids in Michigan. *Environ. Entomol.* 21, 409-416.

Landis, D. A., and Marino, P. C. (1998). Landscape structure and extra-field processes: impact on management of pests and beneficials. *In* "Handbook of Pest Management." (J. Ruberson, ed.), Marcel Dekker Inc. NY. (in press).

LaSalle, J. (1993). Parasitic Hymenoptera, biological control and biodiversity. *In* "Hymenoptera and Biodiversity." (J. LaSalle, and I. D. Gauld, eds.), pp. 197216. C.A.B International. Oxon, U.K.

LaSalle, J., and Gauld, I. D. (1993). "Hymenoptera and Biodiversity." C.A.B International. Oxon, U.K.

Liss, W. J., Gut, L. J., Westigard, P. H., and Waren C. E. (1986). Perspectives on arthropod community structure, organization and development in agricultural crops. *Annu. Rev. Entomol.* 31, 455-478.

Mahr, D. L., and Ridgway, N. M. (1993). "Biological Control of Insects and Mites: An Introduction to Beneficial Natural Enemies and Their Use in Pest Management." North Central Regional Extension Publication No. 481. Madison, WI.

Marino, P. C., and Landis, D. A. (1996). Effects of landscape structure on parasitoid diversity and parasitism in agroecosystems. *Ecol. Appl.* 6, 276-284.

McLaughlin, A., and Mineau, P. (1995). The impact of agriculture on biodiversity. *Agric. Ecosys. Environ.* 55, 201-212.

Merriam, G. (1988). Landscape dynamics in farmland. *Trends Ecol. Evol.* 3, 16-20.

Mensah, R. K., and. Madden, J. L. (1993). Development and application of an integrated pest management programme for the psyllid, *Ctenarytaina thysanura* on *Boronia megastigma* in Tasmania. *Entomol. Exp. Appl.* 66, 59-74.

Mohyuddin, A. I. (1991). Utilization of natural enemies for the control of insect pests of sugar-cane. *Insect Sci. Applic.* 12, 19-26.

Murdoch, W. W. (1994). Population regulation in theory and practice. *Ecology* 75, 271-287.

Murdoch, W. W., and Briggs, C. J. (1996). Theory for biological control: recent developments. *Ecology* 77, 2001-2013.

Murdoch, W. W., Chesson J., and Chesson, P. L.. (1985). Biological control in theory and practice. *Amer. Nat.* 125, 344-366.

Murdoch, W. W., Swarbuck, S. L., Luck, R. F., Walde, S., and Yu, D. S. (1996). Refuge dynamics and metapopulation dynamics: an experimental test. *Amer. Nat.* 147, 424-444.

National Research Council. (1996) " Ecologically-Based Pest Management: Solutions for a New Century." Board on Agriculture, Committee on Pest and Pathogen Control

Through Management of Biological Control Agents and Enhanced Cycles and Natural Processes. National Academy Press. Washington, DC.

Olszak, R. W. (1991). The relations between the aphids and parasitoids occurring on apple trees and on six species of shrubs. *In* "Biology and Impact of Aphidophaga." (L. Polgar, R. J. Chambers, A. F. G. Dixon, and I. Hodek, eds.), pp. 61-65. SPB Academic. The Hague.

Olszak, R. W. (1994). Parasitoids associated with different aphid species, their effectiveness and population dynamics. *Norw. J. Agric. Sci. Suppl.* 16, 362367.

Pickett, C. H., and Bugg, R. L. "Enhancing Biological Control: Habitat Management to Promote Natural Enemies of Agricultural Pests." University of California Press. Berkeley, CA. (in preparation).

Pickett, S. T. A., and White, P. S. (1985). "The Ecology of Natural Disturbance and Patch Dynamics." Academic Press. San Diego, CA.

Powell, W. (1986). Enhancing parasitoid activity in crops. *In* "Insect Parasitoids." (J. Waage, and D. Greathead, eds.), pp. 319- 340. Academic Press. London, U.K.

Price, P. W. (1991). Evolutionary theory of host and parasitoid interactions. *Biol. Cont.* 1, 83-93.

Price, P. W. (1994). Evolution of parasitoid communities. *In* "Parasitoid Community Ecology." (B. A. Hawkins, and W. Sheehan, eds.), pp. 472-491. Oxford University Press. Cambridge, U.K.

Rabb, R. L., Stinner R. E., and van den Bosch, R. (1976). Conservation and augmentation of natural enemies. *In* "Theory and Practice of Biological Control." (C. B. Huffaker, and P. S. Messenger, eds.), pp. 233-254. Academic Press. New York, NY.

Reice, S. R. (1994). Nonequilibrium determinants of biological community structure. *Amer. Sci.* 82, 424-434.

Ryszkowski, L., Karg, J., Margalit, G., Paoletti, M. G., and Zlotin, R. (1993). Aboveground insect biomass in agricultural landscapes of Europe. *In* "Landscape Ecology and Agroecosystems." (R. G. H. Bunce, L. Ryszkowski, and M. G. Paoletti, eds.), pp. 71-82. Lewis Publishers. Boca Raton, FL.

Sedivy, J. (1995). Hymenopterous parasitoids of cereal leaf beetle, *Oulema galleciana* Heyd. *Ochr. Rostl.* 31, 227-235.

Sheehan, W. (1986). Response by specialist and generalist natural enemies to agroecosystem diversification: a selective review. *Environ. Entomol.* 15, 456-461.

Sheehan, W. (1991). Host range patterns of hymenopteran parasitoids of exophytic lepidopteran foliovores. *In* "Insect-Plant Interactions." (E. Bernays, ed.), Vol. III. pp. 209-247. CRC Press. Boca Raton, FL.

Sousa, W. P. (1984). The role of disturbance in natural communities. *Annu. Rev. Ecol. Sys.* 15, 353-391.

Stilling, P. (1990). Calculating the establishment rates of parasitoids in classical biological control. *Amer. Entomol.* 36, 225-230.

Szentkiralyi, F., and Kozar, F. (1991). How many species are there in apple insect communities?: testing the resource diversity and intermediate disturbance hypotheses. *Ecol. Entomol.* 16, 491-503.

Townes, H. (1972). Ichneumonidae as biological control agents. *In* "Proc. Tall Timbers Conf. on Ecol. Anim. Cont. by Hab. Manag." (R. Komarek, ed.), 3, 235-248.

van den Bosch, R., and Telford, A. D. (1964). Environmental modification and biological control. *In* "Biological Control of Pests and Weeds." (P. DeBach, ed.), pp. 459-488. Reinhold. New York, NY.

van Emden, H. F. (1963). Observations on the effects of flowers on the activity of parasitic Hymenoptera. *Entomol. Mon. Mag.* 98, 265-270.

van Emden, H. F. (1965). The role of uncultivated land in the biology of crop pests and beneficial insects. *Sci. Hort.* 17, 1 21-136.

van Emden, H. F. (1990). Plant diversity and natural enemy efficiency in agroecosystems. *In* "Critical Issues in Biological Control." (M. Mackauer, L. E. Ehler, and J. Roland, eds.), pp. 63-80. Intercept. Andover, Hants, U.K.

Wratten, S. D. (1994). Habitat management for biological control: a contribution to agricultural sustainability. *In* "Sustainable Agriculture: Proceedings of a Meeting Held at Orange Agricultural College." (G. M. Gurr, B. J. Baldwin, and B. R. Johnson, eds.), pp. 34-59. The University of Sydney, New South Wales, Australia.

Wratten, S. D., and Thomas, C. F. G. (1990). Farm-scale dynamics of predators and parasitoids in agricultural landscapes. *In* "Landscape Ecology and Agroecosystems." (R. G. H. Bunce, L., Ryszkowski, and M. G. Paoletti, eds.), pp. 219-237. Lewis Publishers. Boca Raton, FL.

Zhao, J. Z., Ayers, G. S., Grafius, E. J., and Stehr, F. W. (1992). Effects of neighboring nectar-producing plants on populations of pest Lepidoptera and their parasitoids in broccoli plantings. *Great Lakes Entomol.* 25, 253-258.

CHAPTER
7

HABITAT ENHANCEMENT AND CONSERVATION OF NATURAL ENEMIES OF INSECTS

David N. Ferro and Jeremy N. McNeil

I. INTRODUCTION

The use of biological control agents is one approach to reduce the undesirable ecological and health problems associated with the overuse of chemical insecticides in agroecosystems. However, indigenous natural enemies often fail to maintain crop damage below economically acceptable levels. This is often due to their inability to cause high mortality during the initial colonization of agroecosystems by insect pests, which may be associated with the patchiness of prey and a number of interacting factors related to foraging behavior and numbers of natural enemies. However, it is not our intent to address these issues, as they have recently been considered by other authors (Wade and Murdoch, 1988; Karieva, 1990). In this chapter we concentrate on the biological information needed to enhance the actual densities of natural enemies, especially in the early part of the season, through habitat management.

Many cropping systems are annuals, so there is a distinct time window during which natural enemies occur in this temporary habitat. However, prior to sowing and following harvest natural enemies must find other resources essential for survival either within the field or in adjacent habitats. These resources could include food (such as pollen, honeydew, or nectar), alternate (often nonpest) arthropod hosts, and suitable sites to overwinter. Weedy fields frequently harbor a greater diversity and abundance of natural enemies (Shelton and Edwards, 1983) that can regulate pest populations. For example, Smith (1976 a,b) found over 4000 cabbage aphids *(Brevicoryne brassicae* (L.)) per m^2 in weed-free Brussels sprout plots, while simultaneously recording negligible populations in weedy plots: attributed to higher densities of anthocorid predators and aphidophagous syrphids. Similarly, intercropping and strip-cutting within the agroecosystem have been considered as a means of modifying the agroecosystem to increase densities of natural enemies (Grossman and Quarles, 1993). However, these approaches are generally unacceptable to most farmers as they are unwilling to have weeds and other plants compete with their crops (but see Chapters 8 and 9).

There is clear evidence that plants outside the cultivated field may provide the necessary resources to increase the impact of natural enemies. For example, Powell (1986) reviewed cases where increased parasitism was attributed to flowers occurring outside cultivated fields, which provided nectar sources for adult parasitoids or alternate hosts at times when pest species were not present in the crop. To date, most research on the enhancement of populations of natural enemies of insect pests has centered on providing greater plant diversity within and just adjacent to the cropping system (Altieri and Whitcomb, 1979; Andow, 1991). However, if biological control specialists wish to maximize the impact of natural enemies through habitat management, we are convinced that it is essential to look beyond the immediate confines of agricultural lands to the uncultivated habitats that separate or surround cultivated fields. In addition, we propose that a more systematic approach be undertaken to obtain a solid understanding of the biology of natural enemies throughout the year and not just during the agricultural growing season. The database should include a thorough understanding of the resources needed by the natural enemies at different times of the year and how these are obtained through movement between different habitats within the agricultural landscape. Once these resources have been identified, it will be necessary to determine the spatial and temporal arrangement of these resources within the agricultural landscape.

II. AGRICULTURAL LANDSCAPE MOSAICS

A landscape is defined as an aggregate of distinct clusters of managed and unmanaged habitats separated by relatively well defined boundaries, especially with respect to vegetation structure (Forman and Godron, 1981). Thus, the landscape is composed of communities or species assemblages surrounded by a matrix with a dissimilar community structure or composition. Within the agricultural landscape, a cultivated field is just one patch. While the crop is of major economic interest as a valued resource it may not be the most important component of the system from the perspective of conservation biological control. Insect species found within the crop patch should be seen as subsets of metapopulations *(sensu* Merriam, 1988), where movement among subpopulations is a key to survival of both pests and their natural enemies (Opdam, 1989). Within this framework, the probabilities of herbivores and their natural enemies establishing in a crop will be inversely proportional to the distance from other subpopulations and the size of the "island" under cultivation (Kindvall and Ahlen, 1992; Tscharntke, 1992; Kruess and Tscharntke, 1994).

Many soil-inhabiting predators need shelter outside of the cropping system to survive, especially in annual cropping systems where fields are heavily cultivated at the end of the growing season. This is particularly true for predators that do not disperse by fling. Polyphagous predators have been shown to reduce populations of aphids infesting arable crops (Edwards *et al.,* 1979; Wratten and Pearson, 1982). A study by Coombes and Sotherton (1986) on the dispersal of carabid and staphylinid adults into cereal fields from field boundaries showed that beetles could be recovered up to 200 m into the fields and that two patterns of dispersal could be distinguished. The carabid *Agonum dorsale* Pont., which does not disperse by flight, gradually colonized deeper into the field from hedgerows, whereas the staphylinid *Tachyporus hypnorum* F., which can migrate by flight, reached peak numbers at the same time at all distances along the transects. Although this study showed the importance of hedgerows in providing overwintering sites for these predators, it also showed the limited dispersal capabilities of these aphid predators.

Thomas and Wratten (1988) sowed the grass *Dactylis glomerata L.* into a raised bank running parallel to rows of the cereal crop to create "'predator conservation strips." This study was followed by another project where several different grasses were sown separately or in combination into raised beds to create "island" habitats for predators within and adjacent to the cropping system (Thomas *et al.,* 1991). They found, within the first year of establishment, predator densities of up to 150 m^{-2} and by the second year up to 1500 m^2. The primary carabid predator was *Demetrias atricapillus* (L.) and the primary staphylinid was *Tachyporus hypnorum (F.).* Although these islands provided a concentration of predators that could move into cereal fields to feed on aphids, it is necessary to determine how to best distribute these islands throughout the cropping system and to identify long-term procedures for maintaining these islands as a resource for natural enemies. These island reservoirs were created and then the population dynamics of the predators and their biological control potential were evaluated without knowing what resources were being provided. We believe an alternative and more direct approach is to study the biology and behavior of natural enemies outside of the crop and then modify the habitat to provide necessary resources (but see Chapter 6).

III.   IDENTIFYING ESSENTIAL RESOURCES: BASES FOR HABITAT MODIFICATION

To illustrate the proposed directed approach we will consider two cases, one with a predator and the other with a parasitoid. Detailed studies of their seasonal biology have identified where the absence of the target pest species results in "gaps" with respect to the availability of resources essential for the survival of natural enemies (see Chapter 3).

The twelve-spotted ladybird beetle *Coleomegilla maculata* (DeGeer), a polyphagous coccinellid (Hodek, 1973), is an important predator of eggs and small larvae of the Colorado potato beetle (Hazzard and Ferro, 1991; Hazzard *et al.,* 1991). Adults overwinter in undisturbed habitats around the edge of fields, usually in aggregations at the base of trees such as willow and poplar (Benton and Crump, 1979), sycamore (Conrad, 1959), and maple (Hazzard *et al.,* 1991). Thus, each spring large numbers of *C. maculata* adults are present in lands adjacent to newly planted potato fields, especially if the field was previously planted with corn. These beetles could eliminate 35-60% of Colorado potato beetle eggs (Hazzard *et al.,* 1991). However, the beetles must find alternate food sources from the time they emerge in mid-April until late May when Colorado potato beetles begin to lay eggs. Furthermore, even though this predator can significantly reduce Colorado potato beetle populations their immature stages are rarely found in potato fields. Thus, ladybeetle adults must exploit other oviposition sites with suitable food resources for their young; these resources often occur outside of the crop. Adults return into the potato fields to feed on second-generation Colorado potato beetle eggs but again must find alternate food and oviposition sites following harvest.

In spring and early summer Lopez and Ferro (unpublished data) observed *C. maculata* adults feeding on pollen of dandelion *Taraxacum officinale* Weber (as previously reported by Solbreck, 1974) and on the pollen of four species of grass *(Bromus tectorum* L.*, Poa annua* L.*, P. trivialis* L., and *Alopecurus myosuroides* Hudson). In addition, at different months during the summer they observed beetles feeding on nectary secretions and pollen of yellow cress *Rorippa palustris* (L.) Besser and on fungal spores on the leaf surface of pigweed *Chenopodiium album* L. Furthermore, adults also fed on aphid colonies occurring on these two plants. This array of nonagricultural plants serve as essential adult food resources both before and after the first oviposition period by the Colorado potato beetle. Thus, if densities of these plants could be increased through the selective application farmscaping techniques (King and Olkowski, 1991), it could increase survivorship of *C. maculata* at different times of the growing season and concentrate populations in close proximity to the potato crop. Prior to entering adult diapause in the fall, *C. maculata* adults feed actively in corn, exploiting both pollen and aphids (Coll and Bottrell, 1991). Further studies on naturally occurring feeding sites exploited in the fall would provide insight into other possible habitat management practices that could help sustain high overwintering populations.

    Cotesia (Apanteles) congregata (Say) is the major parasitoid attacking the tobacco hornworm *Manduca sexta* (L.) and in untreated fields larval parasitism easily exceeds 50% during the second generation of the pest species (McNeil and Rabb, 1973a). Furthermore, the parasitoids developing at this time will enter diapause as prepupae and overwinter within the tobacco agroecosystem (McNeil and Rabb, 1973b). Thus, as more than 50 parasitoids can emerge from one host, these populations represent a considerable reservoir of potential natural enemies for the following year. However, an examination of spring emergence patterns clearly showed that *C. congregata* adults emerge 4-6 weeks prior to the emergence of *M. sexta* (McNeil and Rabb, 1973a), a period considerably longer than estimates of adult parasitoid longevity. Thus, if these populations are to contribute to subsequent generations they must migrate from the agroecosystem and locate alternate hosts within which to complete one generation before subsequently returning to exploit tobacco hornworms. It is therefore essential to identify what nonpest hosts are available and utilized by this parasitoid and to determine if the plant species exploited by these hosts can be integrated into a landscape management scheme to foster this potentially important pool of natural enemies.

    A variation on this approach has been carried out with a certain degree of success with respect to the parasitoid *Anagrus epos* Girault, which attacks eggs of the grape leafhopper *Erythroneura elegantula* Osborn. The grape leafhopper, a major pest of grapes in the San Joaquin Valley of California overwinters as an adult whereas the parasitoid overwinters as an egg within the eggs of a non-economic species, the blackberry leafhopper *Dikrella californica* (Lawson); which occurs throughout the year on wild blackberries *(Rubus* spp.). Doutt *et al. (*1966) reported that vineyards located within 5.6 km of an established blackberry refuge will benefit from the immigration of parasitoids early in the season, whereas those beyond this distance were rarely colonized by wasps until later in the season. More recently, *A. epos* has been reported to overwinter in another leafhopper, *Edwardsiana prunicola* (Edwards) (Kido *et al.,* 1984) and vineyards with French prune (the leafhopper's host) in the vicinity have higher levels of parasitism (Flaherty *et al.,* 1985). Mark-recapture studies have clearly shown that these reservoirs serve as sources of immigrants in spring (Corbett and Rosenheim, 1996).

    For a given cultivated field, the effective size of the agricultural landscape to be included when examining the nonpest alternate hosts exploited by natural enemies as well as planning habitat manipulations to favor biological control agents will vary depending on the insect species under consideration. The landscape of a highly vagile insect will obviously be considerably greater than that of a species with very limited powers of dispersal. The work on the leafhopper parasitoid *A. epos* (Corbett and Rosenheim, 1996) clearly underscores several points concerning the necessity of understanding the movement capabilities of natural enemies. First, as shown in previous studies, the level of colonization is a function of distance from reservoirs. However, more importantly they found that although refuges did contribute to the parasitoids populations in neighboring vineyards, the majority came from sites further away. In fact, the French prune trees acted as a windbreak, thereby favoring the fallout of migrating parasitoids present in the wind stream about the trees. Thus, understanding the dispersal ability of the natural enemies one wishes to conserve will aid in determining the location of noncultivated plant species with respect to the crop.

IV. CONCLUSIONS

A. Landscape Perspective

The effects of diversifying agricultural landscapes on insect pests and their natural enemies are, in and unto themselves, highly variable. Thus, all-encompassing generalizations will be of somewhat limited value in the elaboration of decision-making processes concerning diversification of plant species to enhance the efficacy of parasitoids and predators. As a result many decisions will have to be made on a case by case basis using an understanding of the seasonal biologies of the species under consideration. As noted earlier, most studies have generally considered the crop as the central point of the system and we believe it is time to take a broader perspective, that of landscape ecology (a point also made by Landis and Marino (in press) and in Chapter 6).

We need a much better understanding of the ecology of parasitoids and predators outside of the cultivated habitat, identifying those resources that are necessary for their survivorship and reproduction. We must also determine to what extent populations within the crop contribute to the metapopulation in subsequent years. If these contributions are minor, then investments in habitat management should be oriented specifically to increasing the source populations outside the crop to ensure a greater number of immigrants each year, an action parallel to increasing the dosage of a chemical biocide. However, if the subpopulations within the cropping system contribute significantly to the year-to-year metapopulation dynamics then habitat modifications should not only consider tactics fostering immigration into the crop but also those augmenting the probability of successful emigration when this habitat becomes unsuitable. Such actions could include the addition of plant species to provide alternate hosts and/or food sources, habitats as suitable overwintering sites or the provision of corridors within the cropping system to facilitate movement between the different subcomponents of the metapopulation.

B.   Actions within an IPM Context: Benefits and Constraints

A specific modification of the landscape may prove beneficial with respect to one specific insect pest but as emphasized by Prokopy (1994) any potential actions should be evaluated within the context of a broader integrated management program of the agricultural crop. The reason for caution is that potential benefits may be less than unforeseen costs. For example, while blackberry plants serve as a host plant for alternative hosts of the parasitoid *A. epos* these same plants could be a reservoir for the bacterium responsible for Pierce's disease, a serious disease of grapes (Raju *et al.*, 1983). Thus, an action taken to increase the efficacy of natural enemies could incur losses through increased levels of disease. Intraguild predation (Polis *et al.*, 1989; Rosenheim *et al.*, 1995) may also increase if several species of natural enemies are favored as a result of habitat management. This could result in a lower impact on the target pest, despite a rise in the densities of natural enemies. Another potential problem is with respect to insect pathogens. Roland (1993) analyzed data on the duration of forest tent caterpillar outbreaks in Ontario, Canada and found that the best predictor of outbreaks was the amount of forest edge per km^2. He hypothesized that forest fragmentation may be negatively affecting parasitoids and pathogens that play a major role in dampening outbreaks. Subsequently, Roland and Kaupp (1995) demonstrated reduced transmission of a nuclear polyhedrosis virus at the forest edge compared with the forest interior, possibly due to the negative effect of UV levels on the viability of polyhedra. In general however, these potential problems should not deter research efforts in habitat modification for enhancing the abundance of natural enemies within the agricultural landscape.

Currently IPM practitioners focus on monitoring pest population densities and then recommending control tactics and/or cultural practices to be used. There is no reason the IPM practitioner could not work with growers to create a diversified nursery of resource plants outside of the crop or selectively use herbicides within the agricultural landscape to favor resource plants. This would be an additional cost to the grower and would require a higher level of education for the grower and practitioner. However, federal legislation is mandating further reductions in insecticide use by the year 2000 and this will only happen if we challenge insect pests on all fronts including biological control. For most annual crops, innundative releases of natural enemies are not cost- effective, and if the efficacy of naturally occurring biological control agents are to be maximized it will require habitat modifications favoring conservation of these agents. Thus, researchers must develop precise protocols for IPM practitioners to follow and if growers see the benefits of conserving natural enemies these practices will be accepted. This may require federal support for an area-wide approach. However, we need to start immediately in the acquisition of ecological databases for important natural enemies as these will facilitate the implementation of promising landscape modifications.

REFERENCES

Altieri, M. A., and Whitcomb, W. H. (1979). The potential use of weeds in the manipulation of beneficial insects. *Hort. Sci.* 14, 12-18.

Andow, D. A. (1991). Vegetational diversity and arthropod population response. *Annu. Rev. Entomol.* 36, 561-586.

Benton, A. H., and Crump, A. J. (1979). Observations on aggregation and overwintering in the coccinellid beetle *Coleomegilla maculata (DeGeer). N.Y. Entomol. Soc.* 87, 154-159.

Coll, M., and Bottrell, D. G. (1991). Microhabitat and resource selection of the European corn borer (Lepidoptera: Pyralidae) and its natural enemies in Maryland field corn. *Environ. Entomol.* 20, 526-533.

Conrad, M. S. (1959). The spotted lady beetle, *Coleomegilla maculata* (DeGeer) as a predator of European corn borer eggs. *J. Econ. Entomol.* 52, 843-847.

Coombes, D. S., and Sotherton, N. W. (1986). The dispersal and distribution of polyphagous predatory Coleoptera in cereals. *Ann. Appl. Biol.* 108, 461-474.

Corbett, A., and Rosenheim, J. A. (1996). Impact of a natural enemy overwintering refuge and its interaction with the surrounding landscape. *Ecol. Entomol.* 21, 155-164.

Doutt, R. L., Nakata, J., and Skinner, F. E. (1966). Dispersal of grape leafhopper parasites from a blackberry refuge. *Calif. Agric.* 20, 14-15.

Edwards, C. A., Sunderland, K. D., and George, K. S. (1979). Studies on polyphagous predators of cereal aphids. *J. Appl. Ecol.* 16, 811-823.

Flaherty, D. L., Wilson, L. T., Stern, V. M., and Kido, H. (1985). Biological control in San Joaquin Valley vineyards. *In* "Biological Control in Agricultural IPM Systems." (M. A. Hoy, and D. C. Herzog, eds.). pp. 501-520. Academic Press, New York, NY.

Forman, R. T. T., and Godron, M. (1981). Patches and structural components for a landscape ecology. *BioScience* 31, 733-740.

Grossman, J., and Quarles, W. (1993). Strip intercropping for biological control. *IPM Practitioner* 15, 1-11.

Hazzard, R. V., and Ferro, D. N. (1991). Feeding responses of adult *Coleomegilla maculata* (Coleoptera: Coccinellidae) to eggs of Colorado potato beetle (Coleoptera: Chrysomelidae) and green peach aphids (Homoptera: Aphididae). *Environ. Entomol.* 20, 644-651.

Hazzard, R. V., Ferro, D. N., Van Driesche, R. G., and A. F. Tuttle. (199 1). Mortality to eggs of Colorado potato beetle (Coleoptera: Chrysomelidae) from predation *by Coleomegilla maculata* (Coleoptera: Coccinelidae). *Environ. Entomol.* 20, 841-848.

Hodek, I. (1973). "Biology of Coccinellidae." Academia, Prague, Czechoslovakia.

Karieva, P. (1990). The spatial dimension in pest-enemy interactions. *In* "Critical Issues in Biological Control." (M. Mackauer, L. E. Ehler, and J. Roland, eds.), pp. 213-227. Intercept Ltd., Andover, Hants, U.K.

Kido, H., Flaherty, D. L., Bosch, D. F., and Valero, K. A. (1984). French prune trees as overwintering sites for the grape leafhopper egg parasite. *Amer. J. Enol. Vitic. 35,* 156-160.

Kindvall, O., and Ahlen, I. (1992). Geometrical factors and metapopulation dynamics of the bush cricket, dynamics of the bush cricket, *Metrioptera bicolor* Philippi (Orthoptera: Tettigoniidae). *Cons. Biol.* 6, 520-529.

King, S., and Olkowski, W. (1991). Farmscaping and IPM. *IPM Practitioner* 13, 1-12.

Kruess, A., and Tscharntke, T. (1994). Habitat fragmentation, species loss, and biological control. *Science* 264, 1581-1584.

Landis, D. A., and Marino, P. C. 1998. Landscape structure and extra-field processes: impact on management of pests and beneficials. *In* "Handbook of Pest Management." (J. Ruberson, ed.), Marcel Dekker Inc., New York, NY. (in press).

McNeil, J. N., and Rabb, R. L. (1973a). Life histories and seasonal biology of four hyperparasites of the tobacco hornworm, *Manduca sexta* (Lepidoptera: Sphingidae). *Can. Entomol.* 105, 1041-1052.

McNeil, J. N., and Rabb, R. L. (1973b). Physical and physiological factors in diapause initiation of two hyperparasites of the tobacco hornworm, *Manduca sexta. J Insect Physiol.* 19, 2107-2118.

Merriam, G. (1988). Landscape dynamics in farmland. *Trends Ecol. Evol. 3,* 16-20.

Opdam, P. (1989). Dispersal in fragmented populations. *In* "Species Dispersal in Agricultural Habitats." (R. G. H. Bunce, and D. C. Howard, eds.), pp. 3-17. Belhaven Press. London, U.K.

Polis, G. A., Myers, C. A., and Holt, R. D. (1989). The ecology and evolution of intraguild predation: potential competitors that eat each other. *Annu. Rev. Ecol. Syst.* 20, 297-330.

Powell, W. (1986). Enhancing parasite activity in crops. *In* "Insect Parasitoids." (J. Waage, and D. Greathead, eds.), pp. 319-340. Academic Press, London, U.K.

Raju, B., Goheen, A. C., and Frazier, N. W. (1983). Occurrence of Pierce's disease in plants and vectors in California. *Phytopathol.* 73, 1309-1313.

Roland, J. (1993). Large scale forest fragmentation increases the duration of tent caterpillar outbreak. *Oecologia* 93, 25-30.

Roland, J., and Kaupp, W. F. (1995). Reduced transmission of forest tent caterpillar (Lepidoptera: Lasiocampidae) nuclear polyhedrosis virus at the forest edge. *Environ. Entomol.* 24, 1175-1178.

Rosenheim, J. A., Kaya, H. A., Ehler, L. E., Marois, J. J., and Jaffee, B. A. (1995). Intraguild predation among biological-control agents: theory and evidence. *Biol. Cont.* 6, 303-335.

Shelton, M. D., and Edwards, C. R. (1983). Effects of weeds on the diversity and abundance of insects in soybeans. *Environ. Entomol.* 12, 296-298.

Smith, J. 0. (1976a). Influence of crop background on aphids and other phytophagous insects on Brussels sprouts. *Ann. Appl. Biol.* 83, 1-13.

Smith, J. G. (1976b). Influence of crop background on natural enemies of aphids on Brussels sprouts. *Ann. Appl. Biol.* 83, 15-29.

Solbreck, C. (1974). Maturation of post-hibernation flight behaviour in the coccinellid *Coleomegilla maculata* (DeGeer). *Oecologia* 17, 265-275.

Thomas, M. B., and Wratten, S. D. (1988). Manipulating the arable crop environment to enhance the activity of predatory insects. *Aspects Appl. Biol.* 17, 57-66.

Thomas, M. B., Wratten, S. D., and Sotherton, N. W. (1991). Creation of 'island' habitats in farmland to manipulate populations of beneficial arthropods: predator densities and emigration. *J. Appl. Ecol.* 28, 906-917.

Tscharntke, T. (1992). Fragmentation of *Phragmites* habitats, minimum viable population size, habitat suitability, and local extinction of moths, midges, flies, aphids and birds. *Cons. Biol.* 6, 530-535.

Walde, S. J., and Murdoch, D. D. (1988). Spatial density dependence in parasitoids. *Annu. Rev. Entomol.* 33, 441-466.

Wratten, S. D., and Pearson, J. (1982). Predation of sugar beet aphids in New Zealand. *Ann. Appl. Biol.* 101, 178-181.

CHAPTER
8

SOWN WEED STRIPS: ARTIFICIAL ECOLOGICAL
COMPENSATION AREAS AS AN IMPORTANT
TOOL IN CONSERVATION BIOLOGICAL CONTROL

W. Nentwig, T. Frank, and C. Lethmayer

I. INTRODUCTION

Today our agricultural landscape is so intensively managed that the original
species diversity of many natural habitats has disappeared or became endangered.
Due to the intensive use of herbicides, vegetational diversity (especially that of
flowering species) has become extremely reduced. A similar reduction is observed
among herbivores and predaceous species (Heydemann and Meyer, 1983). The decrease
of floral and faunal diversity in agroecosystems is a consequence of intense management
(Altieri and Letourneau, 1982; Risch, 1987; Kruess and Tscharntke, 1994).
Additionally, high agrochemical input causes many unintended effects such as resistance
in pests, soil erosion, water pollution, and even climate change (Tivy, 1990; Nentwig,
1995). Therefore, not only is nature threatened but humanity is as well.

There are many proposed approaches aimed at making agriculture more
sustainable, reducing the amount of agrochemicals used, and enhancing biodiversity
in agricultural ecosystems. This can be achieved, for example, by organic farming,
crop rotation, small-scale fields, and maintenance of natural areas between
agroecosystems. The latter, sometimes referred to as ecological compensation areas,
consist of separate or interconnected fields which form a network (e.g., Thomas *et
al.,* 1991; Raskin *et al.,* 1992). This buffer zone can act as refuge area and/or dispersal
center and offers many species adequate niches; compensating, at least partly, for
the negative effects of agriculture.

One type of ecological compensation area is a sown weed strip. Though
it is an artificial habitat (as is the agricultural landscape) such strips have many
advantages and offer good opportunities to combine agricultural use and nature

conservation. Sown weed strips are highly attractive for many species and increase biodiversity. Even rare species can occur in intensively managed agricultural landscape if the minimum habitat area they require is available and their specific ecological requirements can be fulfilled, I. e., their host plant (Frank, 1994). In many cases this is possible with a strip-management program such as the one presented in this chapter. Sown weed strips have been intensively investigated in our research group over the past decade and data primarily on ecological aspects are presented here. We will explain how such strips can be established and how they develop. Since biodiversity is a major goal, we will show how sown weed strips enhance the species number of several beneficial arthropod groups without enhancing the abundance of pests. This apparent contradiction is explained by discussing obvious mechanisms which are required for these effects to occur.

II. HOW TO CREATE SOWN WEED STRIPS

One can contend that if a given agricultural area contains sufficient ecological compensation areas, adding more such areas would be unnecessary. However, "enough" is a relative term. We consider the minimum required ecological compensation area (i.e., totally unmanaged (and unused) or only partly managed areas) to be 5 to 10 % of an agricultural landscape. But almost no intensively used landscape offers such an ideal situation. Therefore, additional ecological compensation areas are necessary. We see two options: separating selected areas and allowing normal plant succession to proceed or sowing a given seed mixture to turn succession in a given direction. In extensively used, large-scale landscapes the first possibility may be feasible. In intensively managed small-scale areas, however, the establishment of "semi-natural" compensation areas may be preferred. In these areas plant development is faster, problematic weeds can be avoided more easily, and weeds are more controllable than in free succession areas.

Sown weed strips usually have a width of 3 to 8 m (Heitzmann-Hofmann, 1995; Günter, 1997). They are situated at the border of a field or divide large fields in small parts so that the distance between strips does not exceed 50 to 100 m. The length of the strip depends on the field length and may be in the range of several hundred meters. Weed strips should start or end at other ecological compensation areas such as road sides, field margins, hedges, forest remnants, dry slopes, and so on. Thus, the total system of sown weed strips forms a network of ecological compensation areas which connects several types of natural, semi-natural, and artificial habitats. The fields are embedded into this network and profit from its high diversity (see below).

After careful soil preparation (Günter, 1997) a seed mixture is sown. The composition of this mixture is based on the results of a long screening process in which specific characters of approximately 100 species of wild flowers and cultivated plants are analyzed. Important plant properties have been investigated in monoculture plots of 10 m^2 (3 to 5 replicates) which had been sown within a large wheat field (Heitzmann-Hofmann, 1995). These properties included insect diversity and abundance of beneficial species in patches of the test plant species (Weiss and Stettmer, 1991; Frei and Manhart, 1992), length of flowering period, survival capacity in agricultural soils, longevity in a complex plant mixture as well as the tendency to dispersal into adjacent fields (Heitzmann-Hofmann, 1995). In the last years of field experiments we tried to optimize the seed quantity to reduce the seed costs by allowing a maximum amount of natural reseeding of weeds. In contrast to earlier mixtures, our currently recommended seed mixture avoids *Brassica* species (but still contains *Sinapis alba,* also a Brassicaceae) because of the potential of the former to act as a reservoir for crop pathogens. However, we did not generally avoid plants which harbor potential pest insects (see discussion below).

Our prepared seed mixture now consists of 29 species of wild flowers and cultivated plants (Table 1) and is available from several commercial seed distributors in Switzerland. The selected species cover a wide range of low, middle sized, and tall plants. They include early to late flowering plants as well as annuals, biennials, and perennials. Though this seed mixture has officially been recommended by the Swiss Federal Research Station for Agroecology and Agriculture we do not consider it to be the only useful mixture for sown weed strips. Though it yields satisfying results (see below) it may still be ameliorated, e.g., by changing the seed ratio of given species or by adding and omitting some species. Since the commercial production of wild flower seeds is not cheap there is a particular need to reduce the costs of the wild flower seed mixture.

The seed mixture proposed here is part of a regional concept and is adapted to the Swiss Plateau. It shall not uncritically be transferred to other regions but may be established in other regions after slight modifications with regionally produced seeds. Such field tests and discussions have been started in Germany and Austria. Much simpler seed mixtures have also been tested, and these may consist only of one to four noncrop species (mixtures of clover and grass, *Phacelia,* sunflowers, etc.). Though we think that such reduced mixtures, often only with annual plants, cannot be compared with our multispecies mixture there is still a great deal of research

Table 1. Composition of the seed mixture for weed strips ("University of Berne 1997")[a]

Plant species	g/ha	%
Achillea millefolium	40	0.2
Agrostemma githago	500	2.5
Anthemis tinctoria	10	0.05
Arctium lappa	20	0.1
*Borago officinalis**	200	1.0
Centaurea cyanus	500	2.5
Centaurea jacea	200	1.0
Chrysanthemum leucanthemum	80	0.4
Cichorium intybus	120	0.6
Daucus carota	150	0.75
Dipsacus silvester	5	0.03
Echium vulgare	300	1.5
*Fagopyrum esculentum**	15730	78.65
*Foeniculum vulgare**	200	1.0
Hypericum perforatum	60	0.3
Legousia speculum-veneris	30	0.15
Malva silvestris	80	0.4
*Medicago lupulina**	120	0.6
Melilotus albus	30	0.15
Melilotus officinalis	20	0.1
Oenothera biennis	30	0.15
*Onobrychis viciifolia**	1000	5.0
Origanum vulgare	50	0.25

Table 1-Continued

| Papaver rhoeas | 150 | 0.75 |
Plant species	g/ha	%
Pastinaca sativa	200	1.0
Silene alba	80	0.4
Sinapis alba*	40	0.2
Tanacetum vulgare	5	0.03
Verbascum densiflorum	50	0.25
TOTAL	20000	100.00

[a] Strips contain wild flowers of Swiss origin and *cultivated species (Günter, 1997). Recommended by the Swiss Federal Research Station for Agroecology and Agriculture, suitable for all major crops at amount of 20 kg/ha.

required in order to transfer the system presented here to other climate zones (e.g., Mediterranean countries) or other continents.

Maintenance of the vegetation which grows from such a multispecies seed mixture with additional wild species from the respective seed reservoir in the soil is a difficult task. Experiments on succession usually cover several years and cannot be performed as quickly nor repeated as frequently as desired. Our (admittedly insufficient) results indicate that plant succession in sown strips maintains a high diversity at least in the first 3 years. With respect to the specific site conditions (e.g., soil condition), weedy grasses may be advantaged and become more common under special circumstances. Problematic weeds (e.g., Cirsium sp., Rumex sp.) can be controlled by individual treatments of the plants (Fig. 1). To slow down succession towards grasses and to prevent the dominance of only a few species or the appearance of woody plants, we recommend alternating mowing half the strip every second year, but we also try to minimize maintenance and labor costs (Günter, 1997).

Although our experience with the actual complex seed mixture for weed strips is restricted to only a few years we assume that these ecological compensation areas can stay for many years. When these habitats become less and less diverse some kind of regeneration technique could be applied. This may be accomplished by some kind of a minimum soil treatment or by complete ploughing followed by additional, or new, sowing. So in the long-term, a steady state between young and old sown weed strips will be achieved within a landscape. This provides the type of mosaic in which landscapes naturally regenerate (Remmert, 1991) and comprises an important feature which stabilizes agricultural landscapes.

Extensive evaluations have been performed to optimize the technical aspects of sowing as well as approaches to maintaining and mowing weed strips, their inclusion into crop rotations, and evaluations of their effects on the most common crops (Günter, 1997). Here we will not report on these agronomic aspects but rather focus on more ecological aspects. Sown weed strips not only represent important refugia and dispersal centers, but also offer many additional niches. Thus, we expect a higher diversity of plants and arthropods, some of which may be regarded as beneficials, pests, or as indifferent species.

Our concept of sown weed strips has been developed primarily for use in the temperate zone where increase of productivity is not a major concern. Agricultural yields are already high and actual efforts should not intend to further increase the high production but to stabilize it with respect to an environmentally friendly and sustainable practice. So, we will not give yield data here. Most investigations presented

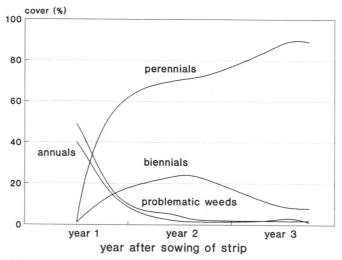

Figure 1. Scheme of the vegetation development in a sown weed strip over time. Based on 3 vegetation samples per year the development has been extrapolated for 4 plant groups in a most successfully developing sown weed strip (Günter, 1997). Sowing occurred in the first year at the end of April, a first cut to reduce spontaneously emerging problematic weeds was performed 6 weeks later, the first vegetation sample was at the end of July. At the end of the second year the vegetation was cut again.

here have been made on a small-scale (1 to 100 m) where soil heterogeneity or uneven fertilizer distribution caused methodological problems. Generally, however, there are no pronounced effects of weed strips on yield neither in a positive nor in a negative direction (Günter, 1997).

III. ENHANCING DIVERSITY OF BENEFICIALS

A.   Spiders

The number of spider species and individuals in sown weed strips and adjacent crops were studied using pitfall traps (Frank and Nentwig, 1995). As a general result, the number of species in the weed strips was always higher than in the adjacent fields. After one winter, the two year old weed strips contained significantly more species than the younger strips. This was likely due to an especially dense and richly structured vegetation in these older weed strips and we assume that this vegetation served as a place of hibernation for many species. Spiders are able to breed and hibernate in great numbers in sown weed strips (Bürki and Hausammann, 1993; Lys and Nentwig, 1994). Densely vegetated Budapest also appeared to be particularly attractive sites for hibernation (Luczak, 1979; Wiedemeier and Duelli, 1993). Most species were found in areas in which cultivated fields were close to sown weed strips (13 m from the strips) than in fields at a greater (50 m) distance to the weed strips.

Similar results were obtained by measuring the number of individuals caught by pitfall traps. The number of individuals also tended to decline with increasing distance to the weed strips. Similarly, Katz et al. (1989) found declining numbers of species and individuals in different crops with increasing distance from a natural meadow. Such results show that semi-natural habitats are able to enrich surrounding crops in terms of number of species and individuals. This can be of great importance for the enhancement of beneficial spiders in fields. Jmhasly and Nentwig (1995) measured the density of spiders in a wheat field bordering weed strips with a vacuum insect net and observed declining densities with increasing distance from weed strips (Fig. 2). In addition to spider densities, web cover (determined by visual observations) was in most cases higher near the weed strips than away from them. Thus, number of web-building spiders declined with increasing distance from the weed strips. A corresponding result was obtained for epigeic spiders studied in the same area.

Frank and Nentwig (1995) found different distribution patterns of spiders in sown weed strips. Some species were almost entirely confined to these strips but rarely occurred elsewhere. For another abundant group of spiders dispersal from

W. Nentwig, T. Frank, and C. Lethmayer

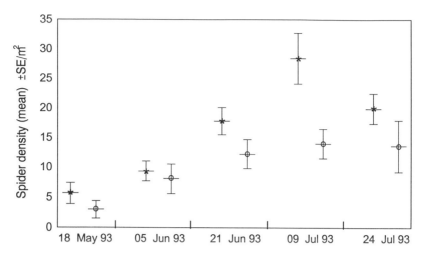

Figure 2. Density estimates (mean ± SE / m²) for spiders in winter wheat at 1.5 m (x) and
13 m (o) from weed strips at different sampling dates (N = 10 samples per site, 9
July: p < 0.05, Mann-Whitney U-test) (Jmhasly and Nentwig, 1995).

weed strips into adjacent fields was observed. In spring, such species were more
abundant in the weed strips but as the season progressed more individuals appeared
in the fields. This distribution pattern is typical for many of the most abundant species
(more than 80% of all specimens) *(Erigone atra* Blackwall, *E. dentipalpis* (Wider),
Oedothorax apicatus Blackwall, *Pardosa agrestis* (Westring), *P. palustris* (L.)) (Frank
and Nentwig, 1995). These species are also dominant in most European agroecosystems.
The importance of spiders as predators of pests in agricultural areas has been frequently
demonstrated (see Nyffeler and Benz, 1987 and Riechert and Bishop, 1990 reviews)
and spiders are considered potentially valuable in the control of pests (Sunderland,
1987). The lack of manipulation and thus the absence of disturbance and the presence
of a complex vegetational structure in strips provide the conditions which favor
colonization by spiders. Thus, the effectiveness of beneficial spiders in crops can
be increased by sown weed strips.

Positive effects of sown weed strips on spider density were not only observed
in crops but also in an apple orchard. Wyss *et al.* (1995) compared the number of
spiders in a strip-managed area of an apple orchard with a control area without sown
weed strips. From spring to autumn, they found more individuals of all spider families
in the strip-managed area of the orchard than in control areas (i.e., with no strips).
In autumn, the number of all web-building spiders observed on the apple trees was
higher in the strip- managed area than in the control area (Fig. 3). Therefore, they
concluded that the larger number of web-building spiders in the strip-managed area
could more efficiently reduce the winged aphids when they returned from their summer

host plants (herbs) to their winter hosts (the apple trees). Since spiders are among the most important predators of aphids and high percentages of their diet is composed of aphids (see review by Nentwig, 1987) this regulation in autumn could lead to an effective reduction of aphids appearing early in the spring of the next year.

B.   Ground beetles

Ground beetles (i.e., Carabidae) are very numerous predacious insects in arable land and therefore are considered important predators (Hance, 1987). Semi-natural habitats (e.g., margin strips as mentioned by Klinger (1987) or Lagerlöf and Wallin (1993) and grassy strips as mentioned by Thomas (1990)) have shown to enhance carabid densities as well as species diversity when compared to the field center. Similar results were obtained on the species diversity of ground beetles in sown weed strips (Frank, 1997). The number of species in the weed strips was, in most cases, higher than in adjacent crops and the number of species in the fields decreased with increasing distance from the weed strips.

Species similarity between weed strip and adjacent field parts (13 m from the strips) was higher than between weed strips and the centers of fields (50 m from the strips) (Frank, 1997). This suggests a particularly intensive interaction in terms of species migration between the sown weed strips and parts of nearby fields. This effect was especially distinct near older weed strips which already served as hibernation sites for many carabid species. A similar trend toward species enrichment in adjacent

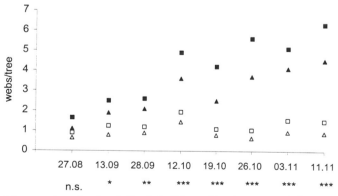

Figure 3. Mean number of webs of all web-building spiders (square symbols) and of Araneidae (triangular symbols) per tree in the strip- managed area (black symbol) and in the control area without strips (open symbol) at 8 sampling dates in autumn 1993 (* $p < 0.05$, ** $p < 0.01$, *** $p < 0.001$, Mann-Whitney U-test) (Wyss et al., 1995).

fields was also observed by Klinger (1987) using margin strips. Lys and Nentwig (1992), studying sown weed strips in another region also found more species of ground beetles in a strip-managed wheat field than in a wheat field without strips. The high species diversity in sown weed strips obtained in different studies indicates the obvious existence of conditions suitable for ground beetles, probably including a richly structured vegetation, associated favorable microclimate, and prey abundance (Zangger *et al.,* 1994). In addition, weed strips provided the beetles protection from deleterious farming operations.

The dispersal ability of ground beetles between semi-natural biotopes and adjacent cultivated fields is important with respect to their potential within integrated pest management. Therefore, movements and densities of the most abundant carabid species in weed strips and an adjacent wheat field were studied using mark-and-recapture techniques (Lys and Nentwig, 1992; Lys *et al.,* 1994). For several species much higher recapture rates, indicating a higher activity, were found in the strip-managed area than in the control area. This higher activity was generally due to a prolongation of the reproductive period in the strip-managed area. Tenerals of *Poecilus cupreus* L. and *Pterostichus melanarius* Illiger have been shown to appear earlier in the season and in greater numbers in the strip-managed area than in the control area. Many more marked adults of these two species moved from the control to the strip-managed area than vice versa, showing a preference of these beetles for the strip-managed area. These two species, and also *Pterostichus anthracinus* Illiger and *Harpalus rufipes* DeGeer, clearly increased their overall activity densities within 3 years in the area where weed strips were available. This effect was also attributed to better nutritional conditions in the strip-managed areas which, in turn, was explained by a higher density of prey.

IV. HERBIVORES: PROMOTING BIODIVERSITY, BUT NOT OF PESTS

A. Aphids

Aphids are rarely considered by scientists in discussions of biodiversity. They are usually only mentioned in connection with their great importance as pest insects of agriculture due to direct damage they cause to plants or their role as vectors of plant pathogens. However, in central Europe, 850 aphid species are known (Müller, 1988) of which only a few species are serious agricultural pests. Since the diversity of aphids is highly dependent on the diversity of the vegetation, a high species number could be expected in sown weed strips. So, it was not astonishing to find a total of

86 aphid species during our studies on strip-managed fields near Berne, Switzerland (Lethmayer, 1995). Though this collection included some species which obviously had been passively transported from other habitats (such as hedges or nearby forests) to the investigated strip-managed area, we discovered a remarkably species-rich insect community living in weed strips. Collected only in a small area (approximately less than ½ ha strips within a much larger landscape) this community represented already one-fifth of the Swiss aphid fauna. Beside the typical agricultural pest species (see below) most of the species turned out to be harmless to agricultural crops, consisting mostly (about 58%) of monophagous species feeding on one species, or species in one plant genus (Table 2). Eight species (about 42% of specimens) are polyphagous and 5 of these feed on crop species: *Aphis fabae* Scopoli (including *Aphis fabae evonymi* Fabricius) on beets and beans, *Macrosiphum euphorbiae* (Thomas) on potatoes, *Metopolophium dirhodum* (Walker) on cereals, and *Dysaphis pyri* (Boyer de Fonscolombe) on pear trees. *Cavariella aegopodii* (Scopoli) can sometimes be harmful to carrots, and *Aphis grossulariae* Kaltenbach can be found either on *Epilobium* sp. as indicated by our result or on *Ribes* sp.

Since some aphid species are important agricultural pests, the question arises whether they are augmented by weed strips or transferred to cultivated fields. This is, however, unlikely for several reasons. First, about 50% of all aphid species are monophagous (or oligophagous) and thus live only on a single host plant. In the case of most noncrop plants this host occurs only in weed strips. So these aphid species will not affect crops directly. However, aphids with host alteration or polyphagous species (some of which are potential pests) will populate crop areas and weed strips as well, when suitable food plants occur. There will also be frequent movement between wild plants in uncultivated areas and cultivated plants (van Emden, 1964). But all these aphids are prey for aphid predators and parasitoids and will probably attract these antagonists. A high density of noncrop aphids and "pest aphids" in the strips represents a host reservoir for specific aphid enemies and facilitates higher populations of aphid predators (by serving as alternative food). This is especially important before and after crop aphids become abundant because the predators and parasitoids will be retained longer in a given landscape and do not need to migrate away. From this perspective weed aphids help to reduce and control crop aphids.

Second, Hausammann (1996a) found that aphid density on crops near weed strips was not increased, indicating that infestation was not more likely near weed strips. Various other studies (Holtz, 1988; Sengonca and Frings, 1988; Wyss, 1995) provided further confirmation with data showing generally lower levels of aphid infestation in an area with weed strips or weedy fields compared to a control area. Additionally, in several of the above mentioned studies the aphid density near weed

Table 2. Aphid species colonizing plants in sown weed strips [a]

FAMILY	HOST PLANT SPECIES	APHID SPECIES	PEST STATUS[b]
Asteraceae	*Centaurea cyanus* L.	*Uroleucon jaceae* (L.)	w
	Tanacetum vulgare L.	*Uroleucon tanaceti* L.	w
		Metopeurum fuscoviride Stroyan	w
	Tripleurospermum inodorum (L.) Schultz-Bipontinus	*Brachycaudus cardui* L.	(p)
	Achillea millefolium L.	*Macrosiphoniella millefolii* (De Geer)	w
		Uroleucon achilleae (Koch)	w
		Brachycaudus cardui L.	(p)
		Aphis fabae Scopoli	p
	Cirsium arvense (L.) Scopoli	*Aphis fabae cirsiiacanthoidis* Scopoli	w
	Cirsium vulgare (Savi) Tenore	*Aphis fabae* Scopoli	p
	Lactuca serriola L.	*Macrosiphum euphorbiae* (Ths.)	p
	Sonchus sp.	*Hyperomyzus lactucae* (L.)	(p)

Family	Plant species	Aphid species	
Apiaceae	Pastinaca saliva L.	Cavariella aegopodii (Scopoli)	(p)
		Cavariella theobaldi (Gill & Bragg)	w
		Aphis fabae Scopoli	p
	Heracleum sp.	Cavariella theobaldi (Gill & Bragg)	w
Caryophyllaceae	Silene alba Miller	Brachycaudus lychnidis L.	w
Papaveraceae	Papaver rhoeas L.	Aphis fabae Scopoli	p
		Hyadaphis foeniculi (Pass.)	w
Rubiaceae	Galium sp.	Aphis fabae evonymi Fabricius	(p)
		Dysaphis pyri (Boyer de Fonscolombe)	p
Polygonaceae	Rumes sp.	Aphis rumicis L.	w
Fabaceae	Medicago sp., Trifolium sp.	Therioaphis trifolii L.	w
Oenotheraceae	Oenothera biennis L.	Aphis fabae Scopoli	p
Onagraceae	Epilobium sp.	Aphis grossulariae Kaltenbach	w

[a] (Frei and Manhart, 1992; Lethmayer, 1995). [b] Symbols: p = potential pest, (p) = occasional pest, w = occurrence only in weed strips.

strips was even lower than in a control field. This has been attributed to the abundance and species richness of aphidophagous predators and parasitic hymenopterans, which are conserved in strip-managed fields.

Finally, apart from the possibility that weed strips offer various potential host plants for several pest species during summer it is also possible that they provide suitable overwintering possibilities. However, most important agricultural aphid pests are host-alternating species and need woody plants (as primary hosts) for hibernation. Well known examples are *Rhopalosiphum padi* (Linaeus) which overwinters on the bird-cherry *(Prunus padus* L.*), Myzus persicae* (Sulzer) on the peach tree *(Persica vulgaris* Miller) or different *Prunus* species, and most *Aphis* species which need various shrubs or trees. Since weed strips do not contain woody plants these noxious aphid species cannot hibernate there.

B.   Phytophagous Beetles

Among phytophagous beetles, the families Chrysomelidae, Nitidulidae, and Curculionidae contain the most abundant and important pest species. Studies on the arthropod fauna of a strip-managed area (Frei and Manhart, 1992; Lethmayer *et al.,* 1997) showed that such a structurally diverse agricultural area with its high floral diversity encourages a high degree of faunal diversity (Table 3). The majority of beetle species in the weed strips, however, consisted of indifferent, harmless, and even rare species. Brassicaceae turned out to be the most attractive host plants to phytophagous beetles, especially for Nitidulidae *(Meligethes aeneus* Fabricius) but also for many Curculionidae *(Ceutorhynchus* spp.) and several Chrysomelidae (for example, *Phyllotreta* spp.). Many species were also found on Fabaceae (mainly on *Medicago* sp. and *Trifolium* sp.). It should be mentioned that our 1997 seed mixture no longer contains *Brassica* species but does include *Sinapis alba* (both in the Brassicaceae). Also, *Trifolium* sp. has been replaced by other soil covering plants because it became too dominant (Table 1). However, *Brassica* sp. and *Trifolium* sp. easily appear within free vegetational succession and are therefore still present.

Although many harmful beetle species also utilize wild plants related to their host crop no pest species occurred in higher numbers in the strips than in the fields (Hausammann, 1996b). The only exception was *M. aeneus* after the flowering time of rape. The main reason for the appearance of pests in the weed strips seems to be the high level of attraction of a suitable crop plant in a field near the strip, i.e. for these beetles the crops are always more attractive than the weed strips. Thus, the latter always have a smaller population. For the same reasons already described for aphids (see above) such low densities, even of pest species, also can be considered

Table 3. Numbers of species of Chrysomelidae and Curculionidae in two years in fields of rape, potato, wheat, in a meadow and in two sown weed strips (Lethmayer *et al.*, 1987).

	Rape	Potato	Wheat	Meadow	Strip 1	Strip 2
Chrysomelidae						
1993	10		10		8	14
1994		11		15	15	14
Curculionidae						
1993	11		10		15	23
1994		14		23	26	16

to be advantageous. In addition, hibernation studies (Bürki and Hausammann, 1993) showed that weed strips offer suitable habitats for a few potential pest species among the weevils (e.g., *Sitona* sp., *Ceutorhynchus* sp.) but never reached an abundance as high as in the natural field boundary. For blossom beetles (Nitidulidae) weed strips provide no adequate hibernation site. Such low populations contain, at the same time, specific enemies such as parasitoids or predators and probably also pathogens.

C.   Other Herbivores

Other investigated phytophagous insect groups were Cicadellidae (Homoptera) and Tenthredinidae (Hymenoptera). Observations on these taxa demonstrated that strip- management plays an important role in maintaining species conservation and biodiversity. Frei and Manhart (1992) collected 12 cicadellid species on 27 plants in weed strips, of which *Zyginidia scutellaris* Herrich-Schaeffer and *Macrosteles laevis* Ribaut (both on different grasses) appeared in highest abundance. Only two species also feed on cultivated plants (i.e., potatoes) but they are not known as transmitters of viral diseases. Most leafhoppers were observed on *Arctium minus* (Hill), *Pastinaca saliva* L., *Onobrychis viciifolia* Scopoli, *Stellaria media* L., and *Knautia arvensis* (L.). All these plants can only exist in untreated places (such as field margins) or in sown weed strips within fields.

Investigations on tenthredenids (sawflies) (Lethmayer, 1995) also confirmed that structurally diverse systems such as weed strips enhance floral and faunal diversity, since the sawfly diversity was highest in weed strips. There were about nine species which found suitable host plants only in the weed strips, but not in the fields; for example, species of *Ametastegia* on Chenopodiaceae or on *Rumex* sp. Pest species such as *Athalia rosae* (L.), *Dolerus* spp., and *Cephus pygmaeus* (L.) can also utilize wild grasses and Brassicaceae in the weed strips. However, they did not make full use of them as alternative developing sites or food resources. As with our other examples, these numerically enhanced populations of herbivores in strips provide important sources of food for predators and polyphagous parasitoids which may be important for the control of crop pests.

V. MECHANISMS

A. Additional Hibernation Sites

Sown weed strips have a high plant diversity with up to 30-40 species in the second year of a strip when censuring on a small-scale of about 100 m. On a larger scale, in 1995 our 18 strips, which covered a total area of 1.4 ha and had a length of more than 3 km, housed at least 120 plant species. This yields a dense and highly diverse vegetation throughout the year, including during winter. In the latter season, suitable hibernation sites are rare (particularly in agricultural fields) and areas like sown weed strips become very attractive. Therefore, arthropods searching for hibernation sites are attracted by these areas and to vegetational structures that provide effective hibernacula, where they may reach high densities. The most important predators in agroecosystems, carabid beetles and spiders, reach much higher hibernation densities and species numbers in sown weed strips than in cereal fields nearby (Table 4). Bürki and Hausammann (1993) found up to 180 coccinellids/m² in the soil under rotting leaves of *Arctium minor* (Hill) but less than 40 beetles under several other plants investigated for comparison and none in adjacent winter wheat fields. Parasitic hymenopterans reached similar high densities under *Symphytum officinale* L. and *Matricaria camomilla* L., but were absent in the soil of a cereal field nearby. These data underline the importance of a permanent vegetation consisting of many different plant species.

B. Increased Performance and Fitness of Predators and Parasitoids

Only a few investigations have focused on the increased predator fitness due to habitat management. The great availability of pollen and nectar attracts high numbers of aphidophagous syrphids. It may be assumed that these food sources

Table 4. Density per m² and number of species of hibernating arthropods in sown weed strips compared to adjacent cereal fields (data from Lys and Nentwig, 1994).[a]

Arthropod Type	Dev. Stage	Mean Density/m²		Total Species Number	
		Sown Weed Strip	Cereal Field	Sown Weed Strip	Cereal Field
Carabidae	Adults	243	55	14	2
	Larvae	87	49	6	3
Staphylinidae		803	85	19	8
Spiders[b]		223	16	5	2

[a] Significance of differences: Carabidae and Araneae: $p < 0.001$, Staphylinidae: $p < 0.01$, Wilcoxon, Mann and Whitney test
[b] In the case of spiders, identification was only to family level, thus the number refers to family numbers.

increase their reproductive potential. However, Salveter (1996) found no relationship between oviposition within wheat fields and the distance to the strips. There was no earlier development of the first syrphid generation and no additional generation after wheat harvest. Due to their great mobility, it is easy for adult syrphids to distribute over large areas. Therefore, effects are not likely to be detectable on a small-scale, but Salveter (1996) concluded that a fitness increase of adults was highly probable.

However, a clear effect on the predator's fitness could be detected in the carabid beetle *Poecilus cupreus* L. This very common species is more active in a system of sown weed strips and adjacent cereal areas than in a large undivided cereal field. The mark and recapture studies of Lys and Nentwig (1992) showed that the beetles migrate between the habitats and always select areas where they find greater prey abundance. Migrating individuals are large and heavy. Thus on average they are in a better nutritional state and produce more eggs (Zangger *et al.*, 1994). This may result in rapid growth; particularly noticeable early and late in the season when quality differences between sown weed strips and wheat fields are highest. Comparable results could also be found for other natural enemies (see below).

If predators and parasitoids find more food and reproduce more their predacious and parasitic impact is increased. This had been demonstrated for many arthropod groups and many types of ecological compensation areas, among them sown weed strips. Jmhasly and Nentwig (1995) found higher predation rates by spiders in close vicinity to such weed strips. Wyss *et al.* (1995) explain the reduction of aphids in orchards by the presence of more orb-weaving spiders, especially of *Araniella* sp., supposedly due to the presence of sown weed strips. Hausammann (1996b) found a parasitization rate of 0.7 to 2.0% among larvae of the rape pollen beetle *Meligethes* sp. (Nitidulidae) by the ichneumonid wasp *Tersilochus heterocerus* Thomas at various locations within a rape field. Parasitization rate, however, increased to 7% close to weed strips this. All these examples show the positive influence of ecological compensation areas such as sown weed strips in conserving predators and parasitoids.

VI. CONCLUSIONS

The positive reaction of many arthropod groups to ecological compensation areas in general, and sown weed strips in particular, shows that there is an urgent "demand" by these animals for such habitats, i.e. for undisturbed, highly structured and perennial areas. Sown weed strips may be a good compromise between extensification of agriculture and nature conservation. In nearly 10 years research on sown weed strips we have shown that such artificial habitats fit perfectly into our (artificial) agricultural landscape. They structure a landscape and increase biodiversity, thus compensating for the ecological monotony of crop fields. At the same time the abundance of potential pest species is not increased, in some species it is even decreased. Therefore, we assume that a sufficient amount of ecological compensation areas (we recommend 5 to 10%) will strongly reduce the necessity for insecticide applications and that it will also represent an important step towards environmentally friendly and sustainable agriculture.

However, there is still a great deal of research needed within the field of ecological compensation areas. We think that it would be worthwhile to compare different types of such conservation areas. Also, long-term studies which include special aspects of maintenance and succession are necessary. Due to the new structures, some species may become new pests (e.g., rodents, slugs, or some insects) and studies of such populations will be important. The network of weed strips will also influence the resource value of a landscape and many populations need to be investigated on a metapopulation level. Finally, there is still limited knowledge about positive and negative interactions between crop fields and conservation areas and we think that research should be intensified. This is also true for mere agronomic aspects (e.g., effects of crop species, yield, pathogens, maintenance practice, etc.), financial aspects (e.g., what is the "value" of compensation areas, or their influence, on the income of farmers?), or even social aspects (e.g., who wants nature conservation, sustainable agriculture, or a special landscape design?).

REFERENCES

Altieri, M. A., and Letourneau, D. K. (1982). Vegetation management and biological control in agroecosystems. *Crop Prot.* 1, 405-430.

Bürki, H.-M., and Hausammann, A. (1993). Überwinterung von Arthropoden im Boden und an Ackerkräutern künstlich angelegter Ackerkrautstreifen. *Agrarökologie* 7, 1-158.

Frank, T. (1994). Ground dwelling and flower visiting arthropods in sown weed strips and adjacent fields. Ph.D Thesis, University of Berne, Switzerland, 86 pp.

Frank, T. (1997). Species diversity of ground beetles (Carabidae) in sown weed strips and adjacent fields. *Biol. Agric. Hort.* (In press).

Frank, T., and Nentwig, W. (1995). Ground dwelling spiders (Araneae) in sown weed strips and adjacent fields. *Acta Oecol.* 16, 179-193.

Frei, G., and Manhart, C. (1992). Nützlinge und Schädlinge an künstlich angelegten Ackerkrautstreifen in Getreidefeldern. *Agrarökologie* 4, 1-14.

Günter, M. (1997). A management plan for sown weed strips in arable land with regard to technical and economical restrictions. University of Berne, Switzerland, Ph.D. thesis (in Preparation).

Hance, T. (1987). Predation impact of carabids at different population densities on the development of *Aphis fabae* in sugar beet. *Pedobiologia* 30, 251-262.

Hausammann, A. (1996a). The effects of weed strip-management on pests and beneficial arthropods in winter wheat fields. *J. Plant Dis. Prot.* 103, 70-81.

Hausammann, A. (1996b). Strip-management in rape crop: is winter rape endangered by negative impacts of sown weed strips? *J. Appl. Entomol.* 120, 505-512.

Heitzmann-Hofmann, A. (1995). Angesäte Ackerkrautstreifen-Veränderungen des Pflanzenbestandes während der natürlichen Sukzession. *Agrarökologie* 13, 1-152

Heydemann, B., and Meyer, H. (1983). Auswirkungen der Intensivkulturen auf die Fauna in den Agrarbiotopen.. *Dt. Rat f. Landschaftspflege* 42, 174-191.

Holtz, F. (1988). Zum Vorkommen von Blattläusen auf Wildpflanzen im Feldrand und im Feldrain. *Mitt. biol. Bundesanst. Land- u. Forstwirtschaft* 247, 77-84.

Jmhasly, P., and Nentwig, W. (1995). Habitat management in winter wheat and evaluation of subsequent spider predation on insect pests. *Acta Oecol.* 16, 389-403.

Katz, E., Duelli, P., and Wiedemeier, P. (1989). Der Einfluss der Nachbarschaft naturnaher Biotope auf Phänologie und Produktion von entomophagen Arthropoden in Intensivkulturen. *Mitt. Dt. Ges. allg. ang. Entomol.* 7, 306-309.

Klinger, K. (1987). Auswirkungen eingesäter Randstreifen an einem Winterweizenfeld auf die Raubarthropodenfauna und den Getreideblattlausbefall. *J. Appl. Entomol.* 104, 47-58.

Kruess, A., and Tscharntke, T. (1994). Habitat fragmentation, species loss, and biological control. *Science* 264, 1581-1584.

Lagerlöf, J., and Wallin, H. (1993). The abundance of arthropods along two field margins with different types of vegetation composition: an experimental study. *Agric., Ecosys., Environ.* 43, 141-154.

Lethmayer, C. (1995). Effects of sown weed strips on pest insects. Ph.D. Thesis, University of Berne, Switzerland, 59 pp.

Lethmayer, C., Nentwig, W., and Frank, T. (1997). Effects of weed strips on the occurrence of noxious coleopteran species (Nitidulidae, Chrysomelidae, Curculionidae). *J. Plant Dis. Prot.* 104, 75-92.

Luczak, J. (1979). Spiders in agrocoenoses. *Polish Ecol. Stud.* 5, 151-200.

Lys, J.-A.., and Nentwig, W. (1992). Augmentation of beneficial arthropods by strip-management. 4. Surface activity, movements and density of abundant carabid beetles in a cereal field. *Oecologia* 92, 373-382.

Lys, J.-A., and Nentwig, W. (1994). Improvement of the overwintering sites for Carabidae, Staphylinidae and Araneae by strip-management in a cereal field. *Pedobiologia* 38, 238-242.

Lys, J.-A., Zimmermann, M., and Nentwig, W. (1994). Increase in activity density and species number of carabid beetles in cereals as a result of strip-management. *Entomol. Exp. Appl.* 73, 1-9.

Müller, F. P. (1988). Aphidina - Blattläuse, Aphiden. *In* "Exkursionsfauna von Deutschland, Wirbellose 2/2." (E. Stresemann, ed.), pp. 87-167. Verlag Volk und Wissen. Berlin, Germany.

Nentwig, W. (1987). The prey of spiders. *In* "Ecophysiology of Spiders." (W. Nentwig, ed.), pp. 249-263. Springer Veriag. Berlin, Germany.

Nentwig, W. (1995). "Humanökologie" Springer Verlag. Berlin, Germany.

Nyffeler, M., and Benz, G. (1987). Spiders in natural pest control: a review. *J. Appl. Entomol.* 103, 321-339.

Raskin, R., Glück, E., and Pflug, W. (1992). Floren- und Faunenentwicklung auf herbizidfrei gehaltenen Agrarflächen -Auswirkungen des Ackerrandstreifenprogramms. *Natur und Landschaft* 67, 7-14.

Remmert, H. (1991). The Mosaic-Cycle Concept of Ecosystems. Springer Verlag. Berlin, Germany.

Riechert, S. E., and Bishop, L. (1990). Prey control by an assemblage of generalist predators: spiders in garden test systems. *Ecology* 71, 1441 -1450.

Risch, J. R. (1987). Agricultural ecology and insect outbreaks. *In* "Insect Outbreaks." (P. Barbosa, and J. C. Schultz, eds.), pp. 217-238. Academic Press. San Diego, CA.

Salveter, R. (1996). Population structure of aphidophagous hoverflies (Diptera: Syrphidae) in an agricultural landscape. Ph.D Thesis, University of Berne, Switzerland. 117 pp.

Sengonca, C., and Frings, B. (1988). Einfluß von *Phacelia tanacetifolia* auf Schädlingspopulationen in Zuckerrübenfeldern. *Pedobiologia* 32, 311-316.

Sunderland, K. D. (1987). Spiders and cereal aphids in Europe. *Bulletin* SROP/WPRS X/1, 82-102.

Thomas, M. B. (1990). The role of man-made grassy habitats in enhancing carabid populations in arable land. *In* "The Role of Ground Beetles in Ecological and Environmental Studies" (N. E. Stork, ed.), pp. 77-85. Intercept. Andover, Hampshire, U.K.

Thomas, M. B., Wratten, S. D., and Sotherton N. W (1991). Creation of "island" habitats in farmland to manipulate populations of beneficial arthropods: predator densities and emigration. *J. Appl. Ecol.* 28, 906-917.

Tivy, J. (1990). "Agricultural Ecology." Longman Scientific and Technical. London, U.K.

van Emden, H. F. (1964). The role of uncultivated land in the biology of crop pests and beneficial insects. *Sci. Hort. 17*, 121-136.

Weiss, E., and Stettmer, C. (1991). Unkräuter in der Agrarlandschaft locken blütenbesuchende Nutzinsekten an. *Agrarökologie* 1, 1-104.

Wiedemeier, P., and Duelli, P. (1993). Bedeutung ökologischer Ausgleichsflächen für die Überwinterung von Arthropoden im Intensivkulturland. *Verh. Ges. Ökol.* 22, 263-267.

Wyss, E. (1995). The effects of weed strips on aphids and aphidophagous predators in an apple orchard. *Entomol. Exp. App.* 75, 43-49.

Wyss, E., Niggli, U., and Nentwig, W. (1995). The impact of spiders on aphid populations in a strip-managed apple orchard. *J. Appl. Entomol.* 119, 473-478.

Zangger, A., Lys, J.-A., and Nentwig, W. (1994). Increasing the availability of food and the reproduction of *Poecilus cupreus* in a cereal field by strip management. *Entomol. Exp. Appl.* 71, 111-120.

CHAPTER
9

HABITAT MANIPULATION AND NATURAL ENEMY
EFFICIENCY: IMPLICATIONS FOR THE CONTROL
OF PESTS

G. M. Gurr, H. F. van Emden, and S. D. Wratten

I. INTRODUCTION

Just as engineering is based on physics so too may agriculture be viewed as a human activity which, fundamentally, must operate in accordance with the principles of a science. This science is ecology. Importantly, however, agricultural systems have developed by trial and error rather than designed in accordance with the principles of their associated science. This is for the simple reason that the practice of agriculture dates back thousands of years before the development of ecology.

Agricultural systems may be classified along a continuum with "traditional" or indigenous systems at one end and high input "western" systems at the other. The former have developed over many hundreds of years of trial and error and may be considered sustainable for the very reason that they have stood the test of time (Altieri, 1991a). Thus, they have "man-made ecological sustainability" (Zadoks, 1993) but not, in most cases, the economic sustainability required to keep pace with rising human populations and desired living standards. In contrast, "western" systems have undergone rapid change in the past 100 years in response to the industrial and green revolutions. Though the latter systems have delivered dramatic productivity increases they are in many cases ecologically unsustainable, especially when applied to marginal environments such as those in much of Australia. For neither of these extremes has ecology been an explicit tool for the shaping of practices. Empirical trial and error has been used in the former; in the case of the latter, a reductionist approach has been taken to challenges such as pest control. The small plot approach of classical agricultural science has often overcome immediate problems by means of a "technical fix" (e.g., pesticides) applicable at the organism-level. However, because of the low number of variables investigated at one time and the usual absence of measurement of ecological factors, this approach has often not recognized shortcomings at the population (e.g., pesticide resistance), community (e.g., pest resurgence), and ecosystem (e.g., groundwater contamination) levels until after the event. Such problems underlie the phases of pest control in cotton defined by Bottrell and Adkisson (1977) as: subsistence, exploitation, crisis, disaster, and recovery, the latter being achieved by a greater recognition of ecological factors in applied pest control.

During the past two decades pest management (with clear linkages to the discipline of population biology) and, more particularly, biological control (with its linkages to community dynamics) are areas of agricultural research which have both drawn upon and made important contributions to ecological theory. Indeed, Waage (1990) has referred to biological control as being "at heart an ecological exercise." Increasing attention has been focused on the nature of the relationship between pest arthropods, crop and noncrop plants, and their physical environment. The emerging discipline of habitat manipulation (or conservation biological control) seeks to manage these relationships to enhance the impact of natural enemies on pest populations. Indeed this approach is one of the key elements in the use of indigenous natural enemies in IPM (van Emden and Peakall, 1966).

Speight (1983) outlined various ways in which agroecosystems may be manipulated to improve pest control and they included: intercropping, use of wild plants in and around crops, and trap cropping. More recently, Perfect (1991) has gone as far as stating that "modern approaches to pest management exploit and promote biodiversity within agricultural systems." While this is probably an overstatement, there is growing activity in this field of research (e.g., recent texts by Boatman, 1994; Glen *et al.,* 1995; Pickett and Bugg, in preparation; and this volume). Unfortunately, however, most attempts to manipulate habitats within agroecosystems to manage pest populations have been intuitive rather than based on careful preliminary research.

Herzog and Funderburk (1986) observed that it is impossible to study all possible combinations of pest, natural enemy, crop, and cultural practice: thus, there is a need for "system-level crop-pest models" to optimize coordination and implementation. This is, however, hampered by the fact that the mechanisms behind interactions between plant diversity, herbivore populations, and natural enemies are barely understood (Power and Karieva, 1990; Wratten *et al.,* in press). Hence, there is a need to develop a better understanding of these mechanisms rather than rushing into intuitive attempts. The first aim of this contribution is, therefore, to examine the extent to which we can now approach a theoretical understanding of the ecological principles which determine the success of habitat manipulation.

A second intention is to temper theory with some practical considerations relating to how habitat manipulation research may be undertaken and translated into practical guidance to farmers. This is important because, although a considerable amount of work has shown *potential* benefits of habitat manipulation, there are relatively few examples where this promise has been successfully translated into practical techniques which are compatible with modern farming practices (reviewed by van Emden, 1990). Further, despite the promise shown to date, some consider that conservation of natural enemies is an approach to biological control which has not received sufficient attention (Dent, 1995). In particular, this approach to pest management has been geographically variable, little work having been conducted in Australia, for example (Gurr, 1994), although in New Zealand, research activity in this area is increasing (White *et al.,* 1995). Consequently, a third aim of this contribution is to address this lack of widespread adoption and to show the potential utility of habitat manipulation strategies for a wide range of agroecosystems.

A balanced approach will be attempted, drawing on relevant ecological theory while keeping the practicalities of a complex (and generally conservative) industry sector in mind. Since habitat manipulation can take various forms, many of which have been the subject of reviews, most recently by van Emden and Dabrowski (1997), this contribution will focus primarily on the provision of nonhost foods for natural enemies. This topic is not comprehensively covered in most other reviews. Using recent examples of this approach from several continents, we aim to exemplify the general potential of habitat modification in relation to improved natural enemy efficiency.

II. ECOLOGICAL PRINCIPLES GUIDING HABITAT MANIPULATION

One critical issue in classical biological control has been the move away from a reliance on an empirical approach (the release of a number of natural enemies which then compete for resources) toward a predictive approach which draws upon the available theoretical framework. In the latter, which releases are preceded by more careful work to determine which agent or agents will give the best level of pest regulation (Ehler, 1990). We believe that habitat management researchers, like those in classical biological control, need to shift from an empirical, hit-and-miss approach to a more rigorous, predictive approach. Such an approach would involve detailed studies of the various organisms and makes use of the imperfect but rapidly developing body of relevant ecological theories outlined below.

A. Diversity and Stability

Traditional agricultural systems have long used diversity to protect against pests and diseases, minimize risk of crop failure, produce a varied diet, and diversify sources of income (Altieri, 1991a). In contrast, the dominant western agricultural systems are characterized by specialization at the expense of diversity (Beus and Dunlap, 1990) and have been termed "fragile," particularly in relationship to pest attack (Altieri and Letourneau, 1982). Altieri (1991b) has argued that pest problems in western agriculture are largely a result of the expansion of crop monoculture at the expense of biodiversity and natural vegetation.

Though the notion that community diversity *leads to* stability was said to have died in the early 1970's (Risch *et al.,* 1983), general agreement remains that diversity generally equates with stability (Perfect, 1991). Despite the fact that many agroecosystems are patently "unsaturated" (*sensu* Brewer, 1979) the pursuit of diversity for its own sake within agroecosystems may be viewed as a "red herring" for two reasons. First, the imperative of agriculture is not *community* stability but the cyclical (usually annual) productivity of crop species which are most commonly early successional stage species and only rarely climax species. *Population* stability, however, is an implicit goal in that keeping pest numbers below an economic threshold is the usual objective (Stern *et al.,* 1959). Second, there is nothing inherently unstable about simple systems (Redfearn and Pimm, 1987 p. 108; Cromartie, 1991), suggesting that pest outbreaks may occur for separate reasons.

What then of the large body of literature relating to how pest control may be achieved by what may be loosely termed "agroecosystem diversity" (e.g., Altieri and Letourneau, 1982; Altieri, 1991b; Altieri, 1994)? In a review of 150 published investigations, Risch *et al.* (1983) found evidence to support the notion that herbivores were less numerous in diverse systems (53% of 198 cases). Another comprehensive review (Andow, 1991a) concluded that herbivorous arthropods were generally less abundant on plants in polycultures, though many species (20.2%) responded in a variable fashion and others (15.3%) were more abundant in polycultures. Vandemeer (1990) provides a good illustration of the unpredictability of the response of even a single pest species *(Spodoptera frugiperda)* to crop diversification in the same country (Nicaragua). In contrast to the findings of other workers (which he reviews) his data showed no lowering of pest attack in a maize-bean intercrop system compared to a maize monoculture.

It is important to stress that most of the studies outlined above concentrated on herbivore population levels, not on their temporal or spatial stability. Yet even the link between diversity and reduced herbivore population level is far from consistent. Risch *et al.* (1983) observed that the mechanisms accounting for reduced herbivore populations in polycultures were rarely studied. Certainly, without knowledge of the mechanisms it is dangerous to ascribe any cause and effect relationship. In relation to mechanisms, Risch (1987) concluded that the many examples of benefits to natural enemies of noncrop vegetation are not sufficient grounds for assuming that it was always a benefit to diversify the agroecosystem by including such vegetation. The mechanisms by which diversity may limit pest populations are discussed in the following section but before this some consideration is given to the negative effects of diversity on herbivore populations.

Way (1977) lists several ways in which decreased diversity can be good for the suppression of pest populations and considered that, at least on a regional (large) scale, diversity exacerbates pest problems. He stated that "fundamentally, most of our pests occur because there is *too much* diversity." Among other things, this diversity provides the alternative resources (food or refuges) which many pest species require for the completion of their life cycles. This disadvantage of diversity may also apply on a smaller scale as shown by Andow and Risch (1985) using plot sizes in the order of 200 m^2. They demonstrated that two different polycultures were less favorable than a corn *(Zea mays* L.) monoculture for the predaceous coccinellid *Coleomegilla maculata* (DeGeer). This was reflected in both a greater abundance of the predator and higher rates of predation of the European corn borer *(Ostrinia nubilalis* (Walker)) in the corn monoculture. This occurred because of a poorer availability of alternative resources (aphids and pollen) in the polycultures in which corn was combined with either beans *(Phaseolus vulgaris* L.) and squash *(Cucurbita maxima* Duchesne) or red clover *(Trifolium pratense* L.). This resulted in an increased migration from such polycultures. Adverse effects may also apply to polycultures even when the diversity adds some feature which natural enemies need (e.g., nectar) because the diversity may disorientate the natural enemy and impede host/prey location (Cromartie, 1991).

The notion that agroecosystems may not necessarily benefit from diversity *per se,* but require only certain elements of diversity which, once identified, could be retained or reintroduced, was voiced over 30 years ago (Way, 1966) and later by van Emden and Williams (1974). Speight (1983) reviewed the negative effects of noncrop vegetation and concluded that uncultivated land with high plant diversity can encourage pests. However, because such vegetation can also provide important resources to natural enemies, selective removal of weed species may lead to a net benefit.

Risch *et al.* (1983) also stated that "careful diversification" can reduce pest numbers and, by separately examining the responses of monophagous and polyphagous pests to diversity, determined that diversity is usually beneficial for the control of monophagous pests. The opposite tended to apply for polyphagous pests (see also Section III, A,1). This may be considered an important step toward understanding the effect of diversity on herbivore populations since the demise of the "diversity-leads-to-stability" dogma. Subsequently, Sheehan (1986) has added to the theoretical framework by explaining that pest control by specialist enemies may be more effective in less diverse agroecosystems. Such an effect may be evident where the concentration of host plants increases the attraction and retention of these enemies and if the lack of patchiness of the vegetation makes location of host/prey easier than in diverse vegetation. However, it is also true that as long as we seek any component of biological control in IPM some uncultivated land will need to be retained in the agroecosystem (van Emden, 1990).

Despite the above advances in theoretical understanding, both Risch (1987) and Altieri (1991b) have commented on the need to better understand the value of environmental heterogeneity and biodiversity in pest outbreaks and Andow (1991a) stated that a theory which predicts when natural enemies will exert significant mortality in polycultures is entirely lacking. Fundamentally, the apparently inconsistent effects of diversity on pest populations are probably a consequence of an over simplistic view of the diversity/stability nexus and a lack of knowledge of the mechanisms involved. In the following section these mechanisms will be examined.

B.   The "Enemies Hypothesis"

In a seminal paper, Root (1973) considered alternative hypotheses for the reduced herbivore populations observed in plots of collards (*Brassica oleracea* L.). He rejected the explanation that natural enemies were favored by the more diverse treatments (the "enemies hypothesis") in favor of a "resource concentration" hypothesis. This held that herbivores were adversely affected in diverse treatments in a more direct manner; the mosaic of vegetation restricted location of, and tenure on, suitable host plants. To a certain extent the subsequent literature has been preoccupied with arbitration between these hypotheses (e.g., Risch *et al.,* 1983; Baliddawa, 1985; Andow, 1991a). Collectively, such studies have tended to support the resource concentration hypothesis, though Risch (1987), while acknowledging the importance of this phenomenon in simple cropping systems, considered the enemies hypothesis more important in explaining pest outbreaks in perennial crops.

Other studies, however, have lent support for the enemies hypothesis even in annual crops. Natural enemies have, for example, been found to be more numerous in corn/soybean (*Glycine max* Merrill) intercrops than in monocultures (Tonhasca, 1993). Thus, there has been a growing consensus that the two hypotheses are probably complementary in many systems (Wratten and van Emden, 1995). However, one mechanism which would limit the degree of complementarity between these hypotheses, and which probably has not received the attention it deserves (Sheehan, 1986) is the degree to which a resource concentration effect may apply to the third trophic level. That is, to what degree are natural enemies (particularly host-specific parasitoids) affected by vegetational diversity (Wickremasinghe and van Emden, 1992). Such effects have recently been investigated for the parasitoid *Pediobius foveolatus* (Crawford) (Coll and Bottrell, 1996). They found that, in the short-term, movement of this specialist natural enemy was similar to that typical of monophagous insect herbivores; immigration was higher in simple habitats and tenure time was shorter in diverse habitats. In the longer-term, however, wasps accumulated in the more complex environment (bean interplanted with tall maize) and this was thought to be due to factors other than host density, such as shade. Clearly the effects of plant diversity on natural enemies may be considerable but complex interactions are likely to be involved (see Chapters 4 and 5). Further detailed studies will be required if the many questions generated by Sheehan's review of this subject are to be resolved.

A further problem associated with diversification has also been suggested by recent observations in New Zealand orchards (Stephens *et al.,* in press) in which buckwheat (*Fagopyrum esculentum* Moench) was planted as a source of nonhost food for natural enemies of pests. The chief effect of this was to increase populations of *Anacharis* sp., a parasitoid of the brown lacewing (*Micromus tasmaniae* Walker), an important predator of pests (Fig. 1). However, in that work rates of parasitism of orchard pest Lepidoptera were almost doubled in the presence of small plots of buckwheat. This suggests that in each situation the impact of habitat manipulation is likely to reflect a "trade-off" between the effects of diversity on the numbers of pests and their natural enemies and the fourth trophic level, the natural enemies of the biological control agents themselves.

Vegetational diversity can exert profound effects on herbivores in agroecosystems but effects also extend to higher trophic levels, and the interactions are complex and not well understood. Both the resource concentration and the enemies hypotheses are important. However, diversifying an agroecosystem can still exacerbate pest damage, where polyphagous pests with specialist natural enemies are important for example. Thus, for habitat manipulation to succeed, a more enlightened approach is required in which the ecology of the organisms is taken into consideration and the circumstances under which natural enemies are most likely to be effective are more clearly understood.

C. Bionomic Strategies

The first attempts to classify the bionomic strategies of organisms were in relation to island biogeography (MacArthur and Wilson, 1967) but the concept was quickly expanded to apply to organisms in other situations. Essentially this held that *r* strategists were good colonizers and typical of ephemeral, unstable habitats whereas *K* strategists were good competitors and typical of stable habitats. Southwood

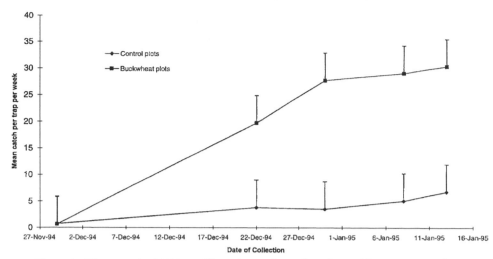

Figure 1.   The mean (and 95% confidence interval) number of parasitic wasps *(Anacharis* sp.) caught in yellow water traps during the 1994-95 trial in areas with and without buckwheat *(Fagopyrum esculentum)*. (From Stephens *et al.* (in press).

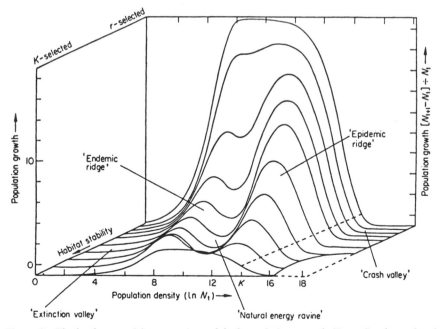

Figure 2.   The landscape of the synoptic model of population growth (From Southwood and Comins (1976, Fig. 1, p. 949))

(1977) applied the bionomic strategy concept to agroecosystems by matching various pest control methods to different bionomic strategies of pests. In relation to biological control, he proposed this would generally be most effective for pests with an intermediate strategy, since extreme r strategists essentially play a game of temporal and spatial "hide and seek" with their enemies, while extreme K strategists have highly evolved behavioral or physical defenses against their enemies. Although Ehler and Miller (1978) have argued that r strategist pests could be controlled by enemies if these also had an r strategy. The matching of biological control with pests with an intermediate strategy has retained overall validity.

In the synoptic model of population dynamics (Southwood and Comins, 1976) the "natural enemy ravine" (Fig. 2) is most pronounced in habitats of intermediate stability and the effect of this is to prevent low pest populations (on the "endemic ridge") from attaining outbreak status (the "epidemic ridge"). Any factor which prevents natural enemies responding to increasing pest populations with appropriate numerical and/or functional responses (e.g., local extinction due to unsuitable physical environment or lack of appropriate nonhost foods for optimal fecundity and host location) will in effect "bridge" the ravine. This may allow pest populations to reach epidemic levels. In this context, the goal of the habitat manipulation becomes clearer, i.e., to make the natural enemy ravine as wide and as deep as possible.

The synoptic model also suggests that attempts to manipulate the habitat to exploit the enemies hypothesis will not be successful for extreme r or extreme K strategist pests and that if such organisms are the target of habitat manipulation strategies these will need to exploit the resource concentration hypothesis. Also, increasing the predator/pest ratio, although intuitively sound, may not result in lower pest populations if the latter are below the natural enemy's oviposition threshold density. That is, the desired aggregative or reproductive numerical response may not occur. If, however, the full reproductive response was not being achieved because the natural enemy's egg productivity was limited by suboptimal pollen, nectar, or alternative host/prey resources then increased diversity can have value (Hickman and Wratten, 1997).

III. PRACTICAL CONSIDERATIONS IN HABITAT MANIPULATION

In this section we will attempt to integrate the guiding ecological principles outlined above with recent published studies and our own current work in an attempt to formulate guidance for future research which may expedite the deployment of habitat manipulation.

A. Choice of Crop Systems for Habitat Manipulation

1. Annual versus perennial crops

Altieri (1991b) has provided a useful classification of agroecosystems which he considered to lie along a hypothetical biodiversity gradient (Fig. 3). The imposition of an "increasing probability for pest build up" gradient is, in the light of the preceding discussion, valid provided that emphasis is placed on the word "probability." Altieri argues that the agroecosystems on the left side of the gradient are more amenable to manipulation since polycultures already contain many of the key factors required by natural enemies and tend to be less frequently disrupted by pesticide applications. It is important to recognize, however, that though contemporary agricultural systems lie along the full extent of this gradient, those which are responsible for the vast bulk of human food production are the grains, vegetables, and other row crops which, even if grown in rotations, are found at the low biodiversity end of the gradient. So, although manipulating polycultures may be easier, we consider the development of strategies which can introduce appropriate diversity into the important (but structurally impoverished) annual crop systems of western agriculture to be the more important goal.

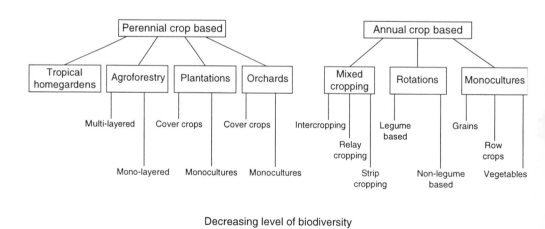

Figure 3. A classification of dominant agricultural agroecosystems on a gradient of diversity and vulnerability to pest outbreak. (From Altieri (1991, Fig. 14.4, p. 170)

There are many examples of successful habitat manipulations in agroecosystems lying in the complex half of Figure 3. Bugg and Waddington (1994), Prokopy (1994), and Gurr *et al.* (1996) have outlined how arthropod pest management in apples can be seen as "evolving" towards such approaches. However, habitat manipulation also has been attempted with considerable success in annual crops (Thomas *et al.,* 1992; White *et al.,* 1995 and Hickman and Wratten, 1997), but not all attempts have been successful. Baliddawa (1985) has encouraged researchers to publish such studies to aid others. Recent examples which have yielded inconclusive or only minor effects on natural enemies include both annual (Puvuk and Stinner, 1992) and perennial (Smith *et al.,* 1996) crops. However, it would be a mistake to consider annual and perennial agroecosystems as mutually exclusive since options exist for diversification of simple crop systems by their integration with the more complex perennial crops. For example, Peng *et al.* (1993) showed how annual and perennial crop systems could be integrated and that the perennial crop (timber and hazel (*Corylus maxima* Mill) trees) could attract and maintain populations of various natural enemy taxa close to the adjacent arable crop (i.e., peas, *Pisum sativum L.*).

Useful guidance on the relative tractability of reducing herbivore populations by diversification in annual and perennial cropping systems has been provided by Andow (1991a) (Table 1). This review of 209 studies of 287 herbivore species indicated that for annual crop systems, a reduction of herbivore populations was more commonly reported than an increase, no change, or variable response to diversification. This generalization was stronger for perennial crops. Importantly, however, dissection of the data to indicate the differential responses of monophagous and polyphagous pests showed that polyphagous pests were likely to be benefited by diverse systems, particularly in perennial crops. Apart from lending support for the resource concentration hypothesis, this finding reinforces the fact that diversification *per se* is no guarantee of reduced pest populations for either annual or perennial crops. Certainly the biology of the pest needs to be taken into consideration and for rational habitat management to occur this needs to be taken into account in choosing the strategy to be deployed.

B.   Choice of Habitat Manipulation Strategy

1.   Overview

One way of viewing habitat manipulation, at least in its more rigorous form, is that it seeks to avoid the hit-and-miss nature of providing diversity *per se* and thereby increases the probability of reducing pest numbers within an agroecosystem. In this respect it is using a knowledge of the biology of the organisms involved and of the ecological principles which determine their interactions with one another (and with the physical environment) to select an appropriate habitat manipulation strategy. As has been elucidated above, the resource concentration hypothesis is well supported by observations and this potentially creates problems. Attempting to exploit this mechanism would require quite fundamental alterations to conventional farming practices, especially those employed in western agriculture. Fundamental alterations are called for as a result of the way in which this hypothesis operates, i.e., by slowing the detection by pests of, and tenure on, host plants by using a confusing mosaic of vegetation. For this to be exploited, polycultures such as strip- and intercropping and cover crops within the primary crops are required. Further reductions in field sizes and the careful introduction of other diversity factors would also be useful.

Clearly, these initiatives will need to consider the risk of exacerbating outbreaks of polyphagous pests. A more practical problem with this approach to diversification is that the strategies constitute radical changes from mainstream western practice and farmers tend to be resistant to such change. However, agricultural practices are constantly evolving so, for example, the removal of land from intensive agriculture

Table 1. The response of monophagous and polyphagous herbivore populations to diversification in annual and perennial crop systems[a]

| | Population density of arthropod species in polyculture compared to monoculture [b] | | | |
	Variable	Higher	No change	Lower
ANNUAL	51	24	31	100
	(24.8)	(11.7)	(15.0)	(48.5)
Monophagous	39	6	27	83
	(25.2)	(3.9)	(17.4)	(53.5)
Polyphagous	12	18	4	17
	(23.5)	(35.3)	(7.8)	(33.3)
PERENNIAL	7	20	5	49
	(8.6)	(24.7)	(6.2)	(60.5)
Monophagous	3	11	4	47
	(4.6)	(16.9)	(6.2)	(72.3)
Polyphagous	4	9	1	2
	(25.0)	(56.3)	(6.2)	(12.5)

[a] From Andow (1991a, Table 3, p. 573.
[a] Percentage of total number of species is in parentheses.

for "set aside" has become a common European response to overproduction. Similarly, undersowing crops with forage plants is now an accepted European solution to the problem of high winter feed costs for cattle (van Emden, 1990). Thus, incorporating some plant diversity into intensive agroecosystems has been attempted by many European farmers but for reasons other than to control pests. Further research is required to ensure that the benefits for pest control are maximized in instances where other factors have led to diversification.

The notion that only certain "types" of diversity are beneficial in terms of pest control favors the likelihood of adoption since the inclusion of specific habitat features which can provide appropriate diversity are likely to be relatively easy to accommodate with the minimum of disruption to normal practice (see Chapter 8). Wratten and van Emden (1995) recently reviewed such habitat management options for enhancing pest control via natural enemy effects. The most widely recognized mechanisms are: food plants to serve as nonhost food sources (e.g., Bugg and Wilson, 1989; Hickman and Wratten, 1997), noncrop vegetation as a habitat for alternative hosts/prey (e.g., Kelly, 1987; Liang and Huang, 1994), and provision of shelter for natural enemies (e.g., Thomas *et al.,* 1992; Tuovinen, 1994). Landis (unpubl. data) has presented guidelines for habitat manipulations in midwestern U.S.A. farming systems. These include alternate hosts and nonhost foods and three categories of shelter: overwintering habitats, within-season habitats, and moderated microclimates. In some instances a single habitat feature may meet all such needs but situations can be envisaged where each may need to be provided by a separate feature.

The provision of nonhost food is one of the most commonly exploited of these mechanisms and this is justified in the context of maximizing the natural enemy ravine of Southwood and Comins (1976). In the remainder of this contribution we focus on this particular approach to habitat manipulation since it is one which has the potential to provide key resources to natural enemies while causing minimal disruption to conventional farming systems, even row crop monocultures.

2. Food plants

For many parasitoids sugars are important in maximizing longevity (Wäckers and Swaans, 1993), fecundity (Hagley and Barber, 1992), and searching activity (Takasu and Lewis, 1995). Not surprisingly then, food availability has been shown to affect parasitism rates (Treacy *et al.,* 1987; Somchoudhury and Dutt, 1988; and others reviewed by Powell, 1986). Thus, availability of nonhost food will be important in many instances for the enemies of pests to mount an optimal numerical and functional response to rising pest numbers and maintain the natural enemy ravine.

The extent to which "honeydew" from aphids and other sucking pests can adequately provide sugars is questionable since such pests are not normally tolerated within crops, particularly where they are vectors of viral diseases. As a consequence, only aphids (preferably nonpestiferous species) on noncrop vegetation will make this resource available. Therefore, provision of nectar by plants is important but availability from crop plants is restricted to those few species which bear extrafloral nectaries (e.g., cotton, *Gossypium* spp, and faba beans *Vicia faba* L.) or those species with entomophilous flowers. Even so, the availability of nectar from the latter may be restricted to discrete periods of flowering which are well into the growing cycle of the crop and therefore not available early in the growing season when it is important that natural enemies colonize successfully (Altieri, 1991b).

Pollen is also an important supplement for the diet of various taxa of entomophagous insects and in the case of adult syrphids is known to be important in egg maturation. This has been exploited as a means of increasing numbers of syrphid adults and eggs within a crop and has led to reduced aphid populations (Hickman and Wratten, 1997). This study employed border plantings of *Phacelia tanacetifolia* Bentham since this plant produces large quantities of pollen. However, despite the fact that it is potentially a good source of nectar (Crane *et al.,* 1984), the nectaries are inaccessible to the short-tongued syrphid adults since the inflorescence has a deep corolla tube. Thus, though this species is known to attract (smaller) parasitic wasps (Holland *et al.,* 1994), syrphid adults would need to meet their requirement for carbohydrate elsewhere, possibly on weeds as observed by Cowgill *et al.* (1993a,b).

That *P. tanacetifolia* is unable to provide all the requirements of the full guild of enemies within the agroecosystem illustrates the importance of the first of three question raised by Andow (1991b): which plants are needed? When and where should they be encouraged to grow? and What tools are available to accomplish answers to the above? Such questions are important for merging theory with practicalities in order to provide guidance to researchers and users of habitat manipulation. In response to these questions, we outline in the following three subsections a number of initiatives which, while not offering a complete solution, may go some way toward the adoption of a less hit-and-miss approach to habitat manipulation.

a. Preliminary microcosm studies   A number of studies have observed the different responses of natural enemies to flowers or other treatments applied in habitat manipulation studies. Differential visitation and use of flowers of different species by adult syrphids have been reported (MacLeod, 1992; Cowgill *et al.,* 1993b). Other studies of quite different habitat management approaches, the provision of overwinter shelter for example, have provided useful guidance on which strategies may be of most benefit to enemies in a habitat manipulation attempt (Thomas *et al.,* 1992).

They are, however, unable to provide more basic biological data on aspects such as fecundity and longevity. These factors have been shown to vary with access to the inflorescences of different species (Idris and Grafius, 1995; Baggen and Gurr, 1998).

Jervis *et al.* (1996) has stated that biological control practitioners should investigate the role of nonhost foods in the survival, fecundity, and searching efficiency of parasitoids before proceeding with a manipulation. In lieu of such information, however, umbelliferous species were recommended since they bear inflorescences with very accessible nectaries. Investigations proposed by Jervis *et al.* (1996) are crucial in the move toward a more rigorous approach to habitat manipulation but such microcosm studies should be expanded to test the response of key pests (as well as the response of natural enemies) to the candidate plants.

Testing the response of pest organisms to prospective habitat manipulation tools is important because pests may be attracted to, and make use of, plants such as mustard (*Brassica hirta* L.) (Matthews-Gehringer *et al.,* 1994) and *P. tanacetifolia* (Wratten and van Emden, 1995). The likelihood of this occurring may be reduced for some pest species by selecting plants which are botanically unrelated to the crop plant. This factor is also likely to be important to avoid it serving as an alternate host for a plant pathogen. However, feeding on the nectar may still occur, particularly by lepidopterans, since the adults of most species freely feed on floral nectar (Kevan and Baker, 1984). Thus, lepidopteran pests are particularly likely to make use of nectar from food plants intended for natural enemies and such an effect has led to the observation of greater numbers of *Pieris rapae* (L.) and *Plutella xylostella* (L.) in plots of *B. oleracea* interplanted with or adjacent to nectar-producing plants (Zhao *et al.,* 1992).

In reviewing habitat management in relation to orchard pest management, Prokopy (1994) commented broadly on the potential for effects to be counterproductive and echoing the message of others (see Section II,A) used the term "selective vegetational diversification" as a way of achieving an optimal balance. One example of the potential for an ambivalent effect of habitat manipulation in orchards applies to the use of rich floral undergrowth as used by Altieri and Schmidt (1985) and Halley and Hogue (1990) to enhance pest control. Although this may increase pest mortality for much of the year there is potential for such a strategy to lead to elevated levels of the tortricid *(Epiphyas postvittana* Walker). Larvae of this pest are known to overwinter on broadleaved weeds on the orchard floor unless appropriate management (grazing, for example) is applied in the winter (Thomas and Burnip, 1993).

Seeking to achieve selective vegetational diversity in a study of the potato moth (*Phthorimaea operculella* Zeller) and its encyrtid parasitoid *Copidosoma koehleri* Blanchard, Baggen and Gurr (1998) used microcosms of various scales ranging from petri dishes to field cages covering entire potato plants. They demonstrated that selectivity could be exploited to prevent the pest from feeding on the food plants. In this instance *Coriandrum sativum* L. conferred an adult longevity significantly greater ($p<0.05$) than those in the control treatment (water) for both the pest and its parasitoid. In contrast, access to the inflorescences of *Borage officinalis* L. led to greater longevity only in *C. koehleri* (Table 2).

These results explain the observations from a field trial in which a strip of flowering plants (mostly *C. sativum)* at one end of a 20 m x 8 m potato plot led to greater rates of parasitism by the egg parasitoid *Copidosoma koehleri* but also to a higher incidence of crop damage in the proximity of the flowers (Baggen and Gurr, 1998). Results from subsequent cage tests strongly suggest that *B. officinalis has* the potential to increase parasitoid activity without allowing the pest to feed and exacerbate crop damage. Such plants, termed "selective food plants", have clear potential to play an important role in habitat management strategies in agroecosystems where nectar feeding pests are important.

The mechanisms by which food plant selectivity may operate include temporal coincidence between nectar availability and insect foraging, differential attractiveness, and morphometric compatibility between the inflorescence and the insect (Jervis *et al.,* 1993). The latter mechanism is thought to be the explanation in the case of *B. officinalis* not conferring any benefit to *P. operculella,* since plants of this genus possess floral architectures which conceal and protect nectaries (Willis, 1973). This is probably one of the easier mechanisms to exploit, since direct comparisons of insect head dimensions and mouth part structure and length can be readily compared with inflorescence characteristics such as depth and diameter of corolla tubes of different plant species. Such studies with bumble bees have shown links between proboscis length and the depth of corollas fed from (Inouye, 1980).

Table 2. The non-selective benefit of coriander (*Coriandrum sativum* L.) to *Phthorimaea operculella* (pest) and *Copidosoma koehleri* (parasitoid) and the selective benefit of borage (*Borage officinalis* L.) to *C. koehleri* only[a]

Treatment[b]	Adult longevity (days)[c]	
	P. operculella	*C. koehleri*
Coriander	13.43 a	9.84 b
Borage	8.76 b	11.96 a
Control	7.76 b	3.80 c

It may also be possible to exploit the differential attractiveness of different nectars to various insect taxa. For example, hexose-rich nectars (such as those common in Compositae) are less attractive to most adult Lepidoptera than sucrose- rich nectars (as commonly produced by extrafloral nectaries) (Rogers, 1985). Studies along these lines will aid the early exclusion of candidate plants likely to have nectar which is attractive to, and accessible by, pests. Fortunately, many important parasitoids are relatively small insects and are not averse to crawling deep into corollas to access nectaries (Jervis *et al.*, 1993), so access should less commonly be a problem than for larger insects. Furthermore, Wäckers and Swaans (1993) observed "flower generalism" in the hymenopteran *Cotesia rubecula,* it being attracted to flowers other that those on which it finds its host (*P. rapae).* They considered that this would be adaptive where host infestation and nectar availability were not synchronized. Since flower generalism in natural enemies (for which Baggen and Gurr (1998) have also presented evidence) is likely to be a common phenomenon, prospects for finding more flower species which meet the needs of beneficial insects, while denying any benefit to pests, appear to be good.

Another important attribute of food plants may be the possession of extrafloral nectaries since these structures are known to be used by natural enemies such as ichneumonids (Bugg *et al.*, 1989) and may have the advantage of making nectar available to many insects which otherwise might not physically be able to access floral nectaries. Indeed, the highly apparent position of such nectaries in relation to the plant surface may conceivably act as a selectivity mechanism, preventing use of long mouthparts by insects such as many lepidopterans. Another advantage of extrafloral nectaries may be to extend the period over which a given plant species produces nectar to beyond that during which it has open flowers.

The potential importance of extrafloral nectaries is illustrated by the complexity of the responses of different insects to cotton isolines without extrafloral nectaries. Though this led to a reduction in attractiveness to pests (e.g., Adjei-Maafo and Wilson, 1983) these isolines were also less attractive to natural enemies. In a study of *Heliothis zea* (Boddie) lower rates of parasitism by *Trichogramma pretiosum* Riley were observed (Treacy *et al.*, 1987). However, increased predation of *Heliothis (Helicoverpa)* larvae resulted in Australia because *Cryptolaemus* ladybirds, which also normally feed from the nectaries, became more carnivorous on the isolines lacking nectaries (Adjei-Maafo, 1980). Such studies suggest that although extrafloral nectar has been said to be less attractive to hymenopteran pollinators (Rogers, 1985), it is utilized by at least some hymenopteran parasitoids and some predators of importance in biological control.

Such effects illustrate the nature of the "trade-off" which Jervis *et al.* (1993) saw between benefits of food plants to pests and to enemies. Unfortunately, the uninformed adoption of proprietary "insectary crop" seed mixes, which are commercially available in the U.S.A., may lead to unforseen benefits to pests and exacerbated crop loss. This is because few of these seed mixes have been critically evaluated (Bugg and Waddington, 1994) so it is quite possible that one or more of the 4-18 plant species typically included in these mixes will provide nectar attractive to and accessible by pests. Such dangers may be avoided by screening of candidate plant species against both natural enemies and pests prior to their inclusion in such seed mixes to ensure that they are "selective food plants" *(sensu* Baggen and Gurr, 1998) to an adequate degree (see Chapter 8).

b. *Criteria-based selection of food plants*   Various reasons have been given for the selection of plant species in habitat manipulation studies. These include: flowering periods which do not coincide with the crop (apples) and divert pollinating insects, *a priori* information that they would provide important resources (such as nectar, pollen, moisture, or shelter), and ability to survive in the selected environment with little maintenance (Matthews-Gehringer, 1984). Others have considered aspects such as cost and availability of seed, competitive ability (with weeds), and early flowering (Wratten and van Emden, 1995). White *et al.* (1995) also stressed the importance of agronomic tractability of plants. Some work in New Zealand has investigated aspects of the phenology of candidate food plants by determining the sowing to flowering periods (Bowie *et al.,* 1995). Such basic screening is important in determining the optimal time for sowing in order to achieve flowering during the period of greatest need. The importance of phenological development has also been reinforced on theoretical grounds. Corbett and Plant (1993) modeled natural enemy movements in relation to interplanted vegetation and determined that this vegetation acted as a source of natural enemies if natural enemies colonized it before crop germination (e.g., wheat undersown with ryegrass (Powell, 1983)), but it acted as a sink for natural enemies if it germinated at the same time as the primary crop.

   A further consideration in relation to choice of food plant species is the risk that it may constitute a threat (i.e., become a noxious weed) (Gurr, 1994), contaminate the primary crop (with its seeds or burrs, for example), poison livestock if accidentally grazed or fed to animals, or serve as alternative host of an important plant pathogen. In this context, we consider that the issue of food plant selection needs to be approached in a more careful and systematic fashion with due regard to the ethical consideration of biological pollution. An approach which may aid researchers considers the various factors of relevance in a given situation and gives each the proper degree of emphasis in the final decision, in a graded-weighted checklist *(sensu* Pearson, 1990). This can be used to review the properties of candidate plants, assign to each criterion a weighting based on its relative importance for the intended environment, and gain a score for the likely suitability of each plant species (Table 3).

c. *"Crop facsimile" studies*   The steps outlined above are seen as useful measures for increasing the likelihood of success from habitat management research while minimizing the expense and degree of risk involved. Despite this, there are instances where the data obtained before moving into farm-scale trials in the intended crop system do not offer a sufficient degree of assurance to the host farmers that there will be a net benefit to them. Further, some crops have such a large suite of pests and these are attended by an equally diverse natural enemy community, that it may be difficult to screen all species in microcosm experiments. In such instances there may be the opportunity to conduct a round of trials in what may be viewed as "facsimile crops."

This approach is being taken with a current program in New South Wales, Australia intending to lessen the intensity of pesticide use in seed lucerne. Because this is a high value crop grown by specialist farmers trials of food plant borders in these were not readily acceptable, particularly since the withholding of insecticides was considered important in order to allow monitoring of arthropods in an undisrupted environment. However, most growers also maintain lucerne stands grown for (less valuable) hay rather than seed. These crops are attacked by most of the same pests species but because higher pest thresholds are tolerated than in seed crops they are rarely treated with insecticides. Consequently these hay crops constitute a facsimile of the seed crops in which habitat manipulation studies have proven acceptable. Provided that biological control benefits can be demonstrated here, persuading growers to try the same approach in their seed crops may be possible.

C.   Spatial Considerations in Habitat Manipulation

Kareiva (1990) has argued that spatial effects need to be considered as well as the behavioral attributes of biological control agents. This is certainly the case in habitat management, since the manipulation feature (e.g., food plant strip or overwintering "island habitats" (*sensu* Thomas *et al.,* 1992)) may occupy only a small proportion of the agroecosystem area, particularly where minimal disruption to conventional farming practice has been an imperative. Wratten and van Emden (1995) pose a number of key questions which are relevant to this issue which can be distilled to the simple inquiry: "enhancement or redistribution; do habitat manipulation features lead only to a localized concentration of existing natural enemies or to greater overall populations?" This question is one which until recently has remained largely unanswered because work has occurred on a relatively small-scale, leading Wratten and van Emden (1995) to stress the need for landscape-level experimentation.

Clearly the mobility of the various natural enemies of importance in a given crop will have a profound effect on this issue. The characteristic form of pollen from *P. tanacetifolia* has been exploited as a way of determining feeding on this plant by syrphid adults caught within crops or as far as 100 m away (Holland *et al.,* 1994). Data for the tachinid *Lixophaga sphenophori* (Villeneuve) is less comprehensive but it has been tracked moving 30 m from food plants (Topham and Beardsley, 1975). Among the large and diverse taxon of the Hymenoptera, mobility may be lower in the case of small species such as *Trichogramma* spp. but for others diffusion rates may exceed 100 m day^{-1} (Corbett and Plant, 1993).

Possibly the study which most closely addresses the issue of whether the provision of food plants leads to local aggregation of natural enemies or actually increases their populations is with hoverflies in British cereal crops (Hickman and Wratten, 1997). In this study, trap catches of adult syrphids (at eight distances from the field boundaries up to 180 m into the crop) were greater in fields with *Phacelia tanacetifolia* borders than in control fields with normal boundary vegetation (Fig 4). Importantly, this field-scale enhancement of adult syrphids translated to a greater numbers of eggs within the *P. tanacetifolia* bordered crops. During a period when third instar syrphid larvae were common significantly fewer aphids were observed in these diversified fields.

In a more recent study on an even larger scale, Marino and Landis (1996) found that percent parasitism in *Pseudaletia unipunctata* (Hearth) was uniform on the scale of individual fields but was greater in "complex" landscapes. It is generally accepted that many parasitoids commute between sites, where host location is the imperative, to others where nonhost feeding occurs. The range over which this commuting occurs will clearly be determined by the mobility of the insect concerned. In the study by Marino and Landis (1996) the higher parasitism rates observed in the complex landscape were thought to result from the greater availability of nectar sources and more favorable microclimates. The absence of any within-field affect may simply reflect the ability of the parasitoids to commute effectively on this scale.

The scale over which habitat manipulation can lead to a redistribution of natural enemies has important implications. If redistribution occurs on a relatively large scale, extending over many fields and farms, those growers practicing the technique could derive a benefit at the expense of more conservative neighbors whose natural enemies may emigrate to where food plants (or other features such as overwintering habitat) are available. These natural enemies may then be disinclined to distribute away from such features and so concentrate their predation/parasitism on pests in adjacent crops. In contrast, if the effects of habitat manipulation on the spatial distribution of natural enemies were confined primarily to the individual field, the aggregation of natural enemies could be deleterious since large sections of the crop could be left "unguarded." Evidence that the latter effect could occur is provided by Bugg *et al.* (1987), who found that the weed *Polygonum aviculare,* a source of nectar, pollen, and alternate prey was so attractive to entomophagous insects that these predators had little inclination to forage within the crop itself. Differences in arthropod activity on a larger scale, though still within fields, have been reported by Landis and Haas (1992). They found that proximity of noncrop habitats was important. In this North American study sections of crop close to wooded edges were consistently associated with high rates of parasitism in the European corn borer *(Ostrinia nubilalis* (Hübner)) (see Chapter 6).

In the above section "selectivity" of food plants, such that they prevented feeding by pests, was discussed on the (unstated) assumption that this was necessary for optimal pest management. However, viewed more closely in the context of spatial movement of pests and their natural enemies, beneficial effects of using nonselective food plants can be envisaged. This would apply if the food plants led to a localized aggregation of pests which in turn led to an enhanced numerical and functional response of their natural enemies. Occurring in close proximity to nonhost foods, such a response would not be inhibited by lack of protein- or calorie-rich foods. Under this scenario

Table 3. The "graded and weighted checklist" approach (Pearson, 1990) used to evaluate *Phacelia tanacetifolia* as a food plant in habitat manipulation of lucerne (*Medicago sativa*) in New South Wales, Australia.[a]

CRITERION	WEIGHTING (1=unimportant to 5=important	ESTIMATED RATING (1 = bad to 5 = good)					SCORE (weighting x rating)
		1	2	3	4	5	
HAZARDS							
Potential weed status	3			*			9
Alternate host for crop pathogen	3				*		12
Livestock toxicity	5					*	25
Product contamination potential	4					*	20
ECONOMIC FACTORS							
Dual crop status	2			*			6
Seed cost (& availability)	2	*					2
Establishment costs	2			*			6
BIOLOGICAL FACTORS							
Pollen production (total/temporal pattern)	4				*		16
Nectar production (total/temporal pattern)	4			*			12
Agronomic compatibility with crop	5				*		20
Vigour/competitiveness with weeds	3			*			9
Perenniality/self sowing annual	1		*				2
						TOTAL =	139[b]

[a] (see text for discussion)
[b] This value to be compared with that for other candidate plant species evaluated on separate sheets.

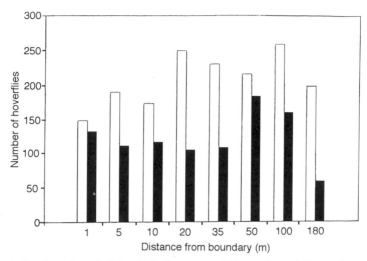

Figure 4. Total number of adult aphidophagous syrphids caught at different distances from
boundaries in fields with (unshaded bars) and without (shaded bars) border strips
of *Phacelia tanacetifolia*. (From Hickman and Wratten (1997).

natural enemies may be able to prevent pest populations from increasing and dispersing
to other areas of the crop and surrounding fields. This hypothesis is currently being
investigated with *P. operculella* in potato crops where it may be encouraged by the
fact that adults of this species tend to colonize crops via the margins.

Addressing issues as challenging as those raised above will require advances
in methodology which will allow the spatial movement of insects to be determined
more readily. The use of pollen grains as markers as described by Hickman *et al.*
(1995) is relatively simple for pollen feeders but probably has little potential for
parasitic Hymenoptera. Although pollen has been recovered from various species,
it is likely that it is consumed indirectly in the nectar and may have originated from
other plants (Jervis *et al.*, 1993). Consequently, the further development and use
of mark, release, recapture techniques using fluorescent or radioactive markers, as
used in recent studies by Corbett and Rosenheim (1996) and Corbett *et al.* (1996),
is important.

IV.   INTEGRATION OF HABITAT MANIPULATION WITH OTHER PEST MANAGEMENT TECHNIQUES

Though classical biological control attempts include some spectacular
successes the success rate is generally viewed as low (Ehler and Hall, 1982) and
certainly leaves room for improvement. One factor which may have contributed to

this level of success is that most classical biological control programs tend to focus on the candidate agents. There is too little emphasis on the environment in which releases are to be made or the aspects of the habitat that may need to be manipulated in order to achieve optimal establishment and pest control. Potentially, habitat manipulation initiatives can enable an exotic natural enemy operating at very low levels for some time to flourish and exert a very considerable impact on pests.

A greater emphasis on the environmental requirements of biological control agents released in an inundative (i.e,. augmentative) manner may therefore also be fruitful. One project currently being planned in New South Wales is to investigate the use of food plant strips between grapevine rows to ensure the maximum possible benefit from *Trichogramma carverae* (Oatman and Pinto) released for the control of *E. postvittana*. Inundative biological control of *E. postvittana* is currently being promoted to grape growers in Australia but at a cost of $A45 per ha for each of the 2 to 3 releases required per season. Thus, *T. carverae* is an input of significant expense, especially since the longevity of unfed adults is less than two days. Under these circumstances provision of nectar-producing flowers within the vineyards has the potential to maximize longevity, hold parasitoids in the desired area, and maximize the return on investment.

One of the greatest challenges which will be faced by those seeking to implement habitat manipulation into western agricultural systems is the tendency for these to employ heavy rates of pesticides. Though habitat manipulation may provide refugia for natural enemies from pesticide treated areas, toxic residues left within the crop may repel or disrupt subsequent activity. Under most circumstances, therefore, use of conventional insecticides will be antagonistic to the aims of habitat manipulation. The further development of narrow spectrum active ingredients (such as insect growth regulators) will be important if habitat manipulation is to function optimally in a variety of agroecosystems (see Chapter 11).

V. CONCLUSIONS

Altieri (1991b) has questioned whether ecology can rescue agriculture when most of the problems are caused by what he terms "the profit motive." A number of recent changes to agriculture have possibly reduced the extent to which the profit motive applies, at least in the short-term and at the scale of individual farms. These include the increasingly vocal "alternative agriculture" movement in many countries (Beus and Dunlap, 1990), the "set aside" scheme in the European Union, and the "Landcare" movement in Australia. Potentially, these will open a window of opportunity for the introduction and integration of ecologically rational pest management strategies. One of the key features of such strategies, particularly in the annual monocultures, will be relatively small, discrete features such as strips of food plants and overwintering shelter. Such structures may be thought of as microecotones; zones where, by the provision of the "right kinds" of physical and floral diversity, pest suppression may be achieved. However, research will need to proceed carefully to ensure that the potential negative effects of diversity, which favor pests or the predator/parasitoid of biological control agents, are minimized. Prospects for the continued and growing use of various forms of habitat manipulation appear good since the benefits may extend beyond pest control to reducing soil erosion and the leaching of agricultural chemicals into surface water, conservation of flora and fauna, and habitat for sporting species such as gamebirds.

Acknowledgments
        We thank Lindsay Baggen for allowing the use of his results and for serving as a productive sounding board for ideas in the development of this work.

REFERENCES

Adjei-Maafo, I. K., and Wilson, L. T. (1983). Factors affecting the relative abundance of arthropods on nectaried and nectariless cotton. *Environ. Entomol.* 12, 349-352.

Adjei-Maafo, I. K. (1980). Effects of nectariless cotton trait on insect pests, parasites and predators with special reference to the effects on the reproductive characters of *Heliothis* spp. Ph.D. Thesis, University of Queensland.

Altieri, M. A. (1991a). Traditional farming in Latin America. *The Ecol.* 21, 93-96.

Altieri, M. A. (1991b). Increasing biodiversity to improve insect pest management in agro-ecosystems. *In* "Biodiversity of Microorganisms and Invertebrates: Its Role in Sustainable Agriculture." (D. L. Hawksworth, ed.), pp. 165-182. CAB International. Wallingford, U.K.

Altieri, M. A. (1994). "Biodiversity and Pest Management in Agroecosystems." Hawthorn Press. New York, NY.

Altieri, M. A., and Letourneau, D. K. (1982). Vegetation management and biological control in agroecosystems. *Crop Prot.* 1, 405-430.

Altieri, M. A., and Schmidt, L. L. (1985). Cover crop manipulation in northern California orchards and vineyards: effects on arthropod communities. *Biol. Agric. Hort.* 3, 1-24.

Andow, D. (1991a). Vegetational diversity and arthropod population response. *Annu. Rev. Entomol.* 36, 561-586.

Andow, D. A. (1991b). Control of arthropods using crop diversity. *In* "CRC Handbook of Pest Management in Agriculture." (D. Pimentel, ed.), Vol. 1, 2nd Edition, pp. 257-283. CRC Press. Boca Raton, FL.

Andow, D. A., and Risch, S. J. (1985). Predation in diversified agroecosystems: relations between a coccinellid predator *Coleomegilla maculata* and its food. *J. Appl. Ecol.* 22, 357-372.

Baggen, L. R., and Gurr, G. M. (1998). The influence of food on *Copidosoma koehleri* (Hymenoptera: Encyrtidae) and the use of flowering plants as a habitat management tool to enhance biological control of potato moth, *Phthorimaea operculella* (Lepidoptera: Gelechiidae). *Biological Control* 11, (in press).

Baliddawa, C. W. (1985). Plant species diversity and crop pest control: an analytical review. *Insect Sci. Appl.* 6, 479-487.

Beus, C. E., and Dunlap, R. E. (1990). Conventional versus alternative agriculture: the paradigmatic roots of the debate. *Rural Sociol.* 55, 590-616.

Boatman, N. D. (1994). "Field Margins: Integrating Agriculture and Conservation." Proc. Symp., University of Warwick, Coventry; 18-20 April 1994. Brit. Crop Prot. Coun. Farnham, U.K.

Bottrell D. G., and Adkisson, P. L. (1977). Cotton insect pest management. *Annu. Rev. Entomol.* 22, 451-481.

Bowie, M. H., Wratten, S. D., and White, A. J. (1995). Agronomy and phenology of "companion plants" of potential for enhancement of insect biological control. *N. Z. J. Crop Hort. Sci.* 23, 423-427.

Brewer, R. (1979). "Principles of Ecology." Saunders, Philadelphia, PA.

Bugg, R. L., and Waddington, C. (1994). Using cover crops to manage arthropod pests of orchards: a review. *Agric., Ecosys., Environ.* 50, 11-28.

Bugg, R. L., and Wilson, L. T. *(1989). Ammi visnaga* (L.) Lamarck (Apiacea): associated beneficial insects and implications for biological control, with emphasis on the bell pepper agroecosystem. *Biol. Agric. Hort.* 6, 241-268.

Bugg, R. L., Ehler, L. E., and Wilson, L. T. (1987). Effect of common knotweed *(Polygonum aviculare)* on abundance and efficiency of insect predators of crop pests. *Hilgardia* 55, 1-53.

Bugg, R. L., Ellis, R. T., and Carlson, R. W. (1989). Ichneumonidae (Hymenoptera) using extra floral nectar of faba bean *(Vicia faba* L.: Fabaceae) in Massachusetts. *Biol. Agric. Hort.* 6, 107-114.

Coll, M., and Bottrell, D. G. (1996). Movement of an insect parasitoid in simple and diverse plant assemblages. *Ecol. Entomol.* 21, 141-149.

Corbett, A., and Plant, R. E. (1993). Role of movement in the response of natural enemies to agroecosystem diversification: a theoretical evaluation. *Environ. Entomol.* 22, 519-531.

Corbett, A., and Rosenheim, J. A. (1996). Quantifying movement of a minute parasitoid, *Anagrus epos* (Hymenoptera: Mymaridae), using fluorescent dust marking and recapture. *Biol. Cont.* 6, 35-44.

Corbett, A., Murphy, B. C., Rosenheim, J. A., and Bruins, P. (1996). Labeling an egg parasitoid, *Anagrus epos* (Hymenoptera: Mymaridae), with rubidium within an overwintering refuge. *Environ. Entomol.* 25, 29-38.

Cowgill, S. E., Wratten, S. D., and Sotherton, N. W. (1993a). The effect of weeds on the numbers of hoverfly (Diptera: Syrphidae) adults and the distribution and composition of their eggs in winter wheat. *Ann. Appl. Biol.* 123, 499-514.

Cowgill, S. E., Wratten, S. D., and Sotherton, N. W. (1993b). The selective use of floral resources by the hoverfly *Episyrphus balteatus* (Diptera: Syrphidae) on farmland. *Ann. Appl. Biol.* 122, 223-231.

Crane, E., Walker, P., and Day, R. (1984). "Directory of Important World Honey Sources." International Bee Research Association. Gerrard Cross, U.K.

Cromartie, W. J. (1991). The Environmental control of insects using crop diversity. *In* "CRC Handbook of Pest Management in Agriculture." (D. Pimentel, ed.), Vol. 1, 2nd Edition, pp. 183-216. CRC Press, Boca Raton, FL.

Dent, D. (1995). "Integrated Pest Management." CAB International. Wallingford, U.K.

Ehler, L. E. (1990). Introduction Strategies in Biological Control of Insects. *In* "Critical Issues in Biological Control." (M. Mackauer, L. E. Ehler, and J. Roland, eds.), pp. 111-134. Intercept, Andover, Hants, U.K.

Ehler, L. E., and Hall, R. W. (1982). Evidence for competitive exclusion of introduced natural enemies in biological control. *Environ. Entomol.* 11, 1-4.

Ehler, L. E., and Miller, J. C. (1978). Biological control in temporary agroecosystems. *Entomophaga* 23, 207-212.

Glen, D. M., Greaves, M. P., and Anderson, H. M. (1995). "Ecology and Integrated Farming Systems." John Wiley and Sons. Chichester, U.K.

Gurr, G. M. (1994). Concluding remarks: landscape ecology and sustainable pest management, *In* "Sustainable Agriculture Workshop, Proceedings of a meeting held at Orange Agricultural College, The University of Sydney, Orange, NSW, 15th-16th April." (G. M. Gurr, B. J. Baldwin, and B. R. Johnston, eds.), pp. 60-64. Orange Agricultural College, Orange, Australia.

Gurr, G. M., Valentine, B. J., Azam, M. N. G., and Thwaite, W. G. (1996). Evolution of arthropod pest management in apples. *Agric. Zool. Rev.* 7, 3569.

Hagley, E. A. C., and Barber, D. R. (1992). Effect of food sources on the longevity and fecundity of *Pholetesor ornigis* (Hymenoptera: Braconidae). *Can. Entomol.* 124, 341-346.

Halley, S., and Hogue, E. J. (1990). Ground cover influence on apple aphid, *Aphis pomi* De Geer (Homoptera: Aphididae), and its predators in a young apple orchard. *Crop Prot.* 9, 221-230.

Herzog, D. C., and Funderburk, J. E. (1986). Ecological bases for habitat manipulation. *In* "Ecological Theory and Integrated Pest Management Practice." (M. Kogan, ed), pp. 217-250. John Wiley and Sons. New York, NY.

Hickman, J. M., and Wratten, S. D. (1997). Use of *Phacelia tanacetifolia* (Hydrophyllaceae) as a pollen source to enhance hoverfly (Diptera: Syrphidae) populations in cereal fields. *J. Econ. Entomol.* 89, 832-840.

Hickman, J. M., Lövei, G. L., and Wratten, S. D. (1995). Pollen feeding by adults of the hoverfly *Melanstoma fasciatum* (Diptera: Syrphidae). *N. Z. J. Zool.* 22, 387-392.

Holland, J. M., Thomas, R. S., and Courts, S. (*1994). Phacelia tanacetifolia* strips as a component of integrated farming. *In* "Field Margins: Integrating Agriculture and Conservation." Proc. Symp. University of Warwick, Coventry, 18-20 April 1994. (N. D. Boatman, ed.), pp. 215-220. Brit. Crop Prot. Coun. Farnham, U.K.

Idris, A. B., and Grafius, E. (1995). Wildflowers as nectar sources for *Diadegma insulare* (Hymenoptera: Ichneumonidae), a parasitoid of diamondback moth (Lepidoptera: Yponomeutidae). *Environ. Entomol.* 24, 1726-1735.

Inouye, D. W. (1980).The effect of proboscis and corolla tube lengths on patterns and rates of flower visitation by bumblebees. *Oecologia* 45, 197-201.

Jervis, M. A., Kidd, N. A., Fitton, M. G., Huddleston, T., and Dawah, H. A. (1993). Flower-visiting by hymenopteran parasitoids. *J. Nat. Hist.* 27, 67-105.

Jervis, M. A., Kidd, N. A. C., and Heimpel, G. E. (1996). Parasitoid adult feeding behaviour and biocontrol - a review. *Biocont. News Inform.* 17, 11-26.

Kareiva, P. (1990). The Spatial Dimension in Pest-Enemy Interactions. In "Critical Issues in Biological Control." (M. Mackauer, L. E. Ehler, and J. Roland, eds.), pp. 231-228. Intercept, Andover, Hants, U.K.

Kelly, G. L. (1987). "Factors affecting the success of *Trissolcus basalis* (Hymenoptera: Scelionidae) as a biological control agent of the green vegetable bug, *Nezara viridula*

(Hemiptera: Pentatomidae)." Ph.D. Thesis, School of Biological Sciences, The University of Sydney, Sydney, Australia.

Kevan, P. C., and Baker, H. G. (1984). Insects on Flowers. *In* "Ecological Entomology." (C. B. Huffaker, and R. L. Rabb, eds.), pp. 607-631. John Wiley and Sons. New York, NY.

Landis, D. A., and Haas, M. J. (1992). Influence of landscape structure on abundance and within field distribution of European corn borer (Lepidoptera: Pyralidae) larval parasitoids in Michigan. *Environ. Entomol.* 21, 409-416.

Liang,W., and Huang, M. (1994). Influence of citrus orchard ground cover plants on arthropod communities in China: A Review. *Agric., Ecosys., Environ.* 50, 29-37.

MacArthur, R. H., and Wilson, E. 0. (1967). "The Theory of Island Biogeography." Princeton University Press, Princeton, NJ.

MacLeod, A. (1992). Alternative crops as floral resources for beneficial hoverflies (Diptera: Syrphidae). *Brighton Crop Prot. Conf. - Pests & Dis.,* 997-1002.

Marino, P. C., and Landis, D. A. (1996). Effect of landscape structure on parasitoid diversity and parasitism in agroecosystems. *Ecol. Appl.* 61, 276-284.

Matthews-Gehringer, D., Lachowski, E., Edwards, S., Wolfgang, S., Schettini, T., Bugg, R. L., and Felland, C. M. (1994). "Herbaceous Ground Covers in a Pennsylvania Apple Orchard: Establishment and Arthropod Assessment 1989-1991." HO-94/02, Rodale Institute Research Center. Kutztown, PA.

Pearson, A. W. (1990). The Management of Research and Development. *In* "Technology and Management." (R. Wild, ed.), pp. 28-41. Nichols Publications. New York, NY.

Peng, R. K., Incoll, L. D., Sutton, S. L., Wright, C., and Chadwick, A. (1993). Diversity of airborne arthropods in a silvoarable agroforestry system. *J. Appl. Ecol.* 30, 551-562.

Perfect, T. J. (1991). Biodiversity and tropical pest management. *In* "The Biodiversity of Microorganisms and Invertebrates: Its Role in Sustainable Agriculture." (D. L. Hawksworth, ed.), pp. 145-148. CAB International. Wallingford, U.K.

Pickett, C. H., and Bugg, R. L. "Enhancing Biological Control: Habitat Management to Promote Natural Enemies of Agricultural Pests." University of California Press. Berkeley, CA. (in preparation).

Powell, W. (1983). The Role of Parasitoids in Limiting Cereal Aphid Populations. *In* "Aphid Antagonists." (R. Cavalloro, ed.), pp. 50-56. Balkema. Rotterdam, The Netherlands.

Powell, W. (1986). Enhancing parasitoid activity in crops. *In* "Insect Parasitoids." (J. Waage, and D. Greathead, eds.), pp. 319-340. Academic Press. London, U.K.

Power, A. G., and Kareiva, P. (1990). Herbivorous insects in agroecosystems. *In* "Agroecology" (C. R. Carroll, J. H. Vandemeer, and P. M. Rosset, eds.), pp. 301-327. McGraw Hill, New York, NY.

Prokopy, R. C. (1994). Integration in orchard pest and habitat management: a review. *Agric., Ecosys. Environ.* 50, 1-10.

Puvuk, D. M., and Stinner, B. R. (1992). Influence of weed communities in corn plantings on parasitism of *Ostrinia nubilalis* (Lepidoptera: Pyralidae) by *Erioborus terebrans* (Hymenoptera: Ichneumonidae). *Biol. Cont.* 2, 312-316.

Redfearn, A., and Pimm, S. L. (1987). Insect Outbreaks and Community Structure. *In* "Insect Outbreaks." (P. Barbosa, and J. C. Schultz, eds.), pp. 99-133. Academic Press. San Diego, CA.

Risch, S. J. (1987). Agricultural Ecology and Insect Outbreaks. *In* "Insect Outbreaks." (P. Barbosa, and J. C. Schultz, eds.), pp. 217-238. Academic Press. San Diego, CA.

Risch, S. J., Andow, D., and Altieri, M. A. (1983). Agroecosystem diversity and pest control: data, tentative conclusions, and new research directions. *Environ. Entomol.* 12, 625-629.

Rogers, C. E. (1985). Extrafloral nectar: entomological implications. *Bull. Entomol. Soc. Amer.* 31, 15-20.

Root, R. B. (1973). Organization of a plant-arthropod association in simple and diverse habitats: the fauna of collards *(Brassica oleracea)*. *Ecol. Monogr. 43,* 95-124.

Sheehan, W. (1986). Response by specialist and generalist natural enemies to agroecosystem diversification: a selective review. *Environ. Entomol.* 15, 456461.

Smith, M. W., Arnold, D. C., Eikenbary, R. D., Rice, N. R., Shiferaw, A., Cheary, B. S., and Carroll, B. L. (1996). Influence of ground cover on beneficial arthropods in pecan. *Biol. Cont.* 6, 164-176.

Somchoudhury, A. K., and Dutt, N. (1988). Evaluation of some flowers as nutritional source of *Trichogramma* spp. *Ind. J. Entomol.* 50, 371-373.

Southwood, T. R. E. (1977). The relevance of population dynamic theory to rest status. *In* "The Origins of Pest, Parasite, Weed and Disease Problems." (J. M. Cherrett, and G. R. Sagar, eds.), pp. 35-54. Blackwell Scientific Publications. Oxford, U.K.

Southwood, T. R. E., and Comins, H. N. (1976). A synoptic population model. *J. Anim. Ecol.* 45, 949-965.

Speight, M. R. (1983). The potential of ecosystem management for pest control. *Agric., Ecosys., Environ.* 10, 183-199.

Stern, V. M., Smith, R. F., van den Bosch, R., and Hagen, K. J. (1959). The integrated control concept. *Hilgardia* 29, 81 -101.

Stephens, M. J., France, C. M., Wratten, S. W., and Frampton, C. M. (1997). Increasing the diversity of an orchard understory to enhance biological control of leafrollers (Lepidoptera: Tortricidae). *Biocont. Sci. Technol.* (in press).

Takasu, K., and Lewis, W. J. (1995). Importance of adult food sources to host searching of the larval parasitoid *Microplitis croceipes*. *Biol. Cont.* 5, 25-30.

Thomas, W. P., and Burnip, G. M. (1993). Lepidopterous insect pests and beneficials - the influence of the understory. *In* " Proceedings of Biological Apple Production Seminar - A Review of Winchmore Research Orchard Trials 1989-1993." (T. P. McCarthy, M. J. Daly, and G. M. Burnip, eds.), pp. 13*19*. *Hort Res. Tech. Rept.* No 9313.

Thomas, M. B., Wratten, S. D., and Sotherton, N. W. (1992). Creation of 'island' habitats in farmland to manipulate populations of beneficial arthropods: predator densities and species composition. *J. Appl. Ecol.* 29, 524-531.

Tonhasca, A. (1993). Effects of agroecosystem diversification on natural enemies of soybean herbivores. *Entomol. Exp. Appl.* 69, 83-90.

Topham, M., and Beardsley, J. W. (1975). An influence of nectar source plants on the New Guinea sugar cane weevil parasite, *Lixophaga sphenophori* (Villeneuve). *Proc. Hawaiian Entomol. Soc.* 22, 145-155.

Treacy, M. F., Zummo, G. R., and Benedict, J. H. (1987). Interactions of host-plant resistance in cotton with predators and parasites. *Agric., Ecosys., Environ.* 13, 151-157.

Tuovinen, T. (1994). Influence of surrounding trees and bushes on the phytoseiid mite fauna on apple orchard trees in Finland. *Agric., Ecosys., Environ.* 50, 39-47.

Vandemeer, J. H. (1990). Intercropping. In "Agroecology." (C. R. Carroll, J. H. Vandemeer, and P. M. Rosset, eds.), pp. 481-516. McGraw Hill. New York, NY.

van Emden, H. F. (1990). Plant diversity and natural enemy efficiency in agroecosystems. *In* "Critical Issues in Biological Control." (M. Mackauer, L. E. Ehler, and J. Roland, eds.), pp. 63-80 Intercept. Andover, Hants, U.K.

van Emden, H. F., and Dabrowski, Z. T. (1997). Issues of biodiversity in pest management. *Insect Sci. Appl.* 15, 605-620.

van Emden, H. F., and Peakall, D. B. (1966). "Beyond Silent Spring: Integrated Pest Management and Chemical Safety." Chapman and Hall. London, U.K.

van Emden, H. F., and Williams, G. F. (1974). Insect stability and diversity in agroecosystems. *Annu. Rev. Entomol.* 19, 455-475.

Waage, J. K. (1990). Ecological Theory and the Selection of Biological Control Agents, *In* "Critical Issues in Biological Control." (M. Mackauer, L. E. Ehler, and J. Roland, eds.), pp. 135-157. Intercept. Andover, Hants, U.K.

Wäckers, F. L., and Swaans, C. P. M. (1993). Finding floral nectar and honeydew in *Cotesia rubecula:* Random or directed? *Proc. Exp. Appl. Entomol.* 4, 67-72.

Way, M. J. (1966). The natural environment and integrated methods of pest control. *J. Appl. Ecol.* 3, 29-32.

Way, M. J. (1977). Pest and Disease Status in Mixed Stands vs. Monocultures: the Relevance of Ecosystem Stability. *In* "The Origins of Pest, Parasite, Weed and Disease Problems." (J. M. Cherrett, and G. R. Sagar, eds.), pp. 127-138. Blackwell Scientific Publications. Oxford, U.K.

White, A. J., Wratten, S. D., Berry, N. A., and Weigmann, U. (1995). Habitat manipulation to enhance biological control of *Brassica* pests by hover flies (Diptera: Syrphidae). *J. Econ. Entomol.* 88, 1171-1176.

Wickremasinghe, M. G. V., and van Emden, H. F. (1992). Reactions of adult female parasitoids, particularly *Aphidius rhopalosiphi,* to volatile chemical cues from the host plants of their aphid prey. *Physiol. Entomol.* 17, 297-304.

Willis, J. C. (1973). "A Dictionary of the Flowering Plants of the World." Eighth Edition (Revised by H. K. Airy Shaw). Cambridge University Press. Cambridge, U.K.

Wratten, S. D., and van Emden, H. F. (1995). Habitat management for enhanced activity of natural enemies of insect pests. *In* "Ecology and Integrated Farming Systems." (D. M. Glen, M. P. Greaves, and H. M. Anderson, eds.), pp. 117-145 John Wiley and Sons. Chichester, U.K.

Wratten, S. D., van Emden, H. F., and Thomas, M. B. Within field and border refugia for the enhancement of natural enemies. *In* "Enhancing Biological Control: Habitat Management to Promote Natural Enemies of Agricultural Pests." (C. H. Pickett, and R. L. Bugg, eds). University of California Press. Berkeley, CA. (in preparation).

Zadoks, J. C. (1993). Crop Protection: Why and How. *In* "Crop Protection and Sustainable Agriculture." (D. J. Chadwick, and J. Marsh, eds.), pp. 48-55. John Wiley and Sons. New York, NY.

Zhao, J. Z., Ayres, G. S., Grafius, E. J., and Stehr, F. W. (1992). Effects of neighbouring nectar-producing plants on populations of pest Lepidoptera and their parasitoids in broccoli plantings. *Great Lakes Entomol.* 25, 253-258.

CHAPTER 10

NATURALLY OCCURRING BIOLOGICAL CONTROLS IN GENETICALLY ENGINEERED CROPS

C. W. Hoy, J. Feldman, F. Gould, G. G. Kennedy,
G. Reed, and J. A. Wyman

I. INTRODUCTION AND OVERVIEW

Genetically engineered crops have entered the marketplace. How will the profound changes in these crops affect the interaction of existing biological control agents with pests? Because many genetically engineered changes in crop cultivars are feasible, a wide range of effects on biological control are possible. This chapter provides a framework for evaluating the likely impact on conservation of biological control agents of genetically engineered changes in crop cultivars. We draw examples both from the literature, most of which describes conventionally developed crop cultivars, and our recent research and experience with transgenic crops that express the *Bacillus thuringiensis* d-endotoxin (*Bt*).

Pest resistant *Bt*-expressing cultivars have been one of the first commercial products derived from genetic engineering technology in agriculture. Insect toxins can affect biological control agents when they feed on toxic plants or on intoxicated prey. Direct effects on biological control agents can also accrue from exposure to other resistance mechanisms expressed on the plant surface; in plant trichome exudates, for example. Even more likely, however, are effects on insect natural enemies through complex modification of the food web. Pests that are targeted by the resistance mechanism are typically reduced in number. A concomitant reduction in pesticide use, however, may result in increased number and diversity of other arthropods in the crop. These changes in availability of various arthropods could have either positive or negative impacts on biological control. The degree to which biological control can be preserved and integrated with genetically engineered pest resistance mechanisms should be considered in the early stages of cultivar development. Widespread concern over pest adaptation to genetically engineered cultivars has accompanied their regulatory approval and commercial release. We suggest that conservation of biological control agents and their potential importance in avoiding the evolution of resistant pests will be a useful consideration in designing new genetically engineered cultivars.

Other genetically engineered changes in crops may result in altered horticultural characteristics which may affect the abundance and influence of biological control agents. Improved tolerance of damage, for example due to viruses transmitted by aphids, changes the criteria for acceptable biological control, making success more likely. Some horticultural characteristics that might be altered, availability of pollen for example, can directly affect populations of biological control agents. Increasing crop tolerance of extreme temperatures or moisture conditions may alter the phenological synchrony between biological control agents and pests. These and other horticultural characteristics will be discussed with regard to their implications for conservation of biological control agents.

The extent to which genetically engineered crop cultivars fit into ecologically based pest management is likely to determine their ultimate acceptance by society as a whole and their longevity in the field. Immediate acceptance by producers, however, may require immediately apparent benefits. We discuss the complexity involved with marketing a genetically engineered cultivar and the importance of natural enemy conservation and biological pest control as a marketing tool relative to other marketing considerations. Our goal in this chapter is to provide a framework for predicting and measuring the impact of crop genetic engineering on conservation of biological control agents, a framework that may help guide future development and use of genetically engineered crop cultivars.

II.   GENETIC ENGINEERING FOR INSECT RESISTANCE

Many plant characters affect natural enemies (parasitoids, predators, pathogens) of insect herbivores (Price *et al.,* 1980; Boethel and Eikenbary, 1986; Barbosa and Letourneau, 1988; Hare, 1992; Dicke, 1996). Consequently, traits conferring resistance can alter the effectiveness of biological control of the resisted pest or other species in the pest complex attacking the crop (Bergman and Tingey, 1979; Duffey and Bloem, 1986; Duffey *et al.,* 1986; Hare, 1992). The ecological impact of resistance traits, not the methods by which they were incorporated into a crop, are responsible for effects on biological control and the conservation of insect natural enemies. Although experience with genetically engineered traits conferring pest resistance is limited, available data indicate a general compatibility between genetically engineered *Bt* expression in crop plants and the natural enemies of both *Bt* susceptible and nonsusceptible pest species (Hoffman *et al, 1*992; Johnson and Gould, 1992; Dogan *et al.,* 1996; Johnson et *al.,* 1997). In contrast, a wealth of information exists on the tritrophic-level effects of conventional plant defenses and that information can provide insight into the potential effects of genetically engineered pest resistance on natural enemies and biological control.

Plant resistance traits that affect natural enemies may do so either negatively or positively. When viewed at the population level, the effects of plant resistance and natural enemies on pest populations can be additive, synergistic, or antagonistic (van Emden, 1966, 1986; Gutierrez, 1986; Hare, 1992). In a recent review, Hare (1992) found that of 16 cases of tritrophic-level interactions involving parasitoids six involved antagonism, two involved synergism, and five involved an additive relationship between plant resistance and potential biological control.

The effects of plant resistance traits on life history parameters of both pests and natural enemies determine biological control success at the population level (Gutierrez, 1986; Hare, 1992). For example, although the parasitoid *Pediobius foveolatus* Crawford developed more slowly and suffered higher mortality and reduced reproduction when reared on Mexican bean beetles *(Epilachna varivestis* Mulsant) on resistant soybeans, the population growth rate of *P. foveolatus was* reduced less than that of its host. As a result, the intrinsic rate of increase of the parasitoid was higher relative to its host on resistant than on susceptible cultivars (Kauffman and Flanders, 1985).

The mechanisms by which plant resistance traits affect natural enemies may be complex. The nature of the effect on natural enemies depends as much on the specific attributes of the resistance mechanism as it does on the details of the interaction of the natural enemies, their hosts/prey, and the plants on which their hosts/prey occur (e.g., Kauffman and Kennedy, 1989; Kashyap *et al.,* 1991; Farrar and Kennedy, 1993). Plant resistance traits may affect natural enemies directly. Alternatively, their effect may be indirect, mediated through effects of the resistance trait on insect hosts/prey of the natural enemy.

A.   Direct Effects on Biological Control Agents

Many insect parasitoids and predators have an intimate association with the plants upon which their insect hosts or prey feed (Barbosa and Letourneau, 1988). These plants provide not only a habitat but also behavioral cues that mediate searching behavior and a source of water or nutrition for some species which feed on floral or extrafloral nectar, pollen, or plant sap (Hagen, 1986: also see Chapters 4, 5, and 9). Plant resistance factors can potentially affect natural enemies directly through any of these avenues.

Plant chemicals produced constitutively or in response to herbivore feeding play an important role in host searching behavior by many entomophagous arthropods (Vinson, 1976; Dicke, 1996; Nordlund *et al.,* 1988; Dicke and Sabelis, 1988; Dicke *et al.,* 1990a,b; Turlings *et al.,* 1990 Vet and Dicke, 1992: also see Chapters 4 and

5). Any pest resistance traits that alter the production or emission of these behaviorally active chemicals may alter the attractiveness of the plant to some species of natural enemies. Depending on the plants, chemicals, and natural enemy species involved the effect may be positive, neutral, or negative. There is no evidence to date that any plant traits altered through genetic engineering to increase pest resistance have directly affected plant attractiveness to natural enemies. The array of plant resistance traits widely observed in the field to date has been limited mainly to proteins like *Bt* endotoxin and viral coat proteins, which might be expected to have little impact on natural enemies. Such effects are possible as additional novel traits conferring resistance (such as allelochemicals) are engineered into plants. Indeed, as we develop a better understanding of the role of chemical cues in the attraction and searching behavior of parasitoids and predators, it may become feasible to engineer plants specifically to increase their attractiveness to natural enemies of important pest species (Bottrell *et al.,* in press).

Predators and parasitoids may be affected by resistance-associated toxins expressed in the nectar, pollen, or sap on which they feed. For example, the generalist predator *Geocoris punctipes* (Say) which normally feeds on plant material to obtain water, had longer developmental time and reduced survival when confined on soybean foliage resistant to velvet bean caterpillar and soybean looper than when confined on susceptible soybean foliage (Rogers and Sullivan, 1986). Similarly, another predaceous hemipteran, *Podisus maculiventris* (Say) which feeds on plant sap as well as insects also experienced reduced growth and prolonged development on resistant soybean foliage (Orr and Boethel, 1986).

The primary toxins responsible for resistance in the transgenic resistant crops produced to date are lepidopteran- and coleopteran-specific *Bt*. There is no evidence that these toxins directly affect insect parasitoids or predators, although much of the evidence comes from experiments in which natural enemies were exposed to commercial sprayable formulations of *Bt*. In laboratory studies, the coleopteran-active *Bt tenebrionis (Btt)* was nontoxic to the predaceous hemipteran *Perillus bioculatus* F. when applied topically or when the predator was maintained on *Btt*-treated potato foliage. Feeding on Colorado potato beetle *Leptinotarsa decemlineata* (Say) larvae that had fed on *Btt*-treated foliage for 20 hours also had no effect on *P. bioculatus* (Hough-Goldstein and Keil, 1991). Similarly, feeding on *Bt tenebrionis (=san diego)-* treated pollen did not cause mortality of adult *Coleomegilla maculata lengi* (Say), an important predator on Colorado potato beetle eggs and larvae and of aphids on potato. However, consumption of *Btt*-treated Colorado potato beetle eggs by *C. maculata* adults was inversely related to *Btt* concentration (Giroux *et al.,* 1994). In field studies, however, Hilbeck (1994) did not detect any effect of *Btt* foliar sprays on predation on Colorado potato beetle eggs by *C. maculata* in potato.

B.   Indirect Effects on Biological Control Agents

Indirect effects of resistance traits on natural enemies are those mediated through the phytophagous insects which serve as hosts or prey. These effects can result from changes in vulnerability of phytophagous insects to natural enemies resulting from altered behavior or development rates on resistant plants or through changes in their suitability as hosts/prey resulting from intoxication or the sequestration of plant-produced toxins (Barbosa and Letourneau, 1988; Barbosa et *al.*, 1991; Hare, 1992; Dicke, 1996). Transfer of plant-produced toxins to natural enemies can be affected by prey behavioral responses to the toxins in complex ways. For example, development time in the coccinellid aphid predator *Eriopis connexa* Germar was shorter when its aphid prey fed on plants with high rather than low DIMBOA levels and longest at intermediate DIMBOA levels (Martos *et al.,* 1992). Studies with artificial diet demonstrated that DIMBOA increased development time of *E. connexa* and DIMBOA concentrations in aphids were higher when they fed on intermediate than on low plant concentrations. Aphid feeding, however, was inhibited at high plant concentrations which led to lower DIMBOA concentrations in the aphids and their increased suitability as prey (Martos *et al.,* 1992).

In the case of *Bt* transgenic plants, evidence suggests that indirect effects on natural enemies are likely to occur in some instances. Their significance under field conditions has not been established. However, indirect effects are more likely to be of consequence in situations in which herbivores experience sublethal effects on the transgenic plants than in situations in which they are killed quickly, leaving little opportunity for interactions with natural enemies. Several studies have documented the potential for indirect effects involving transgenic tobacco expressing the lepidopteran-active *Bt* CrylA(b). In a small-scale field trial, transgenic tobacco lines expressing high levels of *Bt* provided excellent control of tobacco hornworm *Manduca sexta* L. and tobacco budworm *Heliothis virescens* (L.), whereas lines expressing lower levels of toxin reduced larval growth and delayed development. During both years of the study, larval parasitization of the *H. virescens/Helicoverpa zea* (Boddie) complex tended to be higher (56%) on plant lines expressing low levels of *Bt* than on lines which did not express the toxin (43%). Although the differences were not statistically significant, the results clearly indicated compatibility between *Bt* resistance and hymenopterous larval parasitoids (Warren *et al.,* 1992).

In another study, Johnson and Gould (1992) estimated that the effects of partial resistance to *H. virescens* in a tobacco line expressing a low level of the CrylA(b) endotoxin and natural enemies, primarily the parasitoid *Campoletis sonorensis* (Cameron), were synergistic: causing 11% more mortality of *H. virescens* larvae than expected if their effects were additive. The synergism appeared to be associated with elevated mortality of parasitized larvae during the first stadium and to a prolonged period of vulnerability to parasitism and predation associated with a slower development rate on the transgenic plants. Additional data indicate that prolonged development of host larvae on the moderately resistant transgenic plants extends the period of vulnerability to parasitism by *C. sonorensis,* which primarily attacks small larvae (Johnson, in press). The result is an increased likelihood that larvae will be in a vulnerable stage when wasp activity is high. In contrast, vulnerability to parasitism by *Cardiochiles nigriceps* Viereck, which attacks host larvae over a broader range of size classes, was less affected on resistant plants.

When *H. virescens* larvae were exposed to *C. sonorensis* adults for only one day, parasitism levels were higher on conventional than on transgenic plants. This may have been due to increased movement and decreased feeding by host larvae on transgenic foliage. The latter might have resulted in a reduction in the tactile and volatile cues from feeding damage that are used in host location by *C. sonorensis* (Johnson, in press). This behavioral effect is likely of limited consequence because in longer term exposures *C. sonorensis* adults found the host larvae on transgenic plants. Nonetheless, it remains possible that in large plantings of transgenic plants, a general reduction in host-finding cues associated with feeding damage to the plant could lead to a general reduction in parasitoid activity (Johnson, in press).

Comparable studies have not been conducted with Colorado potato beetle resistant, transgenic potato plants expressing the coleopteran-active CrylllA endotoxin from *Bt tenebrionis.* However, a recent study (Lopez and Ferro, 1995) indicated that the parasitoid *Myiopharus doryphorae* (Riley) parasitized with equal frequency living Colorado potato beetle larvae that had been exposed to lethal or sublethal doses of *Bt tenebrionis* (CryllIA toxin) and those which had not been exposed to the toxin. However, the parasitoid could not complete its development if the host died prior to pupation. If the host survived to the pupal stage then the parasitoid completed its development. Negative indirect effects could occur in Colorado potato beetles on transgenic plants if larvae survive into the second stadium (when they become acceptable hosts for parasitization) but die prior to pupation. Given the very high level of toxin expression in commercial transgenic potato varieties, the potential for indirect effects on natural enemies of Colorado potato beetle is likely to be minimal because susceptible larvae die very quickly on the resistant plants (Wierenga *et al.,* 1996). Nonetheless, the potential exists for indirect effects mediated by other phytophagous insects developing on the transgenic plants or by *Bt*-resistant Colorado potato beetles.

Available data suggest that interactions involving *Bt* toxin and the green peach aphid *Myzus persicae* (Sulzer), which is an important pest on potato, are unlikely. The green peach aphid feeds normally on transgenic potato varieties expressing the CrylllA toxin (Shieh *et al.,* 1994) and the predator *Hippodamia convergens* (Guerin-Meneville) is unaffected by feeding on green peach aphids reared on transgenic potato (Dogan *et al.,* 1996). Nevertheless, if no Colorado potato beetle eggs are present, ladybird beetles that feed on both beetle eggs and aphids could be negatively affected.

C. Impact on Population-Level Food Web Interactions

Resistant cultivars negatively affect natural enemies through reduced prey population density but affect them positively when pesticide use is reduced. The impact of genetically engineered cultivars on conservation biological control will likely vary according to the arthropod community in a particular cropping system and the changes in pesticide use permitted by the resistance trait. Our recent research on the arthropod community in one of the first genetically engineered pest-resistant crops, *Bt* potato, is but one example but it allows us to examine the potential impacts of genetically engineered resistance to Colorado potato beetle on conservation biological control in this system. In many North American potato production areas Colorado potato beetle can completely destroy the potato crop and thereby drastically alter the plant community. With standard season long use of insecticides for Colorado potato beetle, many species of arthropods cannot survive. In *Bt* potato crops, however, Colorado potato beetles are eliminated and plants survive without insecticides, allowing a very different arthropod community.

1. Diversity of arthropod fauna in *Bt* potato

Initial concerns regarding a potato crop with high *Bt* expression were that it could affect detritivores in addition to targeted herbivores, thereby reducing species diversity and simplifying the food web even in the absence of insecticides. During five years of intensive sampling in *Bt* potatoes, using a variety of sampling methods, over 200 species of arthropods have been collected and identified in Pacific Northwest potato fields (Reed, unpublished data). Because identifications are far from complete, the estimated total is as much as 300 species. Based on current taxonomic separation, the species composition of this arthropod community is: 50% predators or parasites, 34% detritivores, and 16% plant or seed feeders. As might be expected, when these groups are compared by number of individuals (probably total biomass as well) detritivores are most numerous, followed by herbivores, and finally carnivores. Colorado potato beetles and green peach aphids are the two most abundant canopy dwelling herbivores in this region, although Colorado potato beetle is absent in *Bt* potatoes.

Generalist predators (Araneae, Geocoridae, Nabidae, and Anthocoridae) are abundant in the canopy. Numerically, three dwarf spiders (Linyphiidae: Erigone) are the predominate species both on the foliage and in the litter. Though the detritivore component of the canopy is relatively small, it is dominated by Collembola and various fungus beetles that appear to be feeding on mycelia and fruiting bodies on senescing leaves. Detritivores are the most common group in the litter. Collembola represent the greatest biomass but a substantial number of mite, dipteran, and coleopteran species are present. Common predators include spiders (at least 18 species) and Carabidae (at least 13 species) (Reed, unpublished data). The Pacific Northwest *Bt* potato crop clearly contains a diverse community of arthropods including abundant and diverse biological control agents and a diverse community of potential prey.

2. Enhanced control of secondary pests in *Bt* potato

In agricultural systems, the complex of arthropod pest species typically consists of one to several primary pests together with a variable number of secondary or potential pests. Primary pests usually are not controlled by naturally occurring biological control agents, cultural controls, or other preventive measures used in the system and frequently reach economically damaging levels. In contrast, secondary pests are species which are usually naturally controlled and become pests only when that control is disrupted. Disruption of natural control occurs most frequently when broad-spectrum insecticides used to control primary pests also destroy insect natural enemies.

When genetically engineered crop varieties eliminate the need for broad-spectrum insecticidal controls for primary pests, naturally occurring control agents are more likely to suppress secondary pest populations. For the near future, genetically engineered varieties for insect pest resistance probably will provide thorough control of one or more primary pests because this provides the most easily recognized economic returns to growers (see Section IV). In these highly resistant cultivars conservation biological control will most likely accrue from the elimination of insecticide applications and the resulting conservation of insect natural enemies that control secondary pest populations. Reduction of insecticide applications sufficient to conserve biological control agents, however, may only be possible if all of the primary pests in the crop are controlled by the resistance factor or other selective controls.

In addition to the Colorado potato beetle, the potato leafhopper *Empoasca fabae* (Harris) is a serious primary pest in the North Central and Eastern United States (Walgenbach and Wyman, 1985). Potato leafhoppers are not regulated effectively by natural enemies but were frequently controlled indirectly by insecticides targeted at Colorado potato beetle. In transgenic potatoes, potato leafhopper still must be controlled, preferably without disrupting natural control of other pests. A similar situation exists for the aster leafhopper *Macrosteles quadrilineatus* (Stal), the vector of purple top in the North Central United States. In contrast to the leafhoppers, two species of aphids, the green peach aphid and the potato aphid *Macrosiphum euphorbiae* (Thomas), are frequent pests of potato but are susceptible to natural control by many species of predaceous and parasitic insects. Aphids are primary pests where transmission of viruses is of concern and secondary pests in other growing areas.

In Wisconsin and Ohio, *Bt* potato management programs have been developed to control potato leafhopper populations without disrupting natural control of aphids. In large replicated plots, *Bt* potato resistance to Colorado potato beetle was combined with foliar sprays (malathion or dimethoate), systemic organophosphate insecticides (phorate and disulfoton), or both for control of potato leafhoppers. Conservation of aphid natural controls was compared between these and standard potato plots which were managed with season-long applications of pyrethroids and organophosphates for Colorado potato beetle and potato leafhopper control. Populations of insect pests, predators, and parasitoids were closely monitored in each plot. In all cases, Colorado potato beetles were essentially absent in *Bt* potato plots and frequently abundant in the standard insecticide treated plots.

Although most of the predaceous and parasitic taxa were present in both insecticide treated standard and untreated *Bt* potato plots, populations consistently were more numerous in the *Bt* potato plantings. Over four years (1992 to 1995) and five locations in Wisconsin (Wyman, unpublished data) season-long generalist predator (Anthocoridae, Chrysopidae, Coccinellidae, Opiliones-Palpatores and Araneae) population densities were significantly higher (a=0.05) in *Bt* potato plantings than in comparable standard plots with broad-spectrum control. The broad-spectrum insecticides reduced total predator populations by an average of 63.8% in five Wisconsin locations (ranging from 32 to 76%). Similar results were obtained in 1993, in Ohio plots, with an 83% average reduction in canopy-dwelling aphid predators (Coccinellidae, Chrysopidae, Nabidae, Anthocoridae, and Syrphidae) in the insecticide-treated plots compared with untreated *Bt* potato plantings (Hoy, unpublished data). Parasitic Hymenoptera populations (primarily Braconidae and Pteromalidae) were significantly (a = 0.05) higher in *Bt* potato plots over three years (1992-94). Four Wisconsin locations where samples were taken had an average reduction in standard plots of 58.4% (ranging from 41 to 78%). During 1993, in Wisconsin, ground-dwelling general predators (Carabidae, Staphylinidae, Opiliones-Palpatores and Araneae) were reduced by 65% in pitfall traps in standard plots. These data clearly demonstrated that the anticipated conservation of naturally occurring control agents in *Bt* potatoes, in the absence of broad-spectrum insecticides, does occur consistently. Biological control agents also were conserved in 1996 in two Wisconsin commercial plantings (50 to 80 acres) where paired comparisons could be made between *Bt* potato plantings and foliar spray programs. Generalist predators were reduced by 45 and 31% in the standard potato fields.

Biological control of Wisconsin aphid populations was achieved in three of the four years in which these comparisons were made in experimental plantings where aphids were not treated with insecticides. In these cases, aphids were effectively held below threshold in 1992 when populations were low and were significantly higher (a = 0.05) in standard plantings compared with *Bt* potato in 1994 and 1995 when populations were high. In all instances, biological control of aphids did not occur until relatively late in the season. Aphid biological control was insufficient in five commercial *Bt* potato fields monitored in 1996, where populations increased late in the season despite enhanced predator populations. Increased use of broad-spectrum fungicides for late-blight control in 1996 may have reduced the effectiveness of entomopathogenic fungi in commercial fields, removing an important aphid mortality factor.

The population density of aphid predators in the crop canopy during the early phases of aphid population increase was very important for successful aphid control during 1993, in Ohio. In standard potato plots treated with pyrethroids for Colorado potato beetle control, predaceous insects in the potato canopy declined to undetectable levels for a two week period during mid-July. By the time predaceous insects were detected again in late July, aphid population densities had increased from undetectable levels to an average of 253 per m of row. Aphid population density reached an average of over 4000 per m of row by late August despite frequent pyrethroid use in the nontransgenic plots, while predator populations remained at very low or undetectable levels. Meanwhile in untreated *Bt* potato plots, predaceous insects were detectable throughout late July and August and increased from approximately one to four per m of row between late July and early August, when aphids were dispersing from the heavily infested standard potato plots. Aphid population density in these plots did not exceed ten per m of row until the last week of August. Throughout late July and early August the ratio of insect predators to aphids ranged from 0.16 to 0.5 in the *Bt* potato plots but remained at or very close to zero in the pyrethroid treated plots. The conservation of aphid predators in *Bt* potato plots during a critical two week period apparently prevented the aphid outbreak.

Insecticides may be required for potato leafhopper control in transgenic potato crops grown in Ohio and Wisconsin, but we may still conserve biological control agents by restricting insecticide use in time. Malathion and dimethoate were effective in leafhopper control and did not appear to disrupt biological control of aphids if applied before mid-July. Although insect predator populations were temporarily reduced after these applications, aphid populations also were reduced and predator population densities increased again when or before aphid population densities increased. Systemic organophosphates provided good leafhopper control throughout the season, but natural enemy populations in Wisconsin were significantly lower in these plantings, probably because of lower prey densities.

In summary, intensive sampling in the potato crop system has demonstrated that transgenic resistance, by controlling primary pests without insecticides, can preserve biological control agents that suppress secondary pests. Furthermore, in general, arthropod species diversity can increase with the use of transgenic crop varieties. We suggest that the most effective strategy would include engineered resistance as a component of an integrated pest management program. As a caveat, if some of the primary pests still must be controlled by broad-spectrum pesticides then engineered resistance may provide no more benefit to conservation biological control than narrow-spectrum insecticides of the past. Genetically engineered cultivars may be no different from very effective and very specific insecticides to the insect pest. If they are the sole means of controlling primary pests we must consider the threat of resistance evolution in pest populations and the role of biological control in avoiding it.

D. Conservation of Biological Control Agents and Resistance Management

A number of literature reviews emphasize the importance of alternate crops or weeds as sources of nectar for parasitoids and sources of alternate hosts for predators, parasitoids, and pathogens of the target pest (e.g., van Emden and Williams, 1973; Zandstra and Motooka, 1978; Powell, 1986). As the acreage grown to pesticidal transgenic crops increases, and population densities of targeted pests are reduced, other sources of hosts for natural enemies (particularly those specializing on the pests controlled by the resistant crop) will become more critical for maintaining natural enemy populations. In addition to weeds and other crops, any acreage of the target crop that is not planted with a toxin- producing cultivar and is not treated with broad-spectrum insecticides could boost natural enemy population densities.

A cornerstone of resistance management for transgenic crops is a refuge, generally thought of as acreage of a nontransgenic cultivar (or noncrop) that is grown so that there will be a source of toxin-susceptible genotypes of the pest insect to mate with resistant individuals. Because the objective is to produce susceptible pests, these refuges should be treated minimally or not at all with insecticides and so may also be a source of natural enemies that feed on the toxin susceptible pests. Nontransgenic refuges established near transgenic crops for resistance management will, therefore, contribute to the conservation and abundance of insect natural enemies. The area devoted to such refuges and the extent of their contribution to conservation biological control will depend on their contribution to suppressing resistance development. Will the positive effect that a refuge has on biological control agents have any impact on the role of that refuge in slowing the rate at which resistance develops?

The general answer to this question is straightforward. If the extra natural enemies produced in the refuge decrease survival of insect pests in the refuge and transgenic crop equally then these natural enemies are likely to have no net effect on the rate at which resistance develops. If natural enemies have a greater/lesser relative impact on insects in the refuge than in the transgenic crop then they will, in effect, decrease/increase the production of susceptible pest genotypes in the refuge and increase/decrease the rate at which resistance evolves. Gould (1994) used a simple genetic model to estimate that the percentage change in the number of generations until pests evolve resistance in a high dose/refuge system was approximately equal to the percentage change in the impact of the natural enemies in the transgenic crop. For example, compared with an equal effect of natural enemies in the refuge and transgenic crop, twice the natural enemy impact in the transgenic crop resulted in twice the number of generations until resistance developed.

Understanding the impact natural enemies could have in a specific system requires empirical data. Arpaia *et al.* (1997) measured potato beetle egg mass density in standard and *Bt* potato plots of 3.0 and 0.1 per plant, respectively. The average egg survival in the *Bt* potato plots was 13.5% and for the standard plots it was 33.7%;

mortality was attributable to *C. maculata* predation. When these values were used to initiate the genetic model of resistance development, it took 118 generations for resistance to develop compared with 52 generations in the absence of *C. maculata* predation. In a commercial system there would be more natural enemy species and *C. maculata* might have access to other prey species such as aphids that are unaffected by *Bt* toxins. Research in progress is examining the influence of multiple prey on the evolution of resistance (N. Mallampalli and P. Barbosa, pers. comm.). Conservation of biological control agents in refuges, however, apparently can contribute to resistance management; this benefit may in turn encourage growers to plant refuges and conserve biological control agents in them.

Although commercial developers of *Bt*-expressing crops are directing their programs toward producing cultivars with high levels of *Bt* expression for the time being, it is worth briefly examining the interaction of the low dose strategy with pests and natural enemies. As noted in Section II,B, Johnson and Gould (1992) found that parasitism of *H. virescens* larvae was significantly greater when they were developing on tobacco that had a low titer of *Bt* toxin than when they were on standard tobacco. Initially this would result in more parasitoids produced per acre of low dose tobacco than per acre of standard tobacco. From a resistance management perspective, Johnson and Gould (1992) concluded that through the third stadium, in the absence of natural enemies, larvae on *Bt* tobacco had a fitness that was 86.8% of their fitness on standard tobacco. With the addition of natural enemies, the relative fitness of larvae on *Bt* tobacco dropped to 54.6%. When these data were used to initialize a simple genetic model that assumes greater parasitism of susceptible than resistant larvae, resistance was estimated to evolve in 139 generations without natural enemies and in 32 generations when natural enemies were present. From a theoretical perspective, however, Gould *et al.* (1991) demonstrated that natural enemies could increase, decrease, or have no effect on the rate of resistance development. The direction of the impact was highly dependent on ecological and behavioral traits of the natural enemy involved. Furthermore, if natural enemies become more abundant and effective in the low dose or high dose/refuge systems farmers may require fewer of their acres in *Bt*-expressing cultivars. This increased refuge from selection could offset any negative impact that natural enemies could have on resistance development.

III. GENETIC ENGINEERING FOR IMPROVED HORTICULTURAL CHARACTERISTICS

Many traits are under consideration for new genetically engineered cultivars. Although early uses of genetic engineering have focused on dramatic reduction in pest populations, future traits are likely to target other horticultural and agronomic improvements. Nevertheless, changes in horticultural characteristics could affect biological control agents in various ways. In this section we discuss some of the possible changes to come from genetic engineering and how they may affect the conservation of biological control agents.

A.   Improved Tolerance of Pest Damage

An excellent example of improved tolerance to insect damage will be realized with genetically engineered resistance to several potato diseases. Resistance to potato leafroll virus and potato virus Y would greatly increase the tolerance for aphids and the likelihood of successful biological control of aphids (see Section II,B,2). Virus resistance traits are currently being added to *Bt*-expressing cultivars. If resistance to late-blight can be added then the likely reduction of fungicide applications would allow additional control of aphid populations by naturally occurring entomopathogenic fungi. Plants with improved structural integrity to prevent lodging, with increased shoot or leaf growth to permit greater tolerance or which use antifeedants to redistribute herbivore damage, all can permit larger acceptable populations of pests and increase the likelihood that naturally occurring biological control agents can maintain the pest population within higher limits. Further reduction in insecticide use would result and allow additional conservation of natural enemies.

B.   Altered Plant Architecture

Crop breeders have often changed the architecture of cultivars. The impacts of these architectural changes on natural enemies has rarely been carefully studied (see Bottrell *et al.,* in press; Cortesero and Lewis, in press; see Chapters 4 and 5). Kareiva and Sahakian (1990) examined the impact of some radical changes in pea plant architecture on the success of ladybeetle predation of aphids. They found that the ladybeetles, *Coccinella septempunctata and Hippodamia variegata* were more effective at controlling aphids when the aphids were on a leafless pea variety than when the aphids were on a normal leafy pea genotype. They determined that one reason for this was that the lady beetles were more prone to failing off the leafy peas. In a similar study, Grevstad and Klepetka (1992) found that ladybird beetles were more efficient at controlling cabbage aphids on *Brassica* spp with specific architectural properties. Marquis and Whelan (1996) have argued, based on an assessment of the ecological literature, that the architecture of native plants has evolved in part to make it easier for vertebrate and invertebrate predators to catch their prey. Genetic engineers have not begun to manipulate plant architecture, but if and when they do, it will be important to at least consider the potential impacts that this could have on natural enemies.

C.   Altered Plant Surfaces

The interface for most plant-insect interactions occurs at the plant surface and the characteristics of that surface play an important role in mediating insect-plant interactions involving phytophagous species as well as natural enemies (Juniper and Southwood, 1986; Hare, 1992; also see Chapters 4 and 5). Leaf surface waxes influence the effectiveness of predaceous insects by affecting their ability to adhere to and move about on the leaf surface (Eigenbrode and Espelie, 1995). Resistance in cabbage to diamondback moth larvae, for example, was associated with the glossy leaf trait and depends on predation for its full expression (Eigenbrode *et al.*, 1995). Both the mobility of predaceous insects and exposure to predation of diamondback moth first instars were enhanced on the glossy leaves. Because the relationship between the chemical composition of the leaf surface and that of the apoplast is related to the permeability of the plant cuticle, changes in the structure of the cuticle or the surface wax layer can alter exposure of arthropods on the plant surface to arthropod-active plant chemicals of apoplastic origin (Derridj *et al.*, 1996).

Perhaps more than any other plant surface characteristic, plant pubescence influences the performance of natural enemies. In potato, for example, Obrycki and Tauber (1984) reported a direct relationship between the density of glandular trichomes on aphid-resistant clones derived from crosses between *Solanum tuberosum* L. and *S. berthaultii* Hawkes and adverse effects on 11 aphidophagous species in the greenhouse. These effects included a reduction in searching time, a decrease in the distance that newly hatched chrysopid and coccinellid larvae moved, a decrease in oviposition by *Chrysopa oculata* Say, and a decrease in survival of adults of the parasitoid *Aphidiius matricariae* Haliday. These effects were attenuated in the field, however, and the authors concluded that natural enemies could be preserved on potato with moderate glandular trichome densities.

Aphid resistance in these clones is associated with two distinct types of glandular trichomes (Gregory *et al*, 1986). Type A trichomes are short with a membrane-bound gland at the apex. Type B trichomes are longer with a gland at the tip that continuously discharges a viscous exudate. Ruberson *et al.* (1989) found that high densities of type B glandular trichomes entrapped and killed *Edovum puttleri* Grissell, an egg parasitoid of Colorado potato beetle, but that high densities of type A glandular trichomes did not adversely affect the parasitoid. They suggested that *E. puttleri* could be conserved on aphid-resistant plants by selecting for plants with a high abundance of type A and a low abundance of type B trichomes.

The performance of arthropod natural enemies of plant pests in many cases involves an intimate association with the plant surface. Alterations of the plant surface characteristics through genetic engineering or conventional plant breeding can significantly improve their performance. An understanding of the ways in which insect natural enemies interact with the plant surface has helped and will continue to help identify traits that can be modified to enhance biological control.

D. Expanded Range of Growing Conditions

Growing conditions for a crop are typically dictated by the needs of the crop species and its market. Genetically engineered changes in crop varieties that expand or alter the range of growing conditions in which a crop can be grown might be expected to have an impact on conservation of biological control agents in the crop. For example, higher temperatures and light intensity result in higher tomato trichome density and greater entrapment of the predator *Phytoseiulus persimilis* Athias-Henriot than its prey *Tetranychus urticae* Koch (Nihoul, 1993, 1994). Therefore, for glasshouse tomatoes, cultivars that grow and mature faster at cool temperatures would conserve *P. persimilis* by avoiding production under conditions that encourage high trichome density. Negative effects of altered growing conditions also are possible. Under low humidity the developmental potential of *Chrysopa camea* Stephens is the same as at high humidity but that of *Chrysopa rufilabris* Burmeister is greatly reduced (Tauber and Tauber, 1983). Therefore, a drought-tolerant variety which would be grown under more xeric conditions could conserve *C. carnea* but result in reduced populations of *C. rufilabris*.

IV.   CONSERVATION BIOLOGICAL CONTROL AND MARKETING GENETICALLY ENGINEERED CROPS

As growers have become reliant on insecticides they have also become accustomed to fast, effective, and affordable elimination of multiple insect pests from the crop canopy. Losses due to insects have become the exception in many cropping systems whereas rapid and predictable results have become the hallmark of current insecticide programs. Most notably, the broad-spectrum nature of insecticides has reduced the management time necessary for insect control because minimal applications can control entire pest complexes simultaneously.

Selective pesticides and biological materials historically have not occupied a major share of the insecticide market. Higher cost, inconsistent efficacy, and the requirement of more intensive management has been an impediment to the wide-scale adoption of many biological insecticides. Exceptions have occurred in localized areas where specific conditions favored their use over broad-spectrum alternatives. Microbial *Bt* products are used extensively in the production of cole crops in the U.S.A. and pirimicarb is a standard for aphid control in Canada because it has have been more effective against target pests than other broad-spectrum insecticides. Genetically engineered resistant plants may be more broadly adopted by growers than selective biological insecticides because they currently offer the same benefits as broad-spectrum insecticides: simple, effective, and affordable elimination of one or more pests. In fact, multiple resistance traits which confer protection from entire pest complexes, are projected to be more common in the future (Estruch, 1997).

Biological control can increase the pest control spectrum of resistant plants, effectively increasing their market value. For example, potato growers using *Bt* potatoes are likely to conserve aphid and mite biological control agents, gaining additional benefits at no extra cost. In the Pacific Northwest potato system, spider mites are generally recognized as secondary pests that reach damaging levels only after the crop is treated with insecticides for aphid control (Section II, B, 2). The use of Colorado potato beetle and potato leafroll virus resistant Russet Burbank potatoes in this area could eliminate the 50,000 lbs (ai) of Propargite currently applied for mite control, because insecticides targeted at Colorado potato beetle and aphids would no longer be necessary (Schreiber and Guenthner, unpublished data: petition for determination of regulatory status, USDA petition #97-204-01P). Under circumstances such as these conservation biological control will play a major role in pest management with no added effort and perhaps unbeknownst to those growers benefiting from it.

Seed companies may be able to promote the contribution of genetically engineered cultivars to conservation of natural enemies as a product feature or benefit. Marketers of resistant plants, therefore, have an incentive to develop practical recommendations for use of their cultivars in ways that are consistent with conservation biological control. However, the benefits of conservation biological control will be most useful as a sales tool if they are consistent and easy to document. Public agencies and private consultants have and can continue to document the importance of conserving biological control agents. If conserved natural enemies provide variable control results, however, then risk-averse growers may place little value on their benefits. This strict focus on easily implemented and measured control success could be reduced if society rewards producers for environmentally benign production practices that conserve biological control agents, as in the price premiums traditionally commanded for organic crops. Some evidence for this societal support has come from focus groups and interviews; consumers indicated willingness to preferentially purchase genetically modified, insect resistant vegetables. In sustained retail market tests, genetically modified potatoes were purchased in greater number and at a premium price relative to standard potatoes (NatureMark Potatoes, unpublished data).

Although resistant plants can successfully be used in conjunction with broad-spectrum insecticides their full benefit will be realized when they are used as a foundation of biologically based pest management programs. Potato growers across North America used 1.3 fewer systemic or foliar insecticide applications on average in NewLeaf (*Bt* expressing) compared with standard cv. 'Russet Burbank' potatoes during 1996 (NatureMark Potatoes, unpublished data). If insecticides can consistently be reduced and management costs or uncertainty in pest suppression do not overcompensate then insect natural enemies are likely to be conserved. The potential for conservation biological control may increase the appeal and ultimately the value of resistant plants for many growers.

V. CONCLUSIONS

We suggest the following questions to guide estimates of the impact of new genetically engineered cultivars on conservation biological control:

1. Does the cultivar reduce or eliminate insecticide use? The number of times and extent to which insecticide use can be prevented by pest-resistant cultivars is very important in determining the prospects for conservation of biological control agents. If insecticide use is reduced but not eliminated, then the remaining use must avoid or minimize negative impacts on biological control agents. The extent to which broad-spectrum insecticide use can be reduced also determines the extent to which diversity in the arthropod community can be enhanced with both new natural enemies and new prey contributing to biological control.
2. If the new cultivar is grown under different conditions or using different cultural practices than standard cultivars, are these conditions under which biological control agents are conserved?
3. Does the cultivar provide resources such as pollen, nectar, domatia, or improved searching environment needed to conserve biological control agents or do traits conferring resistance to pests also negatively impact natural enemies?

Conservation biological control of at least some pests should be a marketable improvement. In the near term, conservation is most likely to come from the reduction in insecticide use upon which development and sales of these new cultivars largely depend. Future cultivar improvements, however, being added to a system with reduced reliance on insecticides, increased natural enemy populations, and reduced pest outbreak potential may be able to focus on more subtle means of conserving natural enemies. By careful and strategic consideration of potential traits targeted for genetic engineering, and likely management practices for the cultivars possessing those traits, developers of genetically engineered cultivars can contribute substantially to the conservation of biological control agents in crop systems.

REFERENCES

Arpaia, S., Gould F., and Kennedy, G. G. 1997. Potential impact of *Coleomegilla maculata* predation on adaptation of *Leptinotarsa decemlineata* to *Bt*-transgenic potatoes. *Entomol. Exp. Appl.* 82, 91-100.

Barbosa, P., and Letourneau, D. K. (1988). "Novel Aspects of Insect-plant Interactions." John Wiley and Sons. New York, NY.

Barbosa, P., Gross, P., and Kemper, J. (1991). Influence of plant allelochemicals on the performance of the tobacco hornworm and its parasitoid *Cotesia congregata. Ecology* 72, 1567-1575.

Bergman, J. M., and Tingey, W. M. (1979). Aspects of interaction between plant genotypes and biological control. *Bull. Entomol. Soc. Amer.* 25, 275-279.

Boethel, D. J., and Eikenbary, R. D. (1986). "Interactions of Plant Resistance and Parasitoids and Predators of Insects." Ellis Horwood Ltd. Chichester, U.K.

Bottrell, D. G., Barbosa, P., and Gould, F. (1998). Manipulating natural enemies by plant variety selection and modification: a realistic strategy? *Annu. Rev. Entomol.* (in press).

Cortesero, A. M., and Lewis, W. J. Understanding and manipulating plant attributes to enhance biological control. *Biol. Cont.* (in press).

Derridj, S., Wu, B. R., Stammitti, L., Garrec, J. P., and Derrien, A. (1996). Chemicals on the leaf surface, information about the plant available to insects. *Entomol. Exp. Appl.* 80, 197-201.

Dicke, M. (1996). Plant characteristics influence biological control agents, implications for breeding for host plant resistance. *In* "Breeding for Resistance to Insects and Mites." (P. R. Ellis, and J. Freuler, eds.), pp. 72-80. IOBC WPRS Bulletin Vol. XIX. Avignon, France.

Dicke, M., and Sabelis, M. W. (1988). How plants obtain predatory mites as bodyguards. *Neth. J. Zool.* 38, 148-165.

Dicke, M., van Beek, T. A., Posthumus, M. A., Ben Dom, M., Van Bokhoven, H., and De Groot, A. E. (1990a). Isolation and identification of volatile kairomone that affects acarine predator-prey interactions. Involvement of host plant in its production. *J. Chem. Ecol.* 16, 381-396.

Dicke, M., Sabelis, M. W., Takabayashi, J., Bruin, J., and Posthumus, M. A. (1990b). Plant strategies of manipulating predator-prey interactions through allelochemicals, prospects for application in pest control. *J. Chem. Ecol.* 16, 3091-3118.

Dogan, E. B., Berry, R. E., Reed, G. L., and Rossignol, P. A. (1996). Biological parameters of convergent lady beetle (Coleoptera: Coccinellidae) feeding on aphids (Homoptera: Aphididae) on transgenic potato. *J. Econ. Entomol.* 89, 1105-1108.

Duffey, S. S., and Bloem, K. A. (1986). Plant defense-herbivore-parasite interactions and biological control. *In* "Ecological Theory and Integrated Pest Management Practice." (M. Kogan, ed.), pp. 135-183. John Wiley and Sons. New York, NY.

Duffey, S. S., Bloem, K. A., and Campbell, B. C. (1986). Consequences of sequestration of plant natural products in plant-insect-parasitoid interactions. *In* "Interaction of Host Plant Resistance and Parasitoids and Predators of Insects." (D. J. Boethel and R. D. Eikenbary, eds.), pp. 31-60. Ellis Horwood Ltd. Chichester, U.K.

Eigenbrode, S. D., and Espelie, K. E. (1995) Effects of plant epicuticular lipids on insect herbivores. *Annu. Rev. Entomol.* 40, 171-194.

Eigenbrode, S. D., Moodie, S., and Castagnola, T. (1995). Predators mediate host plant resistance to a phytophagous pest in cabbage with glossy leaf wax. *Entomol. Exp. Appl.* 77, 335-342.

Estruch, J. J., Carozzi, N. B., Desai, N., Duck, N. B., Warren, G. W., and Koziel, M.G. (1997). Transgenic plants: an emerging approach to pest control. *Nat. Biotechnol.* 15, 137-141.

Farrar, R. R., Jr., and Kennedy, G. (1993). Field cage performance of two tachinid parasitoids of the tomato fruitworm on insect resistant and susceptible tomato lines. *Entomol. Exp. Appl.* 67, 73-78.

Giroux, S., Cote, J. C., Vincent, C., Martel, P., and Coderre, D. (1994). Bacteriological insecticide M-ONE effects on predation efficiency and mortality of adult *Coleomegilla maculata lengi* (Coleoptera: Coccinellidae). *J. Econ. Entomol.* 87, 39-43.

Gould, F. (1994). Potential and problems with high-dose strategies for pesticidal engineered crops. *Biocont. Sci. Technol.* 4, 451-461.

Gould, F., Kennedy, G. G., and Johnson, M. T. (1991). Effects of natural enemies on the rate of herbivore adaptation to resistant host plants. *Entomol. Exp. Appl.* 58, 1-14.

Gregory, P., Ave, D.A., Bouthyette, P., and Tingey, W.M. (1986). Insect-defensive chemistry of potato glandular trichomes. *In* "Insects and the Plant Surface." (B. Juniper, and T. R. E. Southwood, eds.), pp. 173-184. Edward Arnold Ltd. London, U.K.

Grevstad, F. S., and Klepetka, B. W. (1992). The influence of plant architecture on the foraging efficiencies of a suite of ladybird beetles feeding on aphids. *Oecologia* 92, 399-405.

Gutierrez, A. P. (1986). Analysis of the interaction of host plant resistance, phytophagous and entomophagous species. *In* "Interaction of Host Plant Resistance and Parasitoids and Predators of Insects." (D. J. Boethel and R. D. Eikenbary, eds.), pp. 198-215. Ellis Horwood Ltd. Chichester, U.K.

Hagen, K. S. (1986). Ecosystem analysis, plant cultivars (HPR), entomophagous species and food supplements. *In* "Interaction of Host Plant Resistance and Parasitoids and Predators of Insects." (D. J. Boethel, and R. D. Eikenbary, eds.), pp 151-187. Ellis Horwood Ltd. Chichester, U.K..

Hare, J. D. (1992). Effects of plant variation on herbivore-natural enemy interactions. *In* "Plant Resistance to Herbivores and Pathogens, Ecology, Evolution, and Genetics." (R. S. Fritz, and E. L. Simms, eds.), pp. 278-298. University of Chicago Press. Chicago, IL.

Hilbeck, A. (1994). Ecological interactions of the Colorado potato beetle with its host plants and natural enemies in North Carolina. Ph.D. Thesis, North Carolina State University, Raleigh, NC.

Hoffman, M. P., Zalom, F. G., Wilson, L. T., Smilanick, J. M., Malyj, L. D., Kiser, J., Hilder, V. A., and Barnes, W. M. (1992). Field evaluation of transgenic tobacco containing genes encoding *Bacillus thuringiensis* delta endotoxin or cowpea trypsin inhibitor, efficacy against *Helicoverpa zea* (Lepidoptera: Noctuidae). *J. Econ. Entomol.* 85, 2516-2522.

Hough-Goldstein, J., and Keil, C. B. (1991). Prospects for integrated control of the Colorado potato beetle (Coleoptera: Chrysomelidae) using *Perillus bioculatus* (Hemiptera: Pentatomidae) and various pesticides. *J. Econ. Entomol.* 84, 1645-1651.

Johnson, M. T. Interaction of resistant plants and wasp parasitoids of *Heliothis virescens* (Lepidoptera: Noctuidae). *Environ. Entomol.* (in press).

Johnson, M. T., and Gould, F. (1992). Interaction of genetically engineered host plant resistance and natural enemies of *Heliothis virescens* (Lepidoptera: Noctuidae) in tobacco. *Environ. Entomol.* 21, 586-597.

Johnson, M. T., Gould, F., and Kennedy, G. G. (1997). Effect of natural enemies on relative fitness of *Heliothis virescens* genotypes adapted and not adapted to resistant host plants. *Entomol. Exp. Appl.* 82, 219-230

Juniper, B. and Southwood, T. R. E. (1986). "Insects and the Plant Surface." Edward Arnold Ltd. London, U.K.

Kareiva, P., and Sahakian, R. (1990). Tritrophic effects of a simple architectural mutation in pea plants. *Nature* 345, 433-434.

Kashyap, R. K., Kennedy, G. G., and Farrar, R. R. Jr. (1991). Mortality and inhibition of *Helicoverpa zea* egg parasitism rates by *Trichogramma* in relation to trichome/methyl ketone-mediated insect resistance of *Lycopersicon hirsutum f. glabratum,* accession Pl 134417. *J. Chem. Ecol.* 17, 2381-2395.

Kauffman, W. G., and Flanders, R. V. (1985). Effects of variably resistant soybean and lima bean cultivars on *Pediobius foveolatus* (Hymenoptera: Eulophidae), a parasitoid of the Mexican bean beetle, *Epilachna varivestis* (Coleoptera: Coccinellidae). *Environ. Entomol.* 14, 678-682.

Kauffman, W. G., and Kennedy, G. G. (1989). Relationship between trichome density in tomato and parasitism of *Heliothis* spp. (Lepidoptera: Noctuidae) eggs by *Trichogramma* spp. (Hymenoptera: Trichgrammatidae). *Environ. Entomol.* 18, 698-704.

Lopez, R., and Ferro, D. N. (1995). Larviposition response of *Myiopharus doryphorae* (Diptera, Tachinidae) to Colorado potato beetle (Coleoptera: Chrysomelidae) larvae treated with lethal and sublethal doses of *Bacillus thuringiensis* Berliner subsp. *tenebrionis. J. Econ. Entomol.* 88, 870-874.

Marquis, R. J., and Whelan, C. (1996). Plant morphology and recruitment of the third trophic level, subtle and little-recognized defenses? *Oikos* (Forum) 75, 330-334.

Martos, A., Givovich, A., and Niemeyer, H. M. (1992). Effect of DIMBOA, an aphid resistance factor in wheat, on the aphid predator *Eriopsis connexa* Germar (Coleoptera: Coccinellidae). *J. Chem. Ecol.* 18, 469-479.

Nihoul, P. (1993). Controlling glasshouse climate influences the interaction between tomato glandular trichome, spider mite and predatory mite. *Crop Prot.* 12, 443-447.

Nihoul, P. (1994). Phenology of glandular trichomes related to entrapment of *Phytoseiulus persimilis* Athias-Henriot in the glasshouse tomato. *J. Hort. Sci.* 69, 783-789.

Nordlund , D. A., Lewis, W. J., and Altieri, M. A. (1988). Influences of plant-produced allelochemicals on the host/prey selection behavior of entomophagous insects. *In* "Novel Aspects of Insect-plant Interactions." (P. Barbosa, and D. K. Letourneau, eds.), pp. 65-90. John Wiley and Sons. New York, NY.

Obrycki, J. J., and Tauber, M. J. (1984). Natural enemy activity on glandular pubescent plants in the greenhouse, an unreliable predictor of effects in the field. *Environ. Entomol.* 13, 679-683.

Orr, D. B., and Boethel, D. J. (1986). Influence of plant antibiosis through four trophic levels. *Oecologia* 70, 242-249.

Powell, W. (1986). Enhancing parasitoid activity in crops. *In* "Insect Parasitoids." (J. Waage, and D. Greathead, eds.), pp. 319-40. Academic Press, London, U.K.

Price, P. W., Bouton, C. E., Gross, P., McPheron, B. A., Thompson, J. N., and Weis, A. E.. (1980). Interactions among three trophic levels, influence of plant on interactions between insect herbivores and natural enemies. *Annu. Rev. Ecol. Syst.* 11, 41-65.

Rogers, D. J., and Sullivan, M. J. (1986). Nymphal performance of *Geocoris punctipes* (Hemiptera: Lygaeidae) on pest-resistant soybeans. *Environ. Entomol.* 15, 1032-1036.

Ruberson, J. R., Tauber, M. J., Tauber, C. A., and Tingey, W. M. (1989). Interaction at three trophic levels, *Edovum puttleri* Grissell (Hymenoptera: Eulophidae), the Colorado potato beetle and insect resistant potatoes. *Can. Entomol.* 121, 841-851.

Shieh, J. N., Berry, R. E., Reed, G. L., and Rossignol, P. A. (1994). Feeding activity of green peach aphid (Homoptera: Aphididae) on transgenic potato expressing a *Bacillus thuringiensis* spp. *tenebrionis*-endotoxin gene. *J. Econ. Entomol.* 87, 618-622.

Tauber, M. J., and Tauber, C. A. (1983). Life history traits of *Chrysopa carnea* and *Chrysopa rufilabris* (Neuroptera: Chrysopidae), influence of humidity. *Ann. Entomol. Soc. Amer.* 76, 282-285.

Turlings, T. C. J., Tumlinson, J. H., and Lewis, W. J. (1990). Exploitation of herbivore-induced plant odors by host-seeking parasitic wasps. *Science* 250, 1251-1253.

van Emden, H. F. (1966). Plant insect relationships and pest control. *World Rev. Pest Cont.* 5, 115-123.

van Emden, H. F. (1986). The interaction of plant resistance and natural enemies. Effects on populations of sucking insects. *In* "Interaction of Host Plant Resistance and Parasitoids and Predators of Insects." (D. J. Boethel, and R. D. Eikenbary, eds.), pp. 136-150. Ellis Horwood Ltd. Chichester, U.K.

van Emden, H. F., and Williams, G. F. (1973). Insect stability and diversity in agroecosystems. *Annu. Rev. Entomol.* 19, 455-474.

Vet, L. E. M., and Dicke, M. (1992). Ecology of infochemical use by natural enemies in a tritrophic context. *Annu. Rev. Entomol.* 17, 141-172.

Vinson, S. B. (1976). Host selection by insect parasitoids. *Annu. Rev. Entomol.* 21, 109-134.

Walgenbach, J. G., and Wyman, J. A. (1985). Potato leafhopper (Homoptera: Cicadellidae) feeding damage at various potato growth stages. *J. Econ. Entomol.* 78, 671-675.

Warren, G. W., Carozzi, N. B., Desai, N., and Koziel, M. G. (1992). Field evaluation of transgenic tobacco containing a *Bacillus thuringiensis* insecticidal protein gene. *J. Econ. Entomol.* 85, 1651-1659.

Wierenga, J. M., Norris, D. I., and Whalon, M. E. (1996). Stage specific mortality of Colorado potato beetle (Coleoptera: Chrysomelidae) feeding on transgenic potatoes. *J. Econ. Entomol.* 89, 1047-1052.

Zandstra, B. H., and Motooka, P. S. (1978). Beneficial effects of weeds in pest management - a review. *Proc. Natl. Acad. Sci.* 24, 333-338.

CHAPTER
11

PESTICIDES AND CONSERVATION OF NATURAL ENEMIES IN PEST MANAGEMENT

J. R. Ruberson, H. Nemoto, and Y. Hirose

I. INTRODUCTION

Pesticides have long overshadowed the importance of natural enemies in pest management programs. The high efficacy, easy accessibility, and consistent performance of chemical controls have made them the tool of choice for growers in managing their pest problems. But frequent outbreaks of secondary pests after pesticide applications and the increasing prevalence of pesticide resistance in various pests have pointed out the risks of unilateral reliance on pesticides. The Integrated Pest Management (IPM) concept initially embraced the combined use of natural enemies and pesticides to manage pests (Stern *et al.,* 1959). This idea later evolved to include coordinated use of all possible viable tactics, including pesticides, natural enemies, host plant resistance, cultural controls, and other biologically based methods (Smith *et al.,* 1976). The IPM concept has been promulgated for nearly 40 years yet there are few good examples of its actual realization.

Integration of natural enemies into IPM is considered by many to be implicit in the use of pest thresholds. Presumably natural enemy activity is involved in holding pest populations below the designated thresholds; failure of natural enemies is therefore implicit when pest populations exceed the threshold. Given this reasoning, once the threshold is exceeded there need be little concern for natural enemy populations, since the natural enemy complex was ineffective and is no longer important or relevant. Such tunnel vision disregards the nature of the interactions between pest and natural enemy complexes and contributes to secondary pest outbreaks and the enhancement of pesticide resistance. However, natural enemies and pesticides can be effectively integrated with adequate knowledge of the pesticide to be used and its effects on natural enemy populations (Bartlett, 1964; Newsom *et al.,* 1976; Jepson, 1989a; Croft

1990; Greathead, 1995; Chapter 15). Indeed, the use of selective pesticides and/or rates is often the means of natural enemy conservation most readily available to growers.

II. EFFECTS OF PESTICIDES ON NATURAL ENEMIES

Pesticides exert a wide variety of direct and indirect effects on natural enemies (Waage, 1989; Croft, 1990; Greathead, 1995). Broadly speaking, pesticides can exert two different types of effects on natural enemies (see Croft (1990) for detailed discussion). Lethal effects are expressed as acute or chronic mortality arising from contact with a pesticide (i.e., through one or more of various routes of exposure; see Table 1). Sublethal effects, in contrast, are often chronic and are expressed as some change in the insect's life history attributes, such as its fecundity, developmental time, egg viability, consumption rates, behavior, and so forth. At the population-level the outcomes of these two types of effects may be quite similar or vastly different. These outcomes also may be exacerbated by multiple years of pesticide use (Tolstova and Atanov, 1982) or by closely synchronized pesticide use over large geographic areas. In addition, they may have important long-term consequences on natural enemy populations, particularly for less mobile species that remain in or near fields from year to year. Because of the broad range of possible effects and the confounding of factors in the field, long-term evaluation of pesticide effects on natural enemy populations is a challenge. Nevertheless, there is a great need for studies examining pesticide effects on natural enemy and pest populations on large temporal and spatial scales (Jepson, 1989b).

A. Evaluating the Effects of Pesticides on Natural Enemies

The integration of natural enemies and pesticides relies heavily on the availability of valid information on the effects of pesticides on natural enemies. This information is often a challenge to obtain and even when acquired may not be of general applicability because of differences among crop species and varieties, cropping practices, and pest and natural enemy genotypes (Stevenson and Walters, 1983). Nevertheless, such information is vital for development of effective IPM programs.

A number of efforts have been made to standardize screening protocols for natural enemies, with some very useful results, but more efforts are needed in this area (Croft, 1990; Hassan, 1992). When evaluating pesticides against natural enemies, several issues must be considered.

Table 1.   Issues of importance for designing bioassays evaluating pesticide effects on natural
enemies.

Issues Relevant to Bioassay design	Specific considerations
Selection of natural enemy species	Relative importance within system Representative of natural enemy guilds in system Known susceptibility to other pesticides
Life stages/sexes to be screened	Active, exposed stages that may contact residues Concealed/protected stages Gender-specific variability in susceptibility
Routes of pesticide entry	Direct, topical contact Contact with residues on substrate Inhalation of vapors Ingestion of toxified prey/host tissues Ingestion of toxified plant products (e.g., nectar, pollen, sap)
Life history parameters to evaluate	Survival Longevity Developmental time Fecundity/fertility schedules Consumption/parasitization rates Searching behavior/rate Dispersal ability; movement Respiratory rate Population growth/reduction
Plot size for field screenings	Dispersal capability of natural enemy Proximity of treatments to one another (e.g., risks of drift)
Pesticide formulations and rates	Anticipated uses (e.g., foliar, in-furrow) Range of recommended rates Possible dilution in environment; amount reaching target and target substrate

1. What to evaluate

Pesticide evaluations require a critical decision regarding which life stage and sex are to be evaluated (Table 1). Susceptibility to pesticides often varies considerably among life stages and sexes for arthropods (Bartlett, 1964). Stage-specific variability can be especially pronounced for those natural enemies that have stages protected from the external environment, such as endoparasitoids and some pupae (e.g., chrysopids, hemerobiids, and coccinellids). Data on stage-specific susceptibility can be invaluable for timing pesticide applications to minimize impact on natural enemies. Gender-specific data are important for understanding changes in population structure and population dynamics following pesticide applications.

Another consideration is which parameters to measure in assessing pesticide effects. Survival is the most commonly (and easily) evaluated parameter but a variety of other life history attributes also can be affected, many of which can influence efficacy of natural enemies (Table 1). For example, behavioral modifications by pesticides may greatly reduce the efficacy of natural enemies. Longley and Jepson (1996) noted that aphid parasitoids of the genus *Aphidius* induced significantly fewer English grain aphids *Sitobion avenae* (F.) to leave pyrethroid-treated plants than untreated ones because of reduced parasitoid activity. Abandonment of plants by aphids when attacked by predators or parasitoids can contribute significantly to aphid mortality, in addition to that caused directly by parasitism, by exposing the aphids to adverse soil conditions (McConnell, 1989) or ground predators (Losey and Denno, in press). In this case, monitoring only presence or absence of natural enemies or the presence of parasitized aphids would have underestimated the overall reduction in parasitoid efficacy.

2. Crop system profile

An important consideration is which species to screen. This is a simple issue in systems with only one or a few important pests which are attacked by a small number of natural enemy species. However, this situation is relatively rare (but see Chapter 3). Even when the number of pest species is low the natural enemy complex is typically more diverse than the pest complex. Because of limitations in time and resources often only a few species can be evaluated. Species selected should represent various guilds and a cross section of exposure risks. Those species deemed highly important also should be given priority. Species selection can be complicated further by intraspecific variability in natural enemy responses to pesticides (e.g., Grafton-Cardwell and Hoy, 1985; Rosenheim and Hoy, 1986; Vidal and Kreiter 1995).

In field trials, the size of the plots to be used for evaluation also must be considered. Plot size decisions must incorporate dispersal capabilities of the natural enemies of interest. In many, if not most cases, this information is not known and decisions are based on anecdotal information or simply educated guesses. Nevertheless, the plots must be sufficiently large or isolated to minimize or eliminate pesticide drift across plots and between plot movement of natural enemies. Often accommodating appropriate plot sizes requires reducing replication, making analyses even more difficult.

3.   Methods of assessment

Routes of toxicant entry are key factors in the conclusions that are drawn in evaluations of the impact of pesticides on natural enemies (Table 1). Topical assays are often used because of ease in managing dosages, but these assays are not necessarily valid predictors of field results (Stark *et al.*, 1995). Residue tests, using plant substrates or inert surfaces are widely used and provide valuable information on pesticide effects. Residue tests do not, however, necessarily reflect what will occur in the field. The substrate used and the degree of pesticide coverage within the assay container can strongly influence results. In addition to topical and residual exposure, several other routes of entry can be of particular importance for natural enemies. In greenhouse situations, pesticides may be airborne as fumigants and may contact the arthropods both topically and via inhalation. Toxicant also can move across trophic levels via treated prey or hosts (see Chapters 4 and 5) and can accumulate in predators or parasitoids. For example, the chitin synthesis inhibitor diflubenzuron and the juvenile hormone mimic pyriproxyfen were nontoxic to the pentatomid predator *Podisus maculiventris* (Say) when topically applied (De Clercq *et al.*, 1995). However, when *P. maculiventris* were fed treated prey, considerable mortality resulted (De Clercq *et al.*, 1995). Pesticides also may enter plant tissues and toxify plant resources, such as nectar and plant sap, that are important for nutrition of natural enemies (Stapel *et al.*, 1997). Thus, allowing natural enemies to feed on treated plant tissues may be necessary in some cases. Field screenings that monitor population dynamics implicitly take all pertinent routes of entry into consideration but often are confounded by immigration and emigration of natural enemies and other variables such as weather and pesticide coverage.

Finally, several aspects of the pesticides themselves must be considered when evaluating pesticides. The first is formulation. For example, emulsifiable concentrates differ from granular formulations in the degree and manner to which they will be exposed to arthropods. Similarly, encapsulation of broad-spectrum pesticides can reduce their nontarget toxicity (e.g., Hull, 1979). Thus, testing of a single formulation (e.g., technical grade material) may yield improper or irrelevant results relative to other formulations. Second, the rate at which a pesticide is applied can be important in determining its selectivity. For example, some broad-spectrum insecticides exhibit dose-response curves with natural enemy species; and thus may be more selective at reduced rates. Data should be compiled for a range of rates, rather than a single rate, where possible. Third, the mode of application must be considered. If an insecticide is a granular systemic, it will likely be incorporated into the soil, often at the time of planting, and will not be present on the surface of aboveground structures. Likewise, seed treatments will not directly contact many beneficial arthropods. Topical or residual tests using a technical grade formulation of such an insecticide would provide no data of relevance to field use. In contrast, pesticides that are sprayed directly onto plant surfaces should be evaluated for both residual and topical activity.

4. Objectives of assessment

Obtaining specific data on the effects of pesticides on natural enemies is not always straightforward. It requires a hierarchy of laboratory and field assays to determine risks of various rates of a pesticide to natural enemies and to assess short- and long-term effects on pest and natural enemy populations (Jepson, 1989a; Croft, 1990; Hassan, 1989, 1992). Forecasting results based on a few limited assays can lead to erroneous results, whether conducted in the laboratory or field (Bakker and Jacas, 1995). Field tests are particularly critical as many effects of pesticides cannot be discerned readily under laboratory conditions. Field trials, however, are often difficult to manage and interpret and appropriate temporal and spatial scales can be challenging to attain and replicate. Nevertheless, field tests incorporating long-term monitoring of pest and beneficial populations are vital as the final arbiters for assessing the selectivity of pesticides under actual production conditions. Ultimately, the question reduces to the effects of pesticides on natural enemy efficacy, and this can only be measured in the field.

B. Pesticide Selectivity

Understanding the impact of pesticides on natural enemies relative to the target pest(s) is important for anticipating the outcomes of the use of a pesticide. Pesticide selectivity to beneficial arthropods has been broadly classified into two forms (Metcalf, 1982; Hull and Beers, 1985; Croft, 1990). The first of these is physiological selectivity, i.e., in the presence of a given pesticide rate the pesticide is less toxic to the natural enemy than it is to the target pest. In contrast, at other, usually higher, rates the reverse may occur. Physiological selectivity is thus a function or property of the pesticide itself relative to the physiology of both arthropods.

The second form of selectivity, ecological selectivity, pertains to the means by which the pesticide is used and the domain in which it is used (see Section III,A, below). For example, systemic pesticides may be available only to leaf-feeding herbivores and thereby have little or no effect on many natural enemies that frequent the leaf surface. However, feeding on plant sap and nectar are common among various taxa of natural enemies, such as the Heteroptera (Alomar and Wiedenmann, 1996; Chapter 9) and this behavior can expose these natural enemies to toxicant (Hough-Goldstein and Whalen, 1993). Interactions of the pesticide with the leaf surface and phenological stage of the crop also play a role in ecological selectivity. If the insecticide is readily absorbed into the plant's cuticle it may rapidly enter the plant and thereby quickly decrease residual exposure of arthropods on the plant's surface. A dense plant canopy can also reduce penetration and spare natural enemies located deep in the canopy, even when highly toxic pesticides are applied.

Defining levels of pesticide selectivity to natural enemies is further complicated in systems where the natural enemy complex is large and diverse and where multiple species play important roles. Screening pesticides against such complexes is a challenging and daunting task. Applications of pesticides to crop systems and long-term (at least one growing season) monitoring of pest and natural enemy populations provide valuable information on pesticide effects This is particularly true in systems where the natural enemy complex is poorly defined and where more detailed evaluations are impractical (Nemoto, 1986, 1993, 1995).

III. RELATIVE ROLES OF PESTICIDES AND NATURAL ENEMIES IN IPM

The relative importance of natural enemies and pesticides determines, to a large extent, the relative roles of chemical and biological control (Fig. 1). As Waage (1989) and Greathead (1995) have pointed out, there is an inverse relationship between the importance of natural enemies and that of insecticides. When the natural enemy complex is ineffective pesticides provide one of the few means of controlling the pest. On the other end of the spectrum, where the natural enemy complex is highly effective, pesticides are unnecessary. The intermediate region of this continuum is where IPM can be effectively employed (Fig. 1). Pest species occur at various points

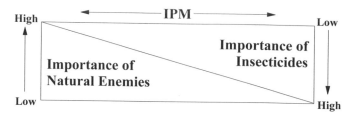

Figure 1. Relative roles of pesticides and natural enemies in agricultural systems. Where natural enemies are ineffective (right), pesticides dominate as pest management tools. In contrast, where natural enemies are effective (left), pesticides are unnecessary. Pests falling into the intermediate zone are candidates for IPM practices integrating natural enemies and pesticides.

along the continuum and management decisions and environmental conditions can shift those points in either direction. In crop systems with diverse pest complexes a wide range of the continuum may be simultaneously represented, making management decisions quite complicated. This complexity and the associated uncertainty can be exploited easily by pesticide distributors and can encourage overuse and abuse of pesticides. Understanding the respective and interactive roles of pesticides and natural enemies provides the basis for developing rational, integrated pest management systems and for effectively integrating these two strategies.

A.   Integrating Pesticides and Natural Enemies

There are several means of integrating natural enemies with pesticides in pest management programs. These involve essentially three areas: use of selective pesticides or rates, temporal separation of pesticide and natural enemies, and spatial separation of pesticide and natural enemies. These areas are each addressed below.

1. Selective pesticides

Use of selective pesticides is the most powerful means of integrating pesticide use and natural enemies (Hull and Beers, 1985). Historically, the pesticide market has been dominated by broad-spectrum, highly disruptive chemicals. Since the advent of the pyrethroid insecticides, more selective compounds have entered the market. Recently, several new classes of pesticides have been developed that exhibit high specificity for target pests but little activity against natural enemies. There are, however, economic constraints limiting the degree to which specificity will be pursued by the pesticide industry. Pesticides that are highly selective have, by definition, narrow target ranges. Thus, the expense of using several selective pesticides versus a single broad-spectrum pesticide can be prohibitive to growers.

Selective pesticides can be particularly valuable in crop systems with multiple pests, especially where potential pests can be wholly suppressed by natural enemies. Such occurrences are common and use of broad-spectrum pesticides in these systems induces secondary pest outbreaks. Such a situation occurs in cotton in the southeastern United States where portions of the region underwent an extensive program to eradicate the cotton boll weevil *Anthonomus grandis* (Boheman). In Georgia, the Boll Weevil Eradication Program (BWEP) was initiated in the fall of 1987 and continued through 1990. Between 1988 and 1990 most of the cotton hectares in the state were treated repeatedly with malathion to kill boll weevils. This same period was characterized by unprecedented populations of the beet armyworm *Spodoptera exigua* (Hübner), a secondary cotton pest which is difficult to control with currently registered insecticides. The massive beet armyworm populations were costly to control. For example, in 1990 the beet armyworm cost Georgia cotton growers $74.00 (US) per acre ($25.9 million statewide) in control costs and crop losses. In subsequent studies, it was demonstrated that a complex of natural enemies is capable of suppressing beet armyworm populations below economic levels and that broad-spectrum insecticides, particularly organophosphates with long residual activity, can impair this complex and release the beet armyworm from its natural enemies (Ruberson *et al.,* 1994a).

This pattern was observed also in California (Eveleens *et al.,* 1973). Malathion, the insecticide used in the BWEP, is a broad-spectrum insecticide that is known to be highly disruptive to beneficial arthropods and to induce secondary pest outbreaks (Ehler and Endicott, 1984; Cohen *et al.,* 1987). Malathion is highly toxic to adult and pupal *Cotesia marginiventris* (Cresson), a key natural enemy of the beet armyworm (Ruberson *et al.,* 1994b). Since completion of the BWEP the use of selective insecticides and rates for control of key pests has become a critical and highly successful element of managing beet armyworms in cotton in the southeastern United States. An enhanced beneficial arthropod complex also pays dividends by suppressing other pests, such as the tobacco budworm *Heliothis virescens* (F.) and cotton bollworm *Helicoverpa zea* (Boddie) and has contributed to dramatic reductions in total pesticide use.

2. Temporal separation

Temporal separation of pesticides and natural enemies can be accomplished in various ways, but is rarely practiced. For example, pesticides can be applied when key natural enemies are absent or when they are present in more tolerant life stages to conserve natural enemies. The success of such an approach relies on detailed life history and phenological data for key natural enemies, so that natural enemy presence and/or population structure can be accurately predicted. The absence of these data for many species, lack of population synchrony, and the need for grower scouting generally have precluded the use of this approach. If natural enemies are to be released against target pests, releases may be conducted after residual toxicity has declined sufficiently to permit natural enemy survival (Malezieux *et al.*, 1992). Temporal synchrony of pesticides and natural enemies also can be reduced by using pesticides with short residual toxicity (Bartlett, 1964).

3. Spatial separation

Spatial separation of natural enemies and pesticides has been practiced in various ways for many years. For example, spot treatment has long been advocated as a means for targeting pests and conserving natural enemies. In practice, however, this method is rarely used because of difficulties associated with delimiting localized pest outbreaks with sufficient precision to permit spot applications. The burgeoning technologies of Geographic Positioning Systems (GPS) and site-specific pest management may provide opportunities for specifically targeting pest infestations within fields (Weisz *et al.*, 1995). Other practices that minimize spatial concurrence of pesticides and natural enemies include treatments of alternate rows (Hull *et al.*, 1983), use of bait formulations that are attractive only to target pests, and the use of trap crops that attract pest populations away from the crop where they can be treated without disrupting the natural enemy complex within the crop (Hokkanen, 1991).

An additional means of integrating pesticides and natural enemies is the use of pesticide-resistant natural enemies in crop systems where pesticide applications are necessary. This area currently is limited to only a few natural enemy species, but there are opportunities for expansion in the future (Beckendorf and Hoy, 1985; Hoy, 1990; Narang *et al.*, 1994). Although use of resistant natural enemies holds promise, grower acceptance likely will be limited until releasing resistant natural enemies is demonstrated to be consistently cost effective. Widespread use of this technique may have some detrimental effects on grower attitudes; it may serve to encourage continued use of broad-spectrum pesticides even as more selective materials become available.

IV. CONCLUSION

Although pesticides and natural enemies have often been viewed as incompatible there are many opportunities to integrate these two elements in comprehensive pest management programs, and this trend will likely increase substantially in the future. In the pesticide industry, there is a growing emphasis on developing pesticides that have minimal environmental impact and that exhibit greater natural enemy selectivity. Similarly, regulatory agencies worldwide are becoming more concerned about the impact of pesticides on natural enemies as a part of the registration process. These developments place a burden on the biological control community and industry to develop standardized, repeatable protocols that adequately and realistically predict risk of pesticide use to nontarget beneficial organisms (Stevenson and Walters, 1983; Aldridge and Carter, 1992; Barrett, 1992). Important steps have been made in this direction (e.g., Carter *et al.,* 1992; Hassan, 1989, 1992), but much work remains to be done before generally acceptable procedures are devised. Further, there is a great need for long-term studies of pesticide impacts on natural enemy faunas to establish the true sustainability of integrated programs.

The widespread use of pesticides will certainly continue as an integral component of many pest management programs, at least in the near future. Ultimately, the accumulated data on pesticide effects on natural enemies must be incorporated into pest management decision making. Despite a large and varied body of information on pesticides and natural enemies very little has actually been utilized in pest management. It is hoped that future trends will move agriculture toward true integration of environmentally sound and sustainable tactics.

REFERENCES

Aldridge, C., and Carter, N. (1992). The principles of risk assessment for non-target arthropods: A UK registration perspective. *In* "Aspects of Applied Biology 31, Interpretation of Pesticide Effects on Beneficial Arthropods." (R. A. Brown, P. C. Jepson, and N. W. Sotherton, eds.), pp. 149-156. Assoc. Applied Biologists. Warwick, U.K.

Alomar, O., and Wiedenmann, R. N. (1996). "Zoophytophagous Heteroptera: Implications for Life History and Integrated Pest Management." Thomas Say Publ. Entomol., Entomol. Soc. Amer. Lanham, MD.

Bakker, F. M., and Jacas, J. A. (1995). Pesticides and phytoseiid mites: Strategies for risk assessment. *Ecotox. Environ. Safety* 32, 58-67.

Barrett, K. L. (1992). The BART Group approach to regulatory testing. *In* "Aspects of Applied Biology 31, Interpretation of Pesticide Effects on Beneficial Arthropods." (R. A. Brown, P. C. Jepson, and N. W. Sotherton, eds.), pp. 165-170. Assoc. Applied Biologists. Warwick, U.K.

Bartlett, B. R. (1964). Integration of chemical and biological control. *In* "Biological Control of Insect Pests and Weeds." (P. DeBach, ed.), pp. 489-511. Chapman and Hall. New York, NY.

Beckendorf, S. K., and Hoy, M. A. (1985). Genetic improvement of arthropod natural enemies through selection, hybridization, or genetic engineering techniques. *In* "Biological Control in Agricultural IPM Systems." (M. A. Hoy, and D. C. Herzog, eds.), pp. 167-187. Academic Press. Orlando, FL.

Carter, N., Oomen, P. A., Inglesfield, C., Samsøe-Peterson, L., de Clercq, R., and Dedryver, C. A. (1992). A European guideline for testing the effects of plant protection products on arthropod natural enemies. *In* "Aspects of Applied Biology 31, Interpretation of Pesticide Effects on Beneficial Arthropods." (R. A. Brown, P. C. Jepson, and N. W. Sotherton, eds.), pp. 157-163. Assoc. Applied Biologists. Warwick, U.K.

Cohen, E., Podoler, H., and El-Hamlauwi, M. (1987). Effects of the malathion-bait mixture used on citrus to control *Ceratitis capitata* (Wiedemann) (Diptera: Tephritidae) on the Florida red scale, *Chrysomphalus aonidum* (L.) (Hemiptera: Diaspididae), and its parasitoid *Aphytis holoxanthus* DeBach (Hymenoptera: Aphelinidae). *Bull. Entomol. Res.* 77, 303-307.

Croft, B. A. (1990). "Arthropod Biological Control Agents and Pesticides." John Wiley and Sons. New York, NY.

De Clercq, P., Vinuela, E., Smagghe, G., and Degheele, D. (1995). Transport and kinetics of diflubenzuron and pyriproxyfen in the beet armyworm *Spodoptera exigua* and its predator *Podisus maculiventris*. *Entomol. Exp. Appl.* 76, 189-194.

Ehler, L. E., and Endicott, P. C. (1984). Effect of malathion-bait sprays on biological control of insect pests of olive, citrus, and walnut. *Hilgardia* 52, 1-47.

Eveleens, K. G., van den Bosch, R., and Ehler, L. E. (1973). Secondary outbreaks of beet armyworm by experimental insecticide applications in cotton in California. *Environ. Entomol.* 2, 497-503.

Grafton-Cardwell, E. E., and Hoy, M. A. (1985). Intraspecific variability in response to pesticides in the common green lacewing, *Chrysoperla carnea* (Stephens) (Neuroptera: Chrysopidae). *Hilgardia* 52, 1-31.

Greathead, D. J. (1995). Natural enemies in combination with pesticides for integrated pest management. *In* "Novel Approaches to Integrated Pest Management." (R. Reuveni, ed.), pp. 183-197. CRC Press. Boca Raton, FL.

Hassan, S. A. (1989). Testing methodology and the concept of the IOBC/WPRS working group. *In* "Pesticides and Non-target Invertebrates." (P.C. Jepson, ed.), pp. 1-18. Intercept. Andover Hants, U.K.

Hassan, S. A. (1992). "Guidelines for Testing the Effects of Pesticides on Beneficial Organisms: Description of Test Methods." IOBC/WPRS Bull. 1992/XV/3.

Hokkanen, H. M. T. (1991). Trap cropping in pest management. *Annu. Rev. Entomol.* 36, 119-138.

Hoy, M. A. (1990). Genetic improvement of arthropod natural enemies: becoming a conventional tactic? *In* "New Directions in Biological Control: Alternatives for Suppressing Agricultural Pests and Diseases." (R. Baker, and P. Dunn, eds.), pp. 405-417. Alan R. Liss. New York, NY.

Hough-Goldstein, J., and Whalen, J. (1993). Inundative release of predatory stink bugs for control of Colorado potato beetle. *Biol. Cont.* 3, 343-347.

Hull, L. A. (1979). Apple, test of pesticides, 1978. *Insectic. Acaric. Tests* 4, 20-22.

Hull, L. A., and Beers, E. H. (1985). Ecological selectivity: Modifying chemical control practices to preserve natural enemies. *In* "Biological Control in Agricultural IPM Systems." (M. A. Hoy, and D. C. Herzog, eds.), pp. 103-122. Academic Press. Orlando, FL.

Hull, L. A., Hickey, K. D., and Kanour, W. W. (1983). Pesticide usage patterns and associated pest damage in commercial apple orchards of Pennsylvania. *J. Econ, Entomol.* 76, 577-583.

Jepson, P. C. (1989a). "Pesticides and Non-target Invertebrates." Intercept. Andover Hants, U.K.

Jepson, P.C. (1989b). The temporal and spatial dynamics of pesticide side-effects on non-target invertebrates. *In* "Pesticides and Non-target Invertebrates." (P.C. Jepson, ed.), pp. 95-128. Intercept. Andover Hants, U.K.

Longley, M., and Jepson, P. C. (1996). Effects of honeydew and insecticide residues on the distribution of foraging aphid parasitoids under glasshouse and field conditions. *Entomol. Exp. Appl.* 81, 189-198.

Losey, J., and Denno R. Positive predator-predator interactions: Enhanced positive predation rates and synergistic suppression of aphid populations. Ecology (in press).

Malezieux, S., Lapchin, L., Pralavario, M., Moulin, J. C., and Fournier, D. (1992). Toxicity of pesticide residues to a beneficial arthropod, *Phytoseiulus persimilis* (Acari: Phytoseiidae). *J. Econ. Entomol.* 85, 2077-2081.

McConnell, J. A. (1989). Incidental mortality of the greenbug, *Schizaphis graminum* (Rondani), on grain sorghum due to adult seven-spotted lady beetle, *Coccinella septempunctata* L., foraging. M.S. Thesis, University of Arkansas, Fayetteville, AR.

Metcalf, R. L. (1982). Insecticides in pest management. *In* "Introduction to Insect Pest Management." (R. L. Metcalf, and W. H. Luckmann, eds.), pp. 217-278. 2nd ed. Wiley Interscience. New York, NY.

Narang, S. K., Bartlett, A. C., and Faust, R. M. (1994). "Applications of Genetics to Arthropods of Biological Control Significance." CRC Press. Boca Raton, FL.

Nemoto, H. (1986). Factors inducing resurgence in the diamondback moth after application of methomyl. *In* "Diamondback Moth Management: Proceedings of the First International Workshop." (N. S. Talekar, and T. D. Griggs, eds.), pp. 387-394. Asian Veg. Res. Dev. Cent. Shanhua, Taiwan.

Nemoto, H. (1993). Resurgence of the Kanzawa spider mite, *Tetranychus kanzawai* Kishida after application of imidacloprid and its countermeasure. *Proc. KantoTosan Plant Prot. Soc.* 40, 245-247. (In Japanese with English summary)

Nemoto, H. (1995). Pest management systems for eggplant arthropods: a plan to control pest resurgence resulting from the destruction of natural enemies. *JARQ* 29, 25-29.

Newsom, L. D., Smith, R. F., and Whitcomb, W. H. (1976). Selective pesticides and selective use of pesticides. *In* "Theory and Practice of Biological Control." (C. B. Huffaker, and P. S. Messenger, eds.), pp. 565-592. Academic Press. New York, NY.

Rosenheim, J. A., and Hoy, M. A. (1986). Intraspecific variation in levels of pesticide resistance in field populations of a parasitoid, *Aphytis melinus* (Hymenoptera: Aphelinidae): the role of past selection pressures. *J. Econ. Entomol.* 79, 1161-1173.

Ruberson, J. R., Herzog, G. A., Lambert, W. R., and Lewis, W. J. (1994a). Management of the beet armyworm (Lepidoptera: Noctuidae) in cotton: Role of natural enemies. *Fla. Entomol.* 77, 440-453.

Ruberson, J. R., Herzog, G. A., Lambert, W. R., and Lewis, W. J. (1994b). Management of the beet armyworm: Integration of control approaches. *Proc. 1994 Beltwide Cotton Prod. Conf.* 2, 857-859.

Smith, R. F., Apple, J. L., and Bottrell, D. G. (1976) The origins of integrated pest management concepts for agricultural crops. *In* "Integrated Pest Management." (J. L. Apple, and R. F. Smith, eds.), pp. 1-16. Plenum Press. New York, NY.

Stapel, J. O., Cortesero, A. M., De Moraes, C. M., Tumlinson, J. H., and Lewis, W. J. (1997). Effects of extrafloral nectar, honeydew, and sucrose on searching behavior and efficiency of *Microplitis croceipes* (Hymenoptera: Braconidae) in cotton. *Environ. Entomol.* (in press).

Stark, J. D., Jepson, P. C., and Mayer, D. F. (1995). Limitations to use of topical toxicity data for predictions of pesticide side effects in the field. *J. Econ. Entomol.* 88, 1081-1088.

Stern, V. M., Smith, R. F., van den Bosch, R. and Hagen, K. S. (1959). The integrated control concept. *Hilgardia* 29, 81-101.

Stevenson, J. H., and Walters, J. H. H. (1983). Evaluation of pesticides for use with biological control. *Agric., Ecosys., Environ.* 10, 201-215.

Tolstova, Y. S., and Atanov, N. M. (1982). Action of chemical substances for plant protection on the arthropod fauna of the orchard. I. Long-term action of pesticides in agrocoenoses. *Entomol. Rev.* 61, 1-14.

Vidal, C., and Kreiter, S. (1995). Resistance to a range of insecticides in the predaceous mite *Typhlodromus pyri* (Acari: Phytoseiidae): inheritance and physiological mechanisms. *J. Econ. Entomol.* 88, 1097-1105.

Waage, J. (1989). The population ecology of pest-pesticide-natural enemy interactions. *In* "Pesticides and Non-target Invertebrates." (P. C. Jepson, ed.), pp. 95-128. Intercept. Andover Hants, U.K.

Weisz, R., Fleischer, S., and Smilowitz, Z. (1995). Site-specific integrated pest management for high-value crops: sample units for map generation using the Colorado potato beetle (Coleoptera: Chrysomelidae) as a model system. *J. Econ. Entomol.* 88, 1069-1080.

CHAPTER
12

CONSERVATION BIOLOGICAL CONTROL OF MOBILE
PESTS: PROBLEMS AND TACTICS

Yoshimi Hirose

I. INTRODUCTION

Many arthropod pests of agriculture move seasonally over wide areas and among crops in diversified agroecosystems (Hirose *et al.,* 1996). In a discussion of their management, Kennedy and Margolies (1985) called these species "mobile pests." Although they did not give any definition of this term, I define mobile pests as polyphagous and multivoltine pests that move freely over large areas and between crops in diversified agroecosystems. Hirose *et al.* (1996) also used the same term, giving a similar description of the phenomenon. They further pointed out that habitats of mobile pests are often ephemeral. As examples of mobile pests, Kennedy and Margolies (1985) cited the two-spotted mite *Tetranychus urticae* Koch, the green peach aphid *Myzus persicae* (Sulzer), European corn borer *Ostrinia nubilalis,* the corn earworm *Helicoverpa zea* (Boddie), the fall armyworm *Spodoptera frugiperda* (J. E. Smith), and cabbage looper *Trichoplusia ni* (Hübner). Additional examples of mobile pests are *Thrips palmi* Karny and soybean bugs, such as *Piezodorus hybneri* (Gmelin), *Riptortus clavatus* (Thunberg), and *Megacopta punctatissima* (Montandon) (Hirose *et al.,* 1996).

The term "migrant pests" (Joyce, 1981) does not seem to be synonymous with the term "mobile pests" because the former includes migratory locusts and planthoppers which often move across great distances. Mobile pests are migratory but the distance over which they move is much shorter than that of typical migrant pests. It can be seen from the above-mentioned examples of mobile pests that most mobile pests are major pests. This may be largely because natural enemies of these pests have difficulty in shifting their habitats to follow the seasonal movement of the pests (Hirose *et al.,* 1996). Little attention has been paid to the relationships between mobile pests and their natural enemies or to the biological control of mobile pests.

Ehler and Miller (1978) discussed relationships between the mobile pests *H. zea* and *T. ni* and their predators in temporary agroecosystems, emphasizing the importance of these predators in their biological control. Unfortunately, their discussion was chiefly limited to relationships on cotton which, among annual crops, is a long season crop. These mobile pests may use only cotton for their three successive generations in a year, because of its long growing season (Stinner *et al.,* 1976 for *H. zea;* and Ehler and van den Bosch, 1974 for *T. ni).* However, one should also consider the situation where mobile pests seasonally shift to different crops. For example, *H. zea* may shift from tobacco to corn in the F1 generation and shift from late corn to cotton in the F3 generation (Table 1). Such pest shifts from crop to crop, or habitat to habitat could have more complicated impacts on the effectiveness of natural enemies as control agents than pest shifts between the same crops in diversified agroecosystems. Most of these potential impacts remain unknown. Clearly, these shifts among mobile pests present potential difficulties in the implementation of conservation biological control (also see Chapter 3). In this chapter, problems associated with the conservation biological control of mobile pests are addressed and its tactics are discussed.

Table 1.   Generalized pattern of habitat shifts by *Helicoverpa zea* in North Carolina[a]

Generation	From	To
Overwinter	Overwintering site	Early corn, Tobacco, Wild hosts
F1	Early corn, Tobacco, Wild hosts	Corn
F2	Corn	Late corn, Soybean, Cotton, Tomato
F3	Late corn, Soybean, Cotton	Tobacco, Soybean, Cotton, Tomato
F4	Tobacco, Soybean, Cotton	Tobacco, Cotton, Tomato Other hosts

[a] (Stinner *et al.,* 1976 and Kennedy and Margolies, 1985).

II. EFFECTIVE NATURAL ENEMIES OF MOBILE PESTS

A. Mobile Natural Enemies

To implement the conservation biological control of mobile pests it is first necessary to know what natural enemies of mobile pests are effective biological control agents. Conservation biological control should not aim to conserve all natural enemies of mobile pests but only those that are effective natural enemies in the field. What characteristics of natural enemies of mobile pests make them effective biological control agents? The following two examples of natural enemies of mobile pests give valuable insights into this question.

Studying egg parasitoids of soybean bugs, such as *Piezodorus hybneri*, *Riptortus clavatus*, and *Megacopta punctatissima*, Hirose *et al.* (1996) found that *Ooencyrtus nezarae* Ishii (Fig. 1), an effective egg parasitoid of these mobile pests, has good colonization ability. Parasitoid females colonize soybean fields in synchrony with an initial increase in host resources. A portion of the females produced by the colonizing females dispersed from fields while their host bugs continued to oviposit. Thus, good colonization ability of *0. nezarae* was associated with a strong tendency to disperse even when hosts were present. Hirose *et al.* (1996) suggested that this type of dispersal of *0. nezarae* females is a strategy for exploiting mobile hosts. The probability of successful parasitization of bugs by *0. nezarae*, an extremely polyphagous parasitoid of bugs inhabiting a variety of habitat, could be higher in habitats other than soybean fields. Indeed, if *0. nezarae* females remain in a habitat until its hosts are gone they could not successfully colonize other habitats including soybean fields at the proper time.

There are early and late maturing soybeans from summer to autumn in any given area. Soybean fields are ephemeral habitats for host bugs and *0. nezarae* because oviposition by bugs in soybean is limited to particular stages of this hosts' food plant (Hirose *et al.*, 1996). Traps placed in open areas isolated from soybean and other host plants of bugs often caught many females of *0. nezarae* (Nobuo Mizutani, unpublished data), indicating that females of this parasitoid move freely over large areas and between crops. Thus, *0. nezarae* as a parasitoid of soybean bugs is an example where effective natural enemies of mobile pests also are mobile.

Another example is that of *Thrips palmi*. It is a multivoltine pest of vegetables, such as eggplant, cucumber, watermelon, muskmelon, and green pepper in Japan. In most areas of Japan this thrips can overwinter only in greenhouses and its overwintered populations disperse from the greenhouses into open fields in July. Its adults are so mobile that the distance traveled from the sources reached 20 km

Figure 1.   A female of the mobile natural enemy *Ooencyrtus nezarae* ovipositing on an egg mass of *Megacopta punctatissima.*

in about 30 days with aid of wind in one case (Matsuzaki *et al.*, 1985). T *palmi,* producing several generations during July to October, repeatedly moves from vegetable plot to vegetable plot in a wide area containing plots of different host vegetables. *Orius* species, such as *0. sauteri* (Poppius), *0. minutus* (Linneaeus), and *0. nagaii* Yasunaga are known to be effective naturally occurring biological control agents of this mobile pest on eggplant (Kawamoto and Kawai, 1988; Nagai *et al.,* 1988; Nagai, 1990, 1991; Kawai and Kawamoto, 1994; Ohno *et al.,* 1995; Takemoto and Ohno, 1996). They are polyphagous predators of thrips, mites, aphids, and lepidopteran eggs.

Dispersal ability of these predators is not exactly known but they are mobile. Although all estimates of diffusion rates for natural enemies by Corbet and Plant (1993) were below 100 m^2/d they stated that it is likely that predatory hemipterans, such as *Orius* species, have diffusion rates of at least 100 m^2/d. Preliminary tests for field release of *0. sauteri* (Y. Hirose, unpublished data) support the contention that it is a highly mobile predator. *Orius* species have shorter generation times than *T. palmi,* suggesting a higher reproductive potential than that of their muitivoltine prey. *Orius* species and other predatory hemipterans characterized by polyphagy have also been reported to be effective naturally occurring biological control agents of mobile pests, such as *Trichoplusia ni* (Ehler, 1977; Ehler and Miller, 1978), *Helicoverpa zea* (van den Bosch et al., 1971; van Steenwyk *et al.,* 1975; Ehler and Miller, 1978), and *H. armigera* (Hübner) (van den Berg and Cock, 1993).

Hirose *et al.* (1996) pointed out that because *0. nezarae* has a shorter generation time than its hosts, it has a high reproductive potential. Nevertheless, the dispersal of a portion of the *0. nezarae* females, in the presence of hosts, resulted in a failure of this parasitoid to increase parasitism of its host during the latter half of its reproductive period. Thus, they mentioned that the failure to increase parasitism as a season progresses may be a problem in naturally occurring control of mobile pests by mobile parasitoids such as *0. nezarae.* However, area wide population studies of mobile pests and their natural enemies are necessary for the evaluation of the effectiveness of mobile natural enemies. The effect of their dispersal and colonization on parasitism or predation of mobile pests should be determined over a wide area containing many pest habitats (Hirose *et al.,* 1996).

B.   Habitat Shifts by Natural Enemies of Mobile Pests

Since mobile pests move from crop to crop or from habitat to habitat seasonally, it is possible that some natural enemies of mobile pests successfully follow these pests to different plant hosts between generations but that others do not. Although habitat shifts by natural enemies of mobile pests rarely have been examined the following two examples of habitat shifts are presented.

Habitat shift by *0. nezarae* (Fig. 2A) is a successful illustration of the adaptive ecology of natural enemies of mobile pests. After hibernation this parasitoid requires a host in a habitat other than crops and thus attacks eggs of *Megacopta punctatissima* on kudzu in May and June. Kudzu is a wild plant found along roadsides and other open areas. Most of the parasitoids emerging from *M. punctatissima* move to early maturing soybean to parasitize eggs of *Riptortus clavatus, Piezodorus hybneri*, and *M. punctatissima* (Takasu and Hirose, 1986), although some parasitoids attack the second generation eggs of *M. puncatatissima* which are no longer abundant on kudzu. Eggs of *R. clavatus* and *P. hybneri* are more abundant than *M. punctatissima* in early maturing soybean. In a field of early maturing soybean in Fukuoka in 1985, parasitism of *R. clavatus* and *P. hybneri* by *0. nezarae* reached 57.4% and 27.2%, respectively (Hirose *et al.,* 1996). In September, *0. nezarae* moves to late maturing soybean to

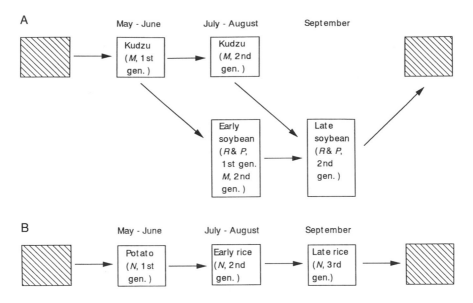

Figure 2.   A schematic representation of habitat shifts by (A) *Ooencyrtus nezarae* and (B) *Trissolcus mitsukurii,* egg parasitoids of mobile phytophagous bugs. The shifts are not based on direct observations on parasitoid movements between habitats but on circumstantial evidence in the field. Open and hatched boxes represent breeding and hibernating habitats, respectively. Arrows represent possible parasitoid movements between habitats. M: *Megacopta puncta,R: Riptortus clavatus,* P: *Piezodorus hybneri,* and N: *Nezara viridula* (L.).

parasitize eggs of these two hosts. More than 80% parasitism of *R. clavatus* occurred in a field of late maturing soybean in Fukuoka in 1984 (Takasu and Hirose, 1985).

Unlike the habitat shift by *O. nezarae,* that by *Trissolcus mitsukurii* (Ashmead) (Fig. 2B) is not completely successful. *T. mitsukurii* is known to parasitize eggs of seven pentatomids and its major host among them is the southern green stink bug *Nezara viridula* (Linneaeus) which damages various crops, such as potato, rice, wheat and soybean, as a mobile pest in Japan. After hibernation, this parasitoid attacks host eggs on potato in May and June. *N. viridura* then shifts its habitat to early planted rice in July and August and the parasitoid also shifts its habitat at this time to attack 2nd generation host eggs. In September, the parasitoid again follows its hosts to late planted rice. Percentage parasitism of *N. viridura* by T. *mitsukurii* in these three habitats (Table 2) indicates that there was a great decrease in the rate of parasitism from potato to early planted rice. Hokyo and Kiritani (1963) ascribed this decrease to differences in dispersal ability between the parasitoid and the host. The above-mentioned descriptions of habitat shifts by *0. nezarae* and *T. mitsukurii* demonstrate that the latter species experiences difficulty in shifting its habitat to follow its hosts; the former species does not. An explanation of the difficulty of the habitat shift made by *T. mitsukurii* from potato to rice was given by Hokyo and Kiritani (1963) but other explanations also are possible. Rice may not be chemically as attractive to foraging *T. mitsukurii* females as is potato or *T. mitsukurii* movement on rice may be physically inhibited. Turner (1983) reported that density and length of soybean trichomes inhibit the movement of foraging females of T. *basalis* (Wollaston), another egg parasitoid of *N. viridura*. Thus, it is possible that difficulties in dispersal ability by T. *mitsukurii* and other natural enemies of mobile pests may be related to chemical and physical properties of their host's food plants and that success in habitat shift by natural enemies of mobile pests depends on the hosts' food plants or host habitats. For example, *Trichogramma* spp., natural enemies of *H. zea*, inflict heavy mortality in corn but not in tobacco due to inhibitory, physical, and chemical properties of the tobacco foliage (Stinner *et al.,* 1976; also see Chapters 4 and 5).

III. CONSERVING NATURAL ENEMIES OF MOBILE PESTS

Stinner and Bradley (1989) considered four major topics in dealing with habitat manipulation to increase effectiveness of predators and parasitoids of mobile pests, such as *Helicoverpa zea* and *Heliothis virescens* (Fabricius): (1) provision of refugia, (2) provision of alternative foods, (3) provision of chemical stimuli, and (4) alteration of plant characters. They reviewed the first two topics for mobile pests, with some reference to the interaction of these two factors with the plant character alterations associated with host plant resistance. Keller and Lewis (1985) also reviewed

Table 2.   Parasitism by *Trissolcus mitsukurii* of *Nezara viridula* in Asso, Wakayama, Japan[a].

Year	Host Generation	Host Habitat	No. of Host Eggs Collected	% Parasitism by *T. mitsukurii*
1961-1962	1st	Potato	5226	51.4
1961-1964	2nd	Early rice	48599	10.9
1961-1962	3rd	Late rice	11555	12.6

[a] (Hokyo, 1970)

provision of chemical stimuli to improve the performance of natural enemies of *H. zea* and *H. virescens* in the field. In a review of the interactions among plant resistance, cultural practices, and biological control Herzog and Funderbank (1985) cited several examples of habitat manipulation to increase effectiveness of natural enemies of mobile pests, such as *H. zea, H. virescens, N. viridula, M. persicae*, and *T. ni*.

Stinner and Bradley (1989) concluded that habitat manipulation, so as to enhance natural enemies, as a strategy for managing mobile pests such as *H. zea* and *H. virescens* offers real potential. However, they reviewed habitat manipulation other than pesticide use. When considering occurrence of multiple pests on one crop, use of selective pesticides should be first recommended to conserve natural enemies of mobile pests. Examples of the use of selective pesticides to conserve these natural enemies are also found in recent reviews of selective pesticide use in biological control (e.g., Hull and Beers, 1985; Mullin and Croft, 1985; Croft, 1990). Another more relevant example is that of the use of selective pesticides to control *Thrips palmi*. Nagai (1990) found that two insect growth regulators (IGR), pyriproxyfen and buprofezin, conserved *Orius* species, predators of *T. palmi*. Using these IGRs, he was successful in controlling *T. palmi, Epilachna vigintioctopunctata* (Fabricius), and *Polyphagotarsonemus latus* (Banks) on eggplants. The IGRs allowed the survival of a large number of *Orius* on eggplants. Similarly, Ohno *et al.* (1995), taking into account the conservation of natural enemies such as species of *Orius,* sprayed pyriproxyfen as a selective insecticide for the control of *T. palmi;* buprofezin as a selective acaricide for the control of *P. latus;* and milbemectine and buprofezin as selective acaricides for the control of *Tetranychus urticae* and *T. kanzawai* Kishida on eggplants. Nagai (1991) suggested dichlorvos as a selective insecticide for the control of the cotton aphid *Aphis gossypii* Glover on eggplants. They obtained successful control and conserved natural enemies. Thus, an integrated pest management system to conserve *Orius* species on eggplant was established using four selective insecticides and acaricides for the control of multiple pests on this crop in Japan (Table 3). *Orius* species are effective naturally occurring biological control agents of *T. palmi* but they can be suppressed by a large scale typhoon which strikes once in several years (Y. Hirose, unpublished data) and *T. palmi* is freed from their suppressive influence. Thus, use of pyriproxyfen for the control of *T. palmi* in such a case is needed.

Although use of selective pesticides offers a real and potent tactic for conserving natural enemies of *T palmi*, such as *Orius* species, this method should be combined with provision of refugia for these natural enemies in order to recruit them into the crop. Close to eggplant plots, *O. sauteri* and *O. minutus* often occur in noncrop habitats such as uncultivated areas with white clover (Takemoto and Ohno, 1996) and coreopsis (Y. Hirose, unpublished data). White clovers and coreopsis bloom in May to July and May to June, respectively, and their flowers infested with high populations of thrips harbor abundant *Orius* species. Thus, these flowers serve as a reservoir for *Orius* species colonizing eggplants infested with *T. palmi* after July. Takemoto and Ohno (1996) reported that *0. nagaii* utilizing agroecosystem plants in the Gramineae as a main habitat were abundant on *T. palmi*-infested eggplants growing adjacent to paddy fields. This suggests the movement of this predator from rice to eggplant. Thus, not only noncrop areas but also crop areas serve as a reservoir for predators of *T. palmi* on eggplant, i.e., *Orius* species. Similarly, in Virginia, corn and alfalfa adjacent to apple orchards apparently serve as a reservoir for *0. insidiosus* throughout the season and thistles harbor large numbers of this predator preying on thrips (McCaffrey and Horsburgh, 1986). Management of mobile natural enemies, such as *Orius* species and *Ooencyrtus nezarae,* in their alternate crop and noncrop habitats in agroecosystems is important to the conservation biological control of mobile pests. The problem of dispersal by mobile natural enemies in the presence of pests might be helped by provision of alternate habitats of mobile natural enemies adjacent to target crops.

Table 3. Selective pesticides to be used in the management of multiple pests on eggplant under naturally occurring biological control of *Thrips palmi by Orius* spp.

Pest	Selective pesticide	Source
Thrips palmi	Pyriproxylen	Nagai (1991) Ohno *et al.* (1995)
Epilachna vigintioctopunctata	Buprofezin	Nagai (1991)
Polyphagotarsonemus latus	Buprofezin	Nagai (1991) Ohno *et al.* (1995)
Tetranychus urticae	Buprofezin Milbemectine	Ohno *et al.* (1995)
T. kanzawai	Buprofezin Milbemectine	Ohno *et al.* (1995)
Aphis gossipii	Dichlorvos	Nagai (1991)

IV. CONCLUSIONS

Effective natural enemies of mobile pests are often mobile natural enemies, although further evaluation is necessary. More data are needed on the effectiveness of mobile natural enemies in regard to their dispersal behavior in the presence of hosts. Mobile natural enemies are characterized by polyphagy, good colonization ability, and high reproductive potential; traits often useful for the natural control of mobile pests in diversified agroecosystems. Difficulty in habitat shifts by natural enemies of mobile pests could be a problem for their conservation biological control. In this context, physical and chemical effects of hosts' food plant on mobile natural enemies are important. Similarly, research on the mechanisms underlying the difficulties that certain natural enemies have in following mobile hosts could lead to enhanced methods of biological control of many major crop pests.

Use of selective pesticides is a practical measure for the enhancement of conservation biological control of mobile pests, particularly where there are multiple pests on the same crop. However, this method should be combined with provision of refugia for natural enemies of mobile pests. These refugia are both noncrop and crop habitats of the natural enemies in a diversified agroecosystem. For example, the management system of *T. palmi* on eggplant (Table 3) should be developed into that for *T. palmi* not only on eggplant but also on other major host crops, such as cucumber, watermelon, and green pepper, in a diversified agroecosystem. In Japan, *T. palmi* usually lives in an area representing a mosaic of many different host and nonhost crops and noncrop areas. Thus, a management system for a mobile pest on a particular crop should be incorporated into the management system for the mobile pest on its major different host crops in the agroecosystem, which in turn, contain noncrop and crop habitats of its mobile natural enemies.

Acknowledgments

I thank Dr. P. Barbosa (University of Maryland) and Dr. W. A. Jones (USDA, Weslaco, Texas) for their useful comments on the manuscript. I also thank Mr. N. Mizutani (Kyushu National Agricultural Experiment Station, Japan) for providing his valuable information on the mobility of *0. nezarae.*

REFERENCES

Corbet, A., and Plant, R. E. (1993). Role of movement in the response of natural enemies to agroecosystem diversification: a theoretical evaluation. *Environ. Entomol.* 22, 519-531.

Croft, B. A. (1990). "Arthropod Biological Control Agents and Pesticides." John Wiley and Sons. New York, NY.

Ehler, L. E. (1977). Natural enemies of cabbage looper on cotton in the San Joaquin Valley. *Hilgardia* 45, 73-106.

Ehler, L. E., and Miller, J. C. (1978). Biological control in temporary agroecosystems. *Entomophaga* 23, 207-212.

Ehler, L. E., and van den Bosch, R. (1974). An analysis of the natural biological control of *Trichoplusia ni* (Lepidoptera: Noctuidae) on cotton in California. *Can. Entomol.* 106, 1067-1073.

Herzog, D. C., and Funderburk, J. E. (1985). Plant resistance and cultural practice interactions with biological control. *In* "Biological Control in Agricultural IPM Systems." (M. A. Hoy, and D. C. Herzog, eds.), pp. 67-88. Academic Press. London, U.K.

Hirose, Y., Takasu, K., and Takagi, M. (1996). Egg parasitoids of phytophagous bugs in soybean: mobile natural enemies as naturally occurring biological control agents of mobile pests. *Biol. Cont.* 7, 84-94.

Hokyo, N. (1970). Part II. Two egg parasites of the southern green stink bug with special reference to the biological control of the host. *In* "Studies on the Population Ecology of the Southern Green Stink Bug, *Nezara viridura* L. (Heteroptera: Pentatomidae)." (K. Kiritani, and N. Hokyo, eds.), pp. 203-222. Res. Council Secretariat, Min. Agric. For. Tokyo (in Japanese).

Hokyo, N., and Kiritani, K. (1963). Two species of egg parasites as contemporaneous mortality factors in the egg population of the southern green stink bug, *Nezara viridura. Jpn. J. Appl. Entomol. Zool.* 7, 214-227.

Hull, L. A., and Beers, E. H. (1985). Ecological selectivity: modifying chemical control practices to preserve natural enemies. *In* "Biological Control in Agricultural IPM Systems." (M. A. Hoy, and D. C. Herzog, eds.), pp. 103-122. Academic Press. Orlando, FL.

Joyce, R. J. V. (1981). The control of migrant pests. *In* "Animal Migration." (D. J. Aidley, ed.), pp. 209-229. Cambridge University Press. Cambridge, U.K.

Kawai, A., and Kawamoto, K. (1994). Predatory activity of *Orius* spp. and effect on the populations of minute sucking pests occurring on eggplant in open fields. *Bull. Natl. Res. Inst. Veg., Omam. Plants Tea Jpn.* Ser. A, No. 9, 85-101 (in Japanese with English synopsis).

Kawamoto, K., and Kawai, A. (1988). Effect of *Orius* sp. (Hemiptera: Anthocoridae) on the population of several pests on eggplant. *Proc. Assoc. Pl. Prot.* Kyushu 34, 141-143 (in Japanese with English synopsis).

Keller, M. A., and Lewis, W. J. (1985). Behavior-modifying chemicals to increase the efficacy of predators and parasitoids of *Heliothis spp. In* "Proc. Workshop on Biological Control of *Heliothis:* Increasing the Effectiveness of Natural Enemies." (E. G. King, and R. D. Jackson, eds.), pp. 449-467. Far Eastern Regional Office, USDA, New Delhi, India.

Kennedy, G. C., and Margolies, D. C. (1985). Mobile arthropod pests: management in diversified agroecosystems. *Bull. Entomol. Soc. Amer.* 31, 21-27.

Matsuzaki, T., Ichikawa, K., Kusakawa, K, and Ogawa, H. (1985). Some factors affecting the seasonal prevalence of *Thrips palmi* Karny on eggplant in the field and greenhouse. *Bull. Kochi Inst. Agr. For. Sci.* 17, 15-24 in Japanese).

McCaffrey, J. P., and Horsburgh, R. L. (1986). Biology of *Orius insidiosus* (Heteroptera: Anthocoridae): a predator in Virginia apple orchards. *Environ. Entomol.* 15, 984-988.

Mullin, C. A., and Croft, B. A. (1985). An update on development of selective pesticides favoring arthropod natural enemies. *In* "Biological Control in Agricultural IPM Systems." (M. A. Hoy, and D. C. Herzog, eds.), pp. 123-150. Academic Press. Orlando, FL.

Nagai, K. (1990). Suppressive effect of *Orius* sp. (Hemiptera: Anthocoridae) on the population density of *Thrips palmi* Karny (Thysanoptera: Thripidae) in eggplant in an open field. *Jpn. J. Appl. Entomol. Zool.* 34, 109-114 (in Japanese with English synopsis).

Nagai, K. (1991). Integrated control programs for *Thrips palmi* Karny on eggplant *(Solanum melongena* L.) in an open field. *Jpn. J. Appl. Entomol. Zool.* 35, 283-289 (in Japanese with English synopsis).

Nagai, K., Hiramatsu, T., and Henmi, T. (1988). Predatory effect of *Orius* sp. (Hemiptera: Anthocoridae) on the density of *Thrips palmi* Karny (Thysanoplera: Thripidae) on eggplant. *Jpn. J. Appl. Entomol. Zool.* 32, 300-304 (In Japanese with English synopsis).

Ohno, K., Takemoto, H., Kawano, K., and Hayashi, K. (1995). Effectiveness of integrated pest control program for *Thrips palmi* Karny on eggplants: a case study in a commercial field. *Bull. Fukuoka Agr. Res. Cent.* 14, 104-109 (in Japanese with English synopsis).

Sheehan, W. (1986). Response by specialist and generalist natural enemies to agroecosystem diversification: a selective review. *Environ. Entomol.* 15, 456-461.

Stinner, R. E., and Bradley, J. R., Jr. (1989). Habitat manipulation to increase effectiveness of predators and parasites. *In* "Proc. Workshop on Biological Control of *Heliothis:* Increasing the Effectiveness of Natural Enemies." (E. G. King, and R. D. Jackson, eds.), pp. 519-527. Far Eastern Regional Office, USDA. New Delhi, India.

Stinner, R. E., Rabb, R. L., and Bradley, J. R., Jr. (1976). Natural factors operating in the population dynamics of *Heliothis zea* in North Carolina. *Proc. Inter. Congr. Entomol.* 15, 622-642.

Takasu, K., and Hirose, Y. (1985). Seasonal egg parasitism of phytophagous stink bugs in a soybean field in Fukuoka. *Proc. Assoc. Pl. Prot.* Kyushu 31, 127-131 (in Japanese with English synopsis).

Takasu, K., and Hirose, Y. (1986). Kudzu-vine community as a breeding site *of Ooencyrtus nezarae* Ishii (Hymenoptera: Encyrtidae), an egg parasitoid of bugs attacking soybean. *Jpn. J. Appl. Entomol. Zool.* 30, 302-304 (in Japanese with English abstract).

Takemoto, H., and Ohno, K. (1996). Integrated pest management of *Thrips palmi* in eggplant fields, with conservation of natural enemies: effects of the surroundings and thrips community on the colonization of *Orius spp. In* "Proc. Intern. Workshop on the Pest Management Strategies in Asian Monsoon Agroecosystems." (N. Hokyo, and

G. Norton, eds.), pp.235-244. Kyushu National Agricultural Experiment Station, Japan.

Turner, J. W. (1983). Influence of plant species on the movement of *Trissolcus basalis* Wollaston (Hymenoptera: Scelionidae) - a parasite of *Nezera viridula* L. *J. Aust. Entomol. Soc.* 22, 271-272.

van den Berg, H. and Cock, M. J. W. (1993). Stage-specific mortality of *Helicoverpa armigera* in three smallholder crops in Kenya. *J. Appl. Ecol.* 30, 640-653.

van den Bosch, R., Leigh, T. F., Falcon, L. A., Stern, V. M., Gonzales, D., and Hagen, K. S. (1971). The developing program of integrated control of cotton pests in California. *In* "Biological Control." (C. B. Huffaker, ed.), pp. 377-394. Plenum Press. New York, NY.

Van Steenwyk, R. A., Toscano, N. C., Ballmer, G. R., Kido, K., and Reynolds, H. T. (1975). Increases of *Heliothis* spp. in cotton under various insecticide treatment regimes. *Environ. Entomol.* 4, 993-996.

CHAPTER
13

A CONSERVATION APPROACH TO USING ENTOMOPATHOGENIC NEMATODES IN TURF AND LANDSCAPES

E. E. Lewis, J. F. Campbell, and R. Gaugler

I. INTRODUCTION

Entomopathogenic nematodes have received a great deal of attention from researchers and industry since the early 1980s because of their many favorable attributes as biological control agents (Kaya and Gaugler, 1993). As a group, they have an extremely broad host ranges and variation in foraging strategies and host associations which potentially offer the ability to control pest species with diverse life histories (Lewis *et al.*, 1992, 1993; Grewal *et al.*, 1994; Lewis *et al.*, 1996; Campbell and Gaugler, 1997). They are used almost exclusively as a biological insecticide, typically applied at high densities to soil in a homogeneous blanket (e.g., 2.5 billion per hectare is the recommended dosage for many crops) with little concern for the fate of applied organisms. Evaluation of entomopathogenic nematode populations after initial applications is usually undertaken only in terms of pest reduction and is limited to a few days or weeks after application. The reasons underlying success or failure of releases of entomopathogenic nematodes are not obvious but may be explained in terms of nematode population ecology.

Current conservation biological control strategies for entomopathogenic nematodes are limited to avoiding their release onto sites where immediate mortality is likely (e.g., unprotected foliar surfaces) or where they are completely ineffective (such as aquatic environs, soils at temperatures that inactivate nematodes, etc.). Considering the current widespread use of nematodes, it is unfortunate that conservation of natural populations in agroecosystems or the fate of applications, have received almost no attention. The few attempts at inoculative applications have met with some success (Gaugler *et al.*, 1992) but this strategy may be inappropriate for some situations.

Determinations of whether to use inundative or inoculative release strategies need to be made on a case-by-case basis but pests of high value commodities with low economic thresholds will usually be relegated to inundative efforts. Although specialized conservation measures for improving entomopathogenic nematode efficacy outside the soil will be discussed, control of non-soil pests is also limited to inundative strategies because nematode infection behaviors are impaired outside the soil environment (Georgis and Gaugler, 1991).

Entomopathogenic nematodes have been isolated from every inhabited continent, in virtually every type of soil habitat where a concerted effort to find them has been attempted. Isolation records demonstrate the great diversity of habitats exploited by entomopathogenic nematodes (Kaya and Gaugler, 1993). Studies indicate that nematodes have the ability to withstand extreme environmental conditions in some cases but few studies offer details of nematode density, local distribution, effects (if any) on host populations, or even species identity in many cases. Although they are poorly understood, natural populations of nematodes are extremely common; 21.7% of 301 soil samples in New Jersey, U.S.A. contained them (Gaugler *et al.,* 1992). A few epizootic outbreaks have been reported (Sexton and Williams, 1981; Akhurst *et al.,* 1992) but because they occur below the soil surface and are difficult to observe outbreaks may often go unreported (Kaya, 1990).

Our goal in this chapter is to synthesize the available information about the environmental fate of entomopathogenic nematodes and from that synthesis devise some guidelines for their use in conservation biological control. We have drawn on literature from natural and applied populations, inoculative and inundative control efforts, and numerous laboratory studies to try to identify factors that influence entomopathogenic nematode survival, population dynamics, and ultimately their effect on host populations.

II. NEMATODE BIOLOGY

Entomopathogenic nematodes comprise two families, the Steinernematidae and Heterorhabditidae; families that are not closely related phylogenetically but which share similar life histories through convergent evolution (Poinar, 1993). They are soil-inhabiting lethal insect parasites. The only free living stage is a non-feeding, developmentally arrested infective juvenile (a dauer larva) whose sole function is to seek out new hosts and initiate infections. This is the life stage of interest for inundative biological control considerations because it is the only one formulated and applied. For longer-term establishment all life stages are of significance. Portals of entry into the host include natural body openings (e.g., spiracles, mouth, and anus) and through thin cuticle. Once in the insect hemocoel, penetrating nematodes "inoculate" an associated bacterium *Xenorhabdus* or *Photorhabdus* spp. for steinernematids and heterorhabditids, respectively, into the host. The bacteria multiply rapidly typically causing host mortality in 24 to 48 hours, although in some host species and stages death can occur in as quickly as 15 minutes (LeBeck *et al.*, 1993). The nematodes feed upon the bacteria and degrading host cadaver, mature, mate, and produce up to three generations within a single host. Infective steinernematid juveniles develop into amphimictic males and females, but never into hermaphrodites. After penetration into the host, heterorhabditid infective juveniles develop into hermaphrodites, whereas subsequent parasitic generations are males, females, and hermaphrodites. As the nutritional quality of the cadaver is depleted, a new generation of infective juveniles is formed which emerges in search of fresh hosts. Each new infective juvenile carries its associated bacterium stored in the gut.

The nematode may appear to be little more than a biological syringe for its associated bacterium yet the relationship between these organisms is one of classic mutualism. Although parasitic, entomopathogenic nematodes have not severed their nutritional relationship with bacteria. Growth and reproduction are dependent upon conditions established in the host cadaver by the bacterium. The bacterium further contributes anti-immune proteins to assist the nematode in overcoming host defenses and antimicrobials which suppress colonization of the cadaver by competing secondary invaders. Conversely, the bacterium lacks invasive powers and is dependent upon the nematode to locate and penetrate suitable hosts. The relationship between each species of entomopathogenic nematode and its bacterial associate tends to be highly specific. Because the bacterium kills the host rapidly, steinernematids and heterorhabditids do not form the intimate host-parasite relationships characteristic of other insect-nematode infections (e.g., mermithids). Consequently, entomopathogenic nematodes infect hundreds of insect species encompassing most insect orders in laboratory exposures. This remarkable spectrum of insecticidal activity has sparked intense interest in the commercial development of nematodes as biological insecticides.

III. PRESENT USES OF ENTOMOPATHOGENIC NEMATODES INCLUDING CONSERVATION BIOLOGICAL CONTROL

Conservation biological control methods are designed to protect and maintain natural or introduced enemies of pests (Rabb *et al.*, 1976). Conservation biological control with entomopathogenic nematodes can include application practices for nematodes that will either favor survival and/or establishment of introduced individuals. It also may involve crop management practices designed to have a positive effect (or to minimize negative impacts) on existing natural or applied populations or applications of entomopathogenic nematodes. Many of the requirements and limitations of individual infective juvenile entomopathogenic nematodes are well understood and have been reviewed (Georgis and Gaugler, 1991; Kaya and Gaugler, 1993). Indeed, the success of entomopathogenic nematodes as inundative biological insecticides is attributable in large part to addressing these limitations. In contrast, requirements for population level survival are poorly known, despite their obvious importance to biological control (Hominick and Reid, 1990). We submit that the requirements for encouraging the survival and persistence of natural populations are similar to those for long-term establishment of applied nematodes and will be treated as such.

The requirements for individual survival are not identical to those for persistence of populations. The choice of which requirements to fulfill depends upon the aim of the user. The importance of survival time of individual infective juveniles, the density of the application, and the likelihood of long-term establishment depend upon the level of control required. In all pest control situations some nematodes must survive long enough to find and infect the target host, regardless of whether the goal is to control via inundative or inoculative releases. Short-term (augmentative) biological control requires individual survival at high enough densities to decrease the pest population to acceptable levels. Inundative biological control, which depends on a single generation of infective juveniles will usually yield faster and higher levels of control than encouraging multiple generations to establish in an area for inoculative control, but the effect will be temporary. Long-term, multigenerational survival and recycling through hosts is necessary for inoculative release programs, but is not important for a short-term, high dose strategy. Three conditions should be met to make inoculative control efforts for soil pests worthwhile: (1) moderately susceptible pests should be present throughout most of the year, (2) pests should have a high economic threshold level, and (3) soil conditions should be favorable for nematode survival (Kaya, 1990). The requirements for inoculative efforts are met in several pest control arenas, including scarab grubs and mole crickets in turfgrass.

In general, the most effective use of nematodes is limited to soil inhabiting pests. The broad experimental host range reported for entomopathogenic nematodes does not translate into a broad-spectrum of insecticidal activity in the field. Countless field trials between ecologically incompatible entomopathogenic nematode species and targeted pests were attempted because their incompatibility was unanticipated. The natural reservoir for entomopathogenic nematodes is the soil yet sweeping efforts beginning in the 1950s were made to use nematodes as inundative biological control agents against foliage feeding pests, with discouraging results. There are no ecological barriers to infection in a petri dish, whereas nematodes applied onto unprotected foliage are exposed to inactivation from ultraviolet radiation, desiccation, and temperature extremes. Aquatic habitats offer shelter from the latter extremes, but control efforts against black fly and mosquito larvae have failed because entomopathogenic nematodes are unable to swim and thus are poorly adapted to initiate infections in aquatic habitats. A few cryptic insects, particularly wood-boring caterpillars, which inhabit protected microhabitats provide a minor exception to the expectation that entomopathogenic nematodes do well only in hosts in soil.

Steinernematid and heterorhabditid nematodes have been in commercial use as inundative biological control agents of insect pests in the U.S.A., Europe, and Asia since the late 1980s. Nematodes are currently applied against pests of cranberries (such as the black vine weevil and cranberry girdler), citrus (such as the citrus root weevil), turfgrass (such as, the mole cricket, white grubs, and billbugs), household pests (including fleas), artichokes (e.g., the artichoke plume moth), mushrooms (such as sciarid flies), tree fruit (e.g., the peach fruit moth), ornamentals (e.g., the black vine weevil and wood borers), and other insect pests of horticulture, agriculture, home, and garden. In addition, entomopathogenic nematodes have demonstrated a surprising ability to suppress populations of phytoparasitic nematodes infesting turfgrass (Smitley *et al.,* 1992). Nevertheless, biological controls overall comprise a slender 1.3% of the $28 billion dollar global pesticide market (R. Georgis, pers. comm., Thermotrilogy, Columbia, MD). Despite possessing impressive attributes for biological control, entomopathogenic nematode sales for 1995 are estimated at only $15 million (P. Grewal, pers. comm., Ohio State University). Nematode use is restricted to so-called niche markets, particularly where chemical agents are restricted or unavailable. In short, nematodes are not reducing reliance on chemical pesticides to any significant degree.

Inundative control using entomopathogenic nematodes may be far from dead given that more than a dozen small companies in nearly as many countries remain in place. However, the paradigm of commercializing nematodes following the chemical pesticide model is flawed. Most biologicals, including nematodes, fit the chemical model poorly. Chemicals are cheap, stable products that are easy to scale up and use whereas nematodes offer none of the aforementioned advantages. Nematode based insecticides are far more susceptible than chemicals to suboptimal temperature, and changes in soil type, thatch depth, and irrigation frequency (Georgis and Gaugler, 1991). Nematodes are inactivated if stored in hot vehicles, cannot be left in spray tanks for long periods, and are incompatible with some agricultural chemicals. Unused product cannot be applied the following year and different species require different screen sizes in application equipment. Nevertheless, alternatives to the chemical pesticide paradigm such as conservation biological control are worth exploring for nematodes, but are poorly developed.

The use of entomopathogenic nematodes has overwhelmingly involved inundative (augmentative) releases yet the origins and future of the use of entomopathogenic nematodes are in inoculative and conservation biological control. The first species discovered was *Steinernema glaseri* (Steiner) from scarab larvae in 1929. Small turfgrass plot experiments assessing this parasite's colonization ability when introduced into new areas began almost immediately in southern New Jersey (Glaser, 1932; Glaser and Farrell, 1935). Encouraging results coupled with the development of *in vitro* rearing methods (McCoy and Glaser, 1936) led to a massive inoculative control program from 1939 to 1942 aimed at the Japanese beetle *Popillia japonica* Newman, an introduced pest. Billions of infective juveniles were released throughout New Jersey yet the releases were largely unsuccessful; *S. glaseri* were re-isolated only from southern New Jersey (Gaugler et al., 1992). The elimination of bacterial symbionts by the use of antimicrobials in the artificial rearing media and the negative effects of the New Jersey climate on the neotropical *S. glaseri* were likely contributing factors to this failure. Inoculative releases were not reported again for more than 40 years.

The next inoculative release involved *Steinernema scapterisci* Nguyen and Smart in 1985 against an introduced pest, mole crickets of the genus *Scaperiscus*, infesting turfgrass in the southeastern U.S.A. This nematode species established, infected (Hudson *et al.*, 1988), and persisted in target populations (Parkman *et al.*, 1993a) and dispersed from release sites (Parkman *et al.*, 1993b). Although subsequent releases were made throughout Florida their impact has been difficult to assess, in large part due to the difficulty of sampling of mole crickets accurately. Nevertheless, Parkman *et al.* (1993b) conclude that "inoculative release of this nematode is a viable alternative to inundative release." In addition to these two intentional inoculative release programs, there have been indications of nematode recycling following a single inundative release. Klein and Georgis (1992) found that a 60% reduction in *P. japonica* populations one month after inundative treatment with the NC strain of *Heterorhabditis bacteriophora* Poinar had increased to 96% after 8 months. Population reductions of 67 and 100% with the NC and HP88 strains of *H. bacteriophora* were followed by 68% and 93 to 97% reductions in the subsequent scarab generation, more than four months after treatment. Nevertheless, there is a dearth of effort on inoculative releases, particularly concerning building the research base which is fundamental to inoculative releases and conservation biological control with entomopathogenic nematodes.

IV. ECOLOGICAL UNDERPINNINGS

Theoretical support for strategy development for inoculative releases or conservation biological control with entomopathogenic nematodes must begin with understanding their ecology. Conservation biological control seeks to encourage or establish populations of natural enemies of sufficient density to suppress pest populations. What density of entomopathogenic nematodes is required to maintain pest populations at acceptable levels? Over how large a spatial scale can this density be maintained? Over how long a temporal scale can this density be maintained? Attempts to manipulate ecological constraints to maintain nematode populations must take into consideration these questions. The driving forces of entomopathogenic nematode population dynamics will determine to what extent we can manipulate nematode density, persistence, and impact on host populations. In this section we discuss the biotic and abiotic determinants of nematode distribution and how entomopathogenic nematode spatial and temporal distribution regulate their impact on insect populations in several ecosystems.

A.   Biotic and Abiotic Determinants of Survival and Distribution

Understanding what influences spatial distribution is critical to assessing how spatial distribution might be manipulated. Important parameters are host distribution, nematode behavior, and abiotic factors (Campbell *et al.*, 1997). For example, if the patchiness of a population results from limited nematode dispersal, the distribution of abiotic or biotic variables will have a significant impact on the effectiveness of, and approach to, manipulating spatial distribution. Creating an environment that encourages entomopathogenic nematode survival is one way to increase persistence. Optimum soil moisture, temperature, and pore size have long been recognized to favor infective juvenile survival in the soil. Minimizing nematode exposure to desiccation and ultraviolet light, post-treatment irrigation to wash the nematodes into the soil, and applying nematodes near dusk are also important (Selvan *et al.,* (1993). Entomopathogenic nematodes are frequently applied to soil concurrently with other treatments or where previous treatment has occurred. A list of noncompatible materials is usually supplied on product containers. Bednarek and Gaugler (1997) have also found that inorganic nitrogen based fertilizers have a negative impact on entomopathogenic nematodes whereas organic fertilizers actually are associated with increases in nematode populations. An excellent review of abiotic and biotic factors affecting entomopathogenic nematode survival in the field is provided by Baur and Kaya (submitted).

Minimizing deleterious effects of sunlight and desiccation can increase infective juvenile survival outside the soil. For control of leafminers (such as *Liriomyza* spp. and *Chromatomyia syngenesiae* Hardy) high humidity is critical for infective juvenile survival (Hara *et al.,* 1993). Williams and MacDonald (1995) showed that *Steinernema feltiae* (Filipjev) or *H. bacteriophora* applied at a rate of 140 to 160 nematodes cm^{-2} with a liquid nonionic wetting agent significantly reduced leafminer populations when ambient relative humidity was 85-90%. Levels of control were significantly increased for *S. feltiae* when ambient relative humidity was maintained above 90%. Percent leafminer reduction was correlated with the duration of nematode survival on the leaf surface, emphasizing the importance of protecting the nematodes from environmental degradation for as long as possible. Using entomopathogenic nematodes to control foliar pests has the greatest potential for success in areas where physical factors can be controlled. In greenhouses, for example, sunlight can be limited or eliminated for 24 or 48 hours, optimum temperature can be maintained (although this varies with nematode species), and relative humidity can be maintained above 90% for extended periods.

Some environmental factors, such as temperature, are not amenable to manipulation. In this case, selecting a nematode species or strain adapted to the habitat into which they are applied is the best strategy. The ability of individual nematodes to withstand local conditions limits persistence in most cases, especially where they are applied to nonendemic areas, which is often the case with commercially produced nematodes. For example, when applying nematodes in hot, dry climates, the isolate of *H. bacteriophora* from the Negev desert region of Israel (Glazer *et al.,* 1996) would be an ideal candidate. Where the climate is cool, isolates from higher latitudes such as the Umea strain of *S. carpocapsae* isolated from Sweden or Western European isolates of *Heterorhabditis* sp. (Griffin and Downes, 1991) might overcome cool temperature inactivity. Ferguson *et al.* (1995) found that locally isolated strains of *H. bacteriophora* persisted in alfalfa fields in northern New York State, U.S.A. significantly longer than any commercial strains applied. One limitation to this approach is that many nematode strains equipped to withstand climatic extremes are as yet unavailable commercially.

B. Spatial Distribution

1. Field studies

The first step toward identifying the factors that favor natural nematode populations is to characterize areas where nematode densities are high. Then we can measure differences that occur between high density areas and those that lack nematode populations. The goal is to make a target control area ideal for the establishment and maintenance of entomopathogenic nematode populations. Whether or not manipulation based on this type of data will actually enhance nematode populations remains to be seen but analysis of population structure may give some clues as to how we can approach manipulations.

Most studies of the distribution of endemic entomopathogenic nematode populations have been conducted on coarse spatial scales (tens of km^2) (e.g., Akhurst and Bedding, 1986; Hominick and Briscoe, 1990; Griffin *et al,* 1991; Hara et al., 1991; Gaugler *et al.,* 1992; Amarasinghe *et al.,* 1994) and indicate only presence or absence from the site sampled. No information is supplied by these studies concerning either horizontal or vertical distributions of nematodes within a population. Limited data on the horizontal distributions of endemic or released nematode populations on a finer scale (cm^2 to m^2) indicate that there is considerable variation in distribution among species, sites, and sampling times (Akhurst et al., 1992; Stuart and Gaugler, 1994; Cabanillas and Raulston, 1994; Campbell *et al.,* 1997). Vertical distribution of nematode populations has rarely been considered, yet location in the soil profile influences the importance of several environmental factors to nematode survival and to some extent determines host affiliations.

Intensive sampling studies that assessed entomopathogenic nematode field distribution have emphasized the differences in field ecology among species and between genera. Analysis of horizontal distribution on small spatial scales suggests that *H. bacteriophora* is highly aggregated; the population density within patches sometimes approaches 65 infective juveniles/cm^2 soil surface. *Steinernema carpocapsae*

Weiser maintains more evenly distributed and less dense populations (15 to 20 infective juveniles/cm^2 soil surface) than *H. bacteriophora* within a field (Campbell *et al.,* 1995; Campbell *et al.,* 1997; unpublished data). The distribution and abundance of *H. bacteriophora* along transects in several habitat types in southwestern New Jersey were extremely variable among and within transects (Stuart and Gaugler, 1994), suggesting that patches may be on the scale of m^2. Analyses of soil type, soil moisture, turfgrass variety, and *P. japonica* distribution did not adequately explain the distribution pattern of *H. bacteriophora* or *S. carpocapsae.* However, the amount of shade influenced the spatial distribution of an undescribed *Heterorhabditis* species in a citrus orchard in a semi-arid region of Israel (Glazer *et al.,* 1996). Nematode recovery was correlated with percent shade and no nematodes were recovered from areas with no shade. Campbell *et al.* 1997) proposed that differences in foraging behavior and host mobility may be important influences on spatial distribution, especially in relatively homogeneous and permanent habitats like turfgrass. The horizontal distribution of endemic *Steinernema riobravis* Cabanillas, Poinar and Raulston in corn fields was also aggregated; nematodes occurred in 81% of the plots at soil depths of 0 to 10 cm, 75% at soil depths of 10 to 20 cm, and 31% at depths of 20 to 30 cm (Cabanillas and Raulston, 1994), suggesting that the nonuniformity of entomopathogenic nematode distributions is three dimensional.

The vertical distribution of *H. bacteriophora* was uniform throughout the top 8 cm of soil whereas *S. carpocapsae* was recovered predominately near the soil surface. Ferguson *et al.* (1995) also found that most of the *S. carpocapsae* applied to alfalfa fields persisted in the top 0 to 5 cm, but *H. bacteriophora* was recovered to depths of 30 to 35 cm. Additional laboratory studies on dispersal (Reed and Carne, 1967; Moyle and Kaya, 1981; Georgis and Poinar, 1983a), host finding (Alatorre-Rosas and Kaya, 1990; Kaya *et al., 1993*), and behavior (Reed and Wallace, 1965; Kondo and lshibashi, 1986; Campbell and Gaugler, 1993) also show that *S. carpocapsae* is a surface-adapted species and that *H. bacteriophora is* not. Cabanillas and Raulston (1994) found that endemic *S. riobravis* occurred to depths of 30 cm but most were recovered from the upper 20 cm.

2. Consequences of spatial distribution to conservation biological control

Few individual factors have been shown to influence entomopathogenic nematode distribution. Amount of shade and soil moisture have some effect on population structure but other environmental parameters (e.g., host density, soil type, plant species, etc.) seem to have little direct effect. Rather, a complex of dynamic characteristics is likely to determine entomopathogenic nematode population structure. Keeping this in mind, it has been proposed that entomopathogenic nematode populations, due to their patchy spatial distribution, may persist as metapopulations (Baur and Kaya, submitted). Metapopulations are populations of local populations which exhibit a "shifting mosaic" type of dynamics (Levins, 1970). In this metapopulation concept individual local populations are highly vulnerable to extinction, they fluctuate asynchronously, and have relatively little migration among patches. The metapopulation persists if local populations are founded at a rate that is at least as great as their extinction rate. What we know about many species of entomopathogenic nematodes' spatial distribution and dispersal ability suggests that this type of population model may be appropriate. As previously discussed, entomopathogenic nematodes are patchy in distribution. Among-patch movement or new patch establishment are unpredictable and probably depend on phoretic transport of infective juveniles, the movement of infected hosts, or other passive means. However, our understanding of the relative importance of within and between local population processes in entomopathogenic nematodes is still very limited. Other types of population structure may also be important in entomopathogenic nematode populations, e.g., there may be source and sink populations, single contiguous functional populations (maintained if dispersal is high enough), or population structures that vary among species, sites, and scales.

The degree to which nematode populations are aggregated might influence persistence in several ways (Levins 1970). First, when a metapopulation is small and made up of only a few local subpopulations then the probability that they all will go extinct at the same time is increased and metapopulation stability is reduced. Second, if the local subpopulations in a metapopulation are extremely isolated from each other then the maintenance or rescue of subpopulations in distress and the probability of founding new subpopulations decreases: populations are more likely to go extinct. Third, the overall stability of the metapopulation will decrease when it is made up of small local populations because each local subpopulation has an increased probability of extinction as its size decreases. Therefore, entomopathogenic nematode spatial distribution and the factors that influence it are not only important in determining the impact nematodes have on pest populations but also on the persistence of this control.

In turfgrass, when hosts are relatively uniform in distribution nematode search behavior and host mobility appear to be the important factors in determining spatial distribution (Campbell et al., 1997). Environmental conditions are likely to influence nematode distribution in more variable and/or abiotically extreme habitats; either directly by influencing the nematode's ability to survive and reproduce or indirectly by affecting the distribution of hosts. Within local populations the probability of extinction depends on demographic processes and environmental stochasticity. Intra- and interspecific competition may be important in the persistence of populations. High levels of infection within a host can result in reduced offspring production (Selvan et al., 1993). Overlap in the distribution of any two different nematode species, either two endemic species or a released species and a native species, may lead to the competitive exclusion of one of them (Kaya and Koppenhoffer, 1996). Many factors can contribute to the outcomes of competitive interactions including foraging strategy, reproductive strategy, reproductive ability, host range, etc. Ideally, before applying nematodes to an area, extant populations would be assessed. The feasibility of manipulating spatial distribution and the influence of these manipulations on persistence remains to be determined, but applying nematodes uniformly to a field will probably not lead to a sustainable population structure.

Releasing *H. bacteriophora* in a homogeneous distribution effectively controls *P. japonica* in turf (Selvan *et al.* 1993); however, it may not leave any hosts available for future generations of the parasite. Since *P. japonica* is univoltine grubs are available most of the year but adult dispersal for recolonization occurs only once annually. Therefore, long-term persistence of the nematodes is unlikely if the application suppresses host populations "too successfully." High density, point applications might be more amenable to long-term persistence for various reasons which need not be mutually exclusive. First, clumped or patchy distributions are typical of natural populations (Stuart and Gaugler, 1994) and nematodes might be best adapted to such a population structure. Second, using point applications may provide sources from which the nematodes can colonize other areas, while potentially providing refugia for hosts to support future generations of nematodes. Ultimately, inoculative releases of nematodes every few years, and potentially at lower application rates than those presently used, could prove to be an extremely cost-effective method for controlling pests and could foster the increased use of nematodes in this and other systems.

C.   Temporal Distribution

1.   Field studies

Temporal (daily and seasonal) variation in the occurrence, density, and infectivity of entomopathogenic nematodes has an impact on biological control. Several studies have addressed temporal variation in nematode populations but generalizations based on these studies should be made with care. Most studies have been conducted in untilled systems; either temperate perennial agroecosystems (such as turfgrass, orchards, and sugar cane) or nonagricultural systems (e.g., forests and roadsides). Few studies have been conducted in unstable ecosystems, such as annual crops or in the tropics. A confounding factor in the assessment of populations in the field is the difference between infectivity *in situ* and infectivity in the laboratory.

Soil samples collected from the field typically are brought to the laboratory, warmed, and moistened before they are baited with hosts susceptible to entomopathogenic nematodes. In other words, the ideal conditions for infection presented in the laboratory may not reflect conditions in the field so those nematodes successfully baited in the laboratory may not have been infective in the field. Correlations of nematode patterns with host patterns are usually only inferred because the soil is usually baited with *Galleria mellonella* L., which is highly susceptible to infection by some species (e.g., *S. carpocapsae),* but not others (e.g., *S. scapterisci).*

A further complicating factor is that the variation in spatial scales on which population persistence is measured can be important, especially if entomopathogenic nematode population structure is based on metapopulation models, as discussed previously (Baur and Kaya, submitted).

Temporal patterns vary among species of entomopathogenic nematodes. Surprisingly, studies from temperate climates show that *Steinernema* species persist throughout the year with no apparent seasonal periodicity. For example, Hominick and Briscoe (1990) saw no seasonal pattern to the recovery of *S. feltiae* from soils collected from sites in roadsides, hedgerows, pasture, and woodland in England, but did find a great deal of variation among sampling times within sites. Campbell *et al.* (1995) studied the seasonal and daily patterns of endemic populations of *S. carpocapsae* in turfgrass in New Jersey. Again, nematode populations had no seasonal pattern in prevalence or density. *S. carpocapsae* was recovered from about 35% of the sections at an average density (in positive samples) of 17 nematodes/cm^2 throughout the sampling period. However, nematodes were not sampled during the winter, nor was infectivity determined under field conditions. In a semi-arid region of Israel, Glazer *et al.* (1996) recovered *Heterorhabditis* sp. in a citrus orchard at higher densities during the winter, when temperatures were cooler and moisture higher than summer. *Steinernema feltiae* populations do vary seasonally in strawberries (Vainio and Hokkanen, 1993) as a result of the cycles of the crop and its insect pests.

Seasonal patterns of *H. bacteriophora* recovery from soil samples show less temporal stability than for *S. carpocapsae* (Campbell *et al.,* 1995). Natural founding populations of *H. bacteriophora* tended to be recovered in June and July during periods when adults of a potential host, *P. japonica,* were active. On average, *H. bacteriophora* was recovered in 8% or less of the samples but where it occurred it was recovered at a higher average density (63 nematodes/cm^2) than *S. carpocapsae.* However, the variability of nematode density was extremely high and most of the samples had nematode densities far below those typically applied when they were used for inundative biological control (i.e., 25 nematodes/cm^2). Reports of epizootics with *Heterorhabditis* species also suggest high temporal variability in nematode recovery (Poinar *et al., 1987*; Akhurst *et al.,* 1992).

The persistence of released nematodes, when it has been monitored, has followed a fairly consistent pattern. Within days of release, population levels drop to a low level (usually less than 20% of the original dosage), which may or may not persist by recycling (Gaugler, 1988; Kaya, 1990; Bedding *et al., 1993*; Currant, 1993; Baur and Kaya, submitted). Long-term persistence of populations requires a combination of infective juvenile survival and recycling in hosts but the relative importance of these components is difficult to assess. Several studies show that long-term persistence of entomopathogenic nematodes occurs. However, the degree of adaptation of released nematodes to local conditions exerts a great influence on their success in the environment. For example, Ferguson *et al.* (1995) released two *H. bacteriophora* isolates, *S. carpocapsae,* and an undescribed *Steinernema* sp. into three New York state alfalfa fields. All isolates except one *H. bacteriophora* were originally isolated from the same region. All isolates persisted for six months without any detectable reduction in recovery, but after 24 months only the locally isolated strain of *H. bacteriophora* was present in high density. Infection in the field was correlated with mean soil temperature and infectivity increased as temperatures rose above 15°C.

2. Consequences of temporal distribution to conservation biological control

The temporal and spatial distribution of hosts is an important factor in the long-term persistence of entomopathogenic nematode populations. Because we know little about the breadth of the natural host range for most species, our ability to assess this is limited. Long-term persistence relies on either the continuous presence of hosts in which the nematodes can recycle or a physiological mechanism of the nematode that allows persistence over periods without hosts. Host distribution is likely to influence the distribution of species such as *S. scapterisci,* which are host specific. On the other hand, generalist species may be more stable, especially if alternative hosts are available during periods when the pest populations are absent. For example, adding alternate hosts to greenhouse planters containing black vine weevils increased *H. bacteriophora* and *S. feltiae* persistence (Burlando *et al.,* 1993). Perhaps field sites where alternate hosts exist naturally, or can be supplied, will provide the necessary conditions for long- term establishment of entomopathogenic nematodes. More likely, however, is the possibility of providing alternate hosts in a confined growing system, such as in a greenhouse. This strategy could potentially increase persistence by providing hosts and increase the effective dose of nematodes, since a single infected *G. mellonella* larva may yield up to 500,000 infective juveniles within 10 days of infection.

V. NEMATODE INTERACTIONS WITH NATURAL HOST POPULATIONS

Laboratory studies have indicated broad host ranges for most entomopathogenic nematode species, but these are undoubtedly overestimates of natural host ranges (Gaugler, 1988; Kaya and Gaugler, 1993). Information on host range is limited because most new species and strains have been isolated using soil baiting methods with *G. mellonella,* which provides no clues as to natural host utilization. Finding infected insects in the field is uncommon. Isolations of entomopathogenic nematodes from natural hosts, which have been summarized by Peters (in press), suggest that *S. carpocapsae* has a broad natural host range among insects associated with the soil surface, including most Lepidoptera. This is consistent with other studies concerning foraging mode and vertical distribution. *Steinernema feltiae* and *H. bacteriophora* also appear to infect a broad range of hosts. However, other species, e.g., *S. glaseri, Steinernema kushidai* Mamiya, and *S. scapterisci*, appear to have narrower natural host ranges, as they are limited to perhaps a single family of hosts.

Expectations of establishment of *Heterorhabditis* and *Steinernema* species differ. In many cases, nematode induced mortality rates are steady and moderate (i.e., 10-30%). However, occasional epizootics in host populations have been reported, as has been noted. Interestingly, long-term steady host mortality rates tend to be associated with *Steinernema,* whereas all reported epizootic outbreaks have been attributed to *Heterorhabditis* nematodes. Georgis and Hague (1981) found natural populations of *S. carpocapsae* in a larch forest in England to cause a constant 10% level of nematode infection of larch sawfly *(Cephalcia lariciphila* Wachtl.) prepupae. *Steinernema kraussei* killed 24 to 27% of the false spruce webworm *(Cephalcia abietis* (L.)) throughout the year (Mrácek, 1986). *Steinernema riobravis,* endemic to the lower Rio Grande valley in Texas, U.S.A., infected prepupae and pupae of the fall armyworm *Spodoptera frugiperda* (J. E. Smith) and *Helicoverpa zea* (Boddie), the corn earworm (Raulston *et al.,* 1992). In corn fields, the level of parasitized *H. zea* and *S. frugiperda* averaged 34.2 and 24.2%, respectively, over five years where hosts were present.

In contrast, Sexton and Williams (1981) found that *Heterorhabditis sp.* distribution in a lucerne field in Victoria, Australia tended to be patchy (nematodes were only recovered from one portion of the field). Where the nematode was present, the white fringed beetle *Graphognathus leucoloma* (Boheman) occurred at lower densities and dead weevils were recovered. In turfgrass, the presence of *H. bacteriophora* was correlated with reduced *P. japonica* larvae populations (Campbell *et al.,* 1995). In another study by these authors, there was a negative correlation between the length of time a section of a turf field harbored *H. bacteriophora* and the density of *P. japonica* larvae (Campbell *et al.,* unpublished). Akhurst *et al.* (1992) reported a *Heterorhabditis* species epizootic in scarabaeid larvae in sugar cane. Nematodes were recovered from 65% of the soil samples and infection rates of scarabs ranged from 26 to 100%. In a California coastal headlands ecosystem, Strong *et al.* (in press) found that root damage by the ghost moth caterpillar, *Hepialus californicus* (Behrens) is a major cause of bush lupine mortality. *Heterorhabditis hepialus* Stock, Gardner and Strong kills a large proportion of the *H. californicus* caterpillars on lupine roots and appears to be specific to this insect species (Strong *et al.,* in press). They proposed that *H. hepialus* was the primary below-ground mortality factor to ghost moth caterpillars in this system. Increased prevalence of the nematode was correlated with reduced entry of the ghost moth caterpillar into the lupine roots and with reduced lupine bush mortality.

In contrast to the examples described above, the interactions between natural or introduced populations of entomopathogenic nematodes and hosts remains obscure in many cases. Bathon (1996) reports that applications of *S. feltiae, Heterorhabditis megidis,* and an undescribed *Heterorhabditis* sp. applied to beech forest, pine forest edge, orchard, and wheat fields generally had little impact on arthropods, as measured using emergence traps. However, there were some minimal impacts on nonpest species including some Coleoptera and Diptera, and a great deal of variation in the impact among the sites in the study. Campbell *et al.* (1995) found that although *S. carpocapsae* was extremely prevalent in turf grass it did not have a detectable impact on *P. japonica* larvae or on mobile surface arthropod population densities. Here, the lack of interaction was attributed to *S. carpocapsae's* ambushing foraging strategy, which is incompatible with the fossorial life history of *P. japonica* grubs (Lewis *et al.,* 1993).

There are some generalizations to be drawn from these studies that can be applied to many situations where entomopathogenic nematodes are used for biological control. The spatial and temporal dynamics of nematode populations in the wild indicate that expecting nematodes to persist in a field as they are applied (that is, in a homogeneous blanket at high density) is probably unrealistic. We propose that a metapopulation model *(sensu* Levins, 1970) is probably an accurate depiction of wild entomopathogenic nematode populations. It is likely that applications of nematodes rapidly break down to a similar population structure. In areas where nematode density is high they can have a significant impact on host populations. However, high densities of nematodes have not been recorded to persist in patches large enough to satisfy most agricultural needs. To exploit natural entomopathogenic nematode populations and be able to predict the long-term effects of applied nematodes we need to tailor our expectations to the biology of the nematodes.

VI. CONCLUSIONS

The requirements of entomopathogenic nematodes for survival depend upon the expectations of the user. If short-term (within-generation) control is the goal, survival of individual infective juveniles is the sole requirement, and it is necessary only to assure that sufficient nematodes live long enough to find and infect enough hosts. Alternatively, when long-term (inoculative) control is desired host suitability (the ability of host to enable the nematodes to recycle), host distribution, and the duration of host availability must be considered for entomopathogenic nematode population maintenance. Whatever the goal of the control effort, the first step toward conservation biological control must be to monitor the fate of entomopathogenic nematodes in the habitats where they are used.

The nearly ubiquitous distribution of entomopathogenic nematodes in so many diverse habitats suggests that conservation of natural populations should be considered for pest control, especially in habitats that are perennial such as forests, orchards, and turf. The development of entomopathogenic nematodes as commercially viable biological insecticides has relegated ecological considerations to the background until recently. Most research was directed toward mass production, formulation, establishing laboratory host ranges, and conducting field efficacy trials. However, the unexpectedly poor performance of some nematode species in field trials has in many cases been explained by ecological constraints that were not considered previous to their release. Recent emphasis on entomopathogenic nematode ecology, especially in terms of host affiliation, has lead to conservation strategies which did not previously receive serious consideration, especially in the long-term. For example, as a group, entomopathogenic nematodes show astounding diversity in habitat requirements and host affiliations yet within species these requirements and relationships tend to be much more specific. Therefore, only the most general requirements can be considered for all species; most must be employed on a more species- or even strain-specific basis.

The requirements for maximizing infective juvenile survival have been explored in detail in the literature and reviewed here. Abiotic soil conditions to favor entomopathogenic nematode survival include adequate moisture and temperatures warm enough to allow infection. Biotic conditions necessary for infective juvenile survival are less well understood and offer many opportunities for new research directions. Survival strategies employed by nematodes in the wild are poorly understood, including those that enable nematodes to endure extremely hostile environs. Competition for hosts between natural populations and applied nematodes is rarely, if ever, considered. The influence of nematode antagonists (i.e., bacteria, fungi, predaceous nematodes and mites) on applications of entomopathogenic nematodes has been addressed in a few studies but no conclusions have been reached.

Long-term population level survival of entomopathogenic nematodes in the soil is even more difficult to address. The records of long-term survival of applied nematodes indicate that recycling through hosts must have occurred. Records of epizootics suggest that under certain conditions dense populations of entomopathogenic nematodes occur, presumably in response to host abundance. However, the presence of dense populations of acceptable hosts does not seem to be the sole requirement for entomopathogenic nematode epizootics. For example, outbreaks of scarab grubs in turf in the northeastern U.S.A. do not always give rise to dense nematode populations. Studies of entomopathogenic nematode population dynamics reveal only that the population structure is patchy in time and space and that they generally lack seasonality. To establish more complete guidelines for conservation of natural populations we need first to understand the requirements and structure of natural populations.

Conservation practices for entomopathogenic nematodes can potentially decrease costs of control and increase the efficacy and predictability of control for both inundative and inoculative releases. Long-term predictability of entomopathogenic nematode influence on host populations lags far behind short-term predictability. This may be because the fate of applications is poorly understood and the biotic and abiotic factors that influence entomopathogenic nematode population dynamics are seldom considered before their use. It is our hope that discussions in this chapter and other related studies will stimulate interest in identifying the factors influencing entomopathogenic nematode survival, population dynamics, and ultimately their effect on host populations. Such data will facilitate the implementation of conservation biological control.

REFERENCES

Akhurst, R. J., and Bedding, R. A. (1986). Natural occurrence of insect pathogenic nematodes (Steinernematidae and Heterorhabditidae) in soil in Australia. *J. Aust. Entomol. Soc.* 25, 241-244.

Akhurst, R. J., Bedding, R. A., Bull, R. M., and Smith, K. R. J. (1992). An epizootic of *Heterorhabditis* spp. (Heterorhabditidae: Nematoda) in sugar cane scarabaeids (Coleoptera). *Fund. Appl. Nematol.* 15, 71-73.

Alatorre-Rosas, R., and Kaya, H. K. (1990). Interspecific competition between entomopathogenic nematodes in the genera *Heterorhabditis and Steinernema* for an insect host in sand. *J. Invert. Pathol.* 55, 179-188.

Amarasinghe, L. D., Hominick, W. M., Briscoe, B. R., and Reid, A. P. (1994). Occurrence and distribution of entomopathogenic nematodes in Sri Lanka. *J. Helminthol.* 68, 277-286.

Bathon, H. (1996). Impact of entomopathogenic nematodes on non-target hosts. *Biocont. Sci. Technol.* 6, 421-434.

Baur, M. E., and Kaya, H. K. 1998. Persistence of entomopathogenic nematodes. *In* "Environmental Persistence of Entomopathogens and Nematodes." Oklahoma State University. Southern Cooperative Series Bulletin. (in press)

Bedding, R. A., Akhurst, R. J., and Kaya, H. K. (1993). "Nematodes and the Biological Control of Insect Pests." CSIRO. East Melbourne, Australia.

Bednarek, A., and Gaugler, R. (1997). Compatibility of soil amendments with entomopathogenic nematodes. *J. Nematol.* 29, 220-227.

Burlando, T. M., Kaya, H. K., and Timper, P. (1993). Insect-parasitic nematodes are effective against black bine weevil. *Calif. Agric.* 47, 16-18.

Cabanillas, H. E., and Raulston, J. R. (1994). Evaluation of the spatial pattern of *Steinernema riobravis* in corn plots. *J. Nematol.* 26, 5-31.

Campbell, J. F., and Gaugler, R. (1997). Inter-specific variation in entomopathogenic nematode foraging strategy: dichotomy or variation along a continuum? *Fund. Appl. Nematol.* 20, 393-398.

Campbell, J. F., and Gaugler, R. (1993). Nictation behaviour and its ecological implications in the host search strategies of entomopathogenic nematodes (Heterorhabditidae and Steinernematidae). *Behaviour* 126, 155-169.

Campbell, J. F., Lewis, E. E., Yoder, F., and Gaugler, R. (1995). Entomopathogenic nematode (Heterorhabditidae and Steinernematidae) seasonal population dynamics and impact on insect populations in turfgrass. *Biol. Cont.* 5, 598-606.

Campbell J. F., Lewis, E. E., Yoder, F., and Gaugler, R. (1997). Entomopathogenic nematode (Heterorhabditidae and Steinernematidae) spatial distribution in turfgrass. *Parasitol.* 113, 473-482.

Currant, J. (1993). Post-application biology of entomopathogenic nematodes in soil. *In* "Nematodes and the Biological Control of Insect Pests." (R. Bedding, R. Akhurst, and H. K. Kaya, eds.), pp. 67-77. CSIRO. East Melbourne, Victoria, Australia.

Ferguson, C. S., Schroeder, P. C., and Shields, E. J. (1995). Vertical distribution, persistence, and activity of entomopathogenic nematodes (Nematoda: Heterorhabditidae and Steinernematidae) in alfalfa snout-beetle (Coleoptera: Curculionidae) infested fields. *Environ. Entomol.* 24, 149-158.

Gaugler, R. (1988) Ecological considerations in the biological control of soil-inhabiting insects with entomopathogenic nematodes. *Agric., Ecosys., Environ.* 24, 351-360.

Gaugler, R., Campbell, J. F., Selvan, S., and Lewis, E. E. (1992). Large-scale inoculative releases of the entomopathogen *Steinernema glaseri:* assessment 50 years later. *Biol. Cont.* 2, 181-187.

Gaugler, R., Lewis, E. E., and Stuart, R. (1997). Ecology in the service of biological control: The case of entomopathogenic nematodes. *Oecologia* (in press).

Georgis, R., and Hague, N. G. M. (1981). A neoaplectanid nematode in the larch sawfly *Cephalcia lariciphila* (Hymenoptera: Pamphiliidae). *Ann. Appl. Biol.* 99, 171-177.

Georgis, R., and Poinar, G. 0., Jr. (1983). Effect of soil texture on the distribution and infectivity of *Neoaplectana carpocapsae* (Nematoda: Steinernematidae). *J. Nematol.* 15, 308-311.

Georgis, R., and Gaugler, R. (1991). Predictability in biological control using entomopathogenic nematodes. *J. Econ. Entomol.* 84, 713-20.

Glaser, R. W. (1932) Studies on *Neoaplectana glaseri,* a nematode parasite of the Japanese beetle *(Popillia japonica).* NJ Dept. Agric. Circ. No. 211. 34 pp.

Glaser, R. W., and Farrell, C. C. (1935). Field experiments with the Japanese beetle and its nematode parasite. *J. NY Entomol. Soc.* 43, 345-371.

Glazer, I., Kozodoi, E., Salame, L., and Nestel, D. (1996). Spatial and temporal occurrence of natural populations of *Heterorhabditis* spp. (Nematoda: Rhabditida) in a semiarid region. *Biol. Cont.* 6, 130-136.

Grewal, P. S., Lewis, E. E., Gaugler, R., and Campbell, J. F. (1994). Host finding behaviour as a predictor of foraging strategy in entomopathogenic nematodes. *Parasitol.* 108, 207-215.

Griffin, C. T., and Downes, M. J. (1991). Low temperature activity in *Heterorhabditis* sp. (Nematoda: Heterorhabditidae). *Nematologica* 37, 83-91.

Hara, A. H., Gaugler, R., Kaya, H. D., and LeBeck, L. M. (1991). Natural populations of entomopathogenic nematodes (Rhabditida: Heterorhabditidae, Steinernematidae) from the Hawaiian Islands. *Environ. Entomol.* 20, 211-216.

Hara, A. H. , Kaya, H. K., Gaugler, R., LeBeck, L. M., and Mello, C. L. (1993). Entomopathogenic nematodes for biological control of the leafminer *Liriomyza trifolii* (Diptera: Agromyzidae). *Entomophaga* 38, 359-369.

Hominick, W. M., and Briscoe, B. R. (1990). Longitudinal survey of fifteen sites for entomopathogenic nematodes (Rhabditida: Steinernematidae). *Parasitol.* 100, 289-294.

Hominick, W. M., and Reid, A. P. (1990). Perspectives on entomopathogenic nematology. *In* "Entomopathogenic Nematodes in Biological Control." (R. Gaugler, and H. K. Kaya, eds.), pp. 327-345. CRC Press. Boca Raton, FL.

Hudson, W. G., Frank, J. H., and Castner, J. L. (1988). Biological control of mole crickets (Orthoptera: Gryllotalpidae) in Florida. *Bull. Entomol. Soc. Amer.* 34, 192-198.

Kaya, H. K. (1990). Soil ecology. *In* "Entomopathogenic Nematodes in Biological Control." (R. Gaugler, and H. K. Kaya, eds.), pp. 93-116. CRC Press. Boca Raton, FL.

Kaya, H. K. (1993). Contemporary issues in biological control with entomopathogenic nematodes. *Food and Fertilizer Tech. Cent,* Taipei, Taiwan. *Extension Bull.* No. 375, 1-13.

Kaya, H. K., and Gaugler, R. (1993). Entomopathogenic nematodes. *Annu. Rev. Entomol.* 38, 181-206.

Kaya, H. K., and Koppenhoffer, A. M. (1996). Effects of microbial and other antagonistic organisms and competition on entomopathogenic nematodes. *Biocont. Sci. Technol.* 6, 357-371.

Kaya, H. K., Bedding, R. A., and Akhurst, R. J. (1993). An overview of insect-parasitic and entomopathogenic nematodes. *In* "Nematodes and the Biological Control of Insects Pests." (R. A. Bedding, R. J. Akhurst, and H. K. Kaya, eds.), pp. 1-10. CSIRO Publications. East Melbourne, Australia.

Klein, M. G. (1990). Efficacy against soil-inhabiting insect pests. *In* "Entomopathogenic Nematodes in Biological Control." (R. Gaugler, and H. K. Kaya eds.), pp. 195-214. CRC Press. Boca Raton, FL.

Klein, M. G., and Georgis, R. (1992). Persistence of control of Japanese beetle (Coleoptera: Scarabaeidae) larvae with steinernematid and heterorhabditid nematodes. *J. Econ. Entomol.* 85, 727-730.

Kondo, E., and Ishibashi, N. (1986). Nictating behavior and infectivity of entomogenous nematodes, *Steinernema* spp., to the larvae of common cutworm, *Spodoptera litura* (Lepidoptera: Noctuidae), on the soil surface. *Appl. Entomol. Zool.* 21, 553-560.

LeBeck, L. M., Gaugler, R., Kaya, H. K., and Hara A. H. (1993). Host stage suitability of the leaf miner *Liriomyza trifolii* to the nematode *Steinernema carpocapsae. J. Invert. Pathol.* 62, 92-98.

Levins, R. (1970). Extinction. *In* "Lectures on Mathematics in the Life Sciences." (M. Gerstenhaber, ed.), pp. 77-107. Vol 2. Amer. Math. Soc. Providence, RI.

Lewis, E. E., Gaugler, R., and Harrison, R. (1992). Entomopathogenic nematode host finding: Response to contact cues by cruise and ambush foragers. *Parasitol.* 105, 309-315.

Lewis, E. E., Gaugler, R., and Harrison, R. (1993). Response of cruiser and ambusher entomopathogenic nematodes (Steinernematidae) to host volatile cues. *Can. J. Zool.* 71, 765-769.

Lewis, E. E., Ricci, M., and Gaugler, R. (1996). Host recognition behavior reflects host suitability for the entomopathogenic nematode, *Steinernema carpocapsae. Parasitol.* 113, 573-579.

McCoy, E. E., and Glaser, R. W. (1936). Nematode culture for Japanese beetle control. *NJ Agric. Circ.* 265. 10 pp.

Moyle, P. L., and Kaya, H. K. (1981). Dispersal and infectivity of the entomogenous nematode, *Neoaplectana carpocapsae* Weiser (Rhabditida: Steinernematidae). *J. Nematol.* 13, 419-421.

Mrácek, Z. (1986). Nematodes and other factors controlling the sawfly, *Cephalcia abietis* (Pamphiliidae: Hymenoptera), in Czechoslovakia. *For. Ecol. Manag.* 15, 75-79.

Parkman, J. P., Hudson, W. G., Frank, J. H., Nguyen, K. B., and Smart, G. C., Jr. (1993a). Establishment and persistence of *Steinernema scapterisci* (Rhabditida: Steinernematidae) in field populations of *Scapteriscus* spp. mole cricket (Orthoptera: Gryllotalpidae). *J. Entomol. Sci.* 28, 182-190.

Parkman, J. P., Frank, J. H., Nguyen K. B., and Smart, G. C., Jr. (1993b). Dispersal *of Steinernema scapterisci* (Rhabditida: Steinernematidae) after inoculative applications for mole cricket (Orthoptera: Gryllotalpidae) control in pastures. *Biol. Cont.* 3, 226-232.

Peters, A. Natural occurrence of insect infections by *Steinernema and Heterorhabditis* spp. in regard to their natural host range and their impact on insect populations. *Biocont. Sci. Technol.* (in press).

Poinar, G. O., Jr. (1993). Origins and phyletic relationships of the entomophilic rhabditids, *Heterorhabditis* and *Steinernema. Fund. Appl. Nematol.* 16, 333-338.

Poinar, G. O., Jr., Jackson, T., and Klein, M. (*1987*). *Heterorhabditis* sp. n. *megidis* (Heterorhabditidae: Rhabditida), parasitic in the Japanese beetle, *Popillia japonica* (Scarabaeidae: Coleoptera), in Ohio. *Proc. Helminth. Soc. Wash.* 54, 53-59.

Rabb, R. L., Stinner, R. E., and van den Bosch, R. (1976). Conservation and augmentation of natural enemies. *In* "Theory and Practice of Biological Control." (C. B. Huffaker, and P. S. Messenger, eds.), pp 233-254. Academic Press. New York, NY.

Raulston, J. R., Pair, S. D., Loera, J., and Cabanillas, H. E. (1992). Prepupal and pupal parasitism of *Helicoverpa zea and Spodoptera frugiperda* (Lepidoptera: Noctuidae) by *Steinernema* sp. in cornfields in the lower Rio. *J. Econ. Entomol.* 85, 1666-1670.

Reed, E. M., and Carne, P. B. (1967). The suitability of a nematode (DD-136) for the control of some pasture insects. *J. Invert. Pathol.* 9, 196-204.

Reed, E. M., and Wallace, H. R. (1965). Leaping locomotion by an insect parasitic nematode. *Nature* 206, 210-211.

Selvan, S., Campbell, J. F., and Gaugler, R. (1993). Density-dependent effects on entomopathogenic nematodes (Heterorhabditidae and Steinernematidae) within an insect host. *J. Invert. Pathol.* 62, 278-284.

Sexton, S., and Williams, P. (1981). A natural occurrence of parasitism of *Graphognathus leucoloma* (Hoheman) by the nematode *Heterorhabditis sp. J. Austr. Entomol. Soc.* 20, 253-255.

Smitley, D. R., Warner, F. W., and Bird, G. W. (1992). Influence of irrigation and *Heterorhabditis bacteriophora* of plant-parasitic nematodes in turf. *J. Nematol.* 24, 637-641.

Strong, D. R., Kaya, H. K., Whipple, A., Chile, A., Kraig, S., Bondonno, M., Byer, K., and Maron, J. L. Entomopathogenic nematodes: natural enemies of root-feeding caterpillars on bush lupine. *Oecologia.* (in press)

Stuart, R. J., and Gaugler, R. (1994). Patchiness in populations of entomopathogenic nematodes. *J. Invert. Pathol.* 64, 39-45.

Williams, E. C., and MacDonald, 0. C. (1995). Critical factors required by the nematode *Steinernema feltiae* for the control of the leafminers *Liriomyza huidobrensis, Liriomyza bryoniae* and *Chromatomyia syngenesiae, Ann. Appl. Biol.* 127, 329-341.

Vainio, A., and Hokkanen, H. M. T. (1993). The potential of entomopathogenic fungi and nematodes against *Otiorhyncus ovatus* L. and *0. dubius* (Col.: Curculionidae) in the field. *J. Appl. Entomol.* 115, 379-387.

ENVIRONMENTAL MANIPULATION FOR
MICROBIAL CONTROL OF INSECTS

James R. Fuxa

I. INTRODUCTION

Microbial control of insects, i.e., the use of microorganisms or their byproducts to suppress pest populations or damage has a long history (Steinhaus, 1975). There are four basic approaches to microbial control, three of which, short-term insecticide (i.e., augmentation), seasonal colonization (another type of augmentation), and introduction-establishment (classical biological control), include the release or application of entomopathogen units (Fuxa, 1997). The fourth approach is environmental manipulation.

Environmental manipulation for microbial control of insects is the enhancement of entomopathogenic suppression of pest populations by means other than the release of entomopathogen units into the environment. In this approach, the usual agricultural or resource management practices are altered to enhance the activity of an entomopathogen population and reduce the pest population or its damage, without significant interference with the overall management practices. Environmental manipulation also can consist of an intentional elimination or delay of a management practice that would be detrimental to an entomopathogen population, as in the case of a change in fungicide applications to create a better environment for an entomopathogenic fungus. This approach generally is aimed at enhancing aspects of the entomopathogen's life cycle that increase its ability to cause high rates of disease prevalence in nature.

Environmental manipulation is promising and logical because the activity of most entomopathogens is greatly influenced by their environment and because potentially it has low cost to the farmer or resource manager compared to most other methods of control. It has low risk of adverse environmental effects. However, its research and implementation have been very limited, primarily due to poor funding and perhaps due to the necessity of understanding and, in some cases, predicting entomopathogen epizootics.

The purpose of this paper is to discuss the ecological/epizootiological basis for environmental manipulations, review research on this approach, and discuss criteria for choosing insect/entomopathogen systems for research and implementation. The research review will not include adjuvants or formulations used in applying entomopathogens, such as UV-light protectants. Similarly, it will not include release methods, such as application through irrigation systems, that might alter the entomopathogen's environment. It will include methods used before or after application or release in order to enhance entomopathogen activity.

II. BASIS IN EPIZOOTIOLOGY/ECOLOGY

All of the approaches to microbial control of insects tend to focus on critical times or deficiencies in the entomopathogen's life cycle. Introduction and establishment position a pathogen in an ecosystem where it can live and reproduce but did not do so previously. The short-term insecticide and seasonal colonization approaches artificially increase the pathogen population and enhance its transmission by placing it in contact with the insect pest.

Environmental manipulation is no exception to this rule. The weak points or critical times in an entomopathogen's life cycle, as is the case with most parasites, occur primarily during the transfer from one host to another. If it is not transovarially transmitted, the pathogen must survive adverse environmental conditions, somehow be transported (usually with no inherent mobility) from the old host to the new, and penetrate the defenses of the new host insect. All three of these critical points, but primarily the first two, have been targets of attempts at environmental manipulation. This is a logical approach. If large numbers of pathogen units have been produced naturally in an ecosystem as is often the case, a slight change in cultural practices has the potential to enhance their survival or transmission to a new host.

In order for this approach to be successful, a thorough knowledge of ecology and epizootiology is required. Epizootiology can be defined as the study of causes and forms of the mass phenomena of disease at all levels of intensity in a host population (Fuxa and Tanada, 1987). In other words, it is the study of animal disease at the population level. Microbial control is essentially applied epizootiology; an attempt to increase disease levels in a host population. Changing resource management practices to increase disease prevalence requires not only knowledge of the entomopathogen's life cycle but also of the crop or resource management practices, the insect pest(s), and other components of the agroecosystem. For example, a population of an entomopathogenic fungus might be preserved by not spraying a fungicide or enhanced by an early crop planting date. The former would minimally require reliable prediction of the fungal epizootics as well as of the effect on phytopathogens. The latter would require an understanding of the effect of planting date on the microhabitat and, in turn, efficacy of the entomopathogen as well as research on the effect of planting date on other pests and crop parameters so that the change in date does not reduce crop yield.

III. RESEARCH ON ENVIRONMENTAL MANIPULATION

Research on environmental manipulation has focused on four areas. The first is improved transport from the pathogen reservoir, usually the soil, to a site such as a leaf surface where the insect host can come into contact with the entomopathogen. Except for certain nematodes and a few fungi which have limited searching ability, entomopathogens depend on passive transport by abiotic or biotic agents while outside the host insect (Andreadis, 1987). From the perspective of any given pathogen unit this passive transport process has a very low probability of success. The second area is improvement in persistence of the entomopathogen at the site where it contacts the insect host. Almost all entomopathogens are harmed quickly by sunlight, low moisture, or chemical pesticides while they are on an exposed surface (Benz, 1987). The third area is overall growth of the entomopathogen population, which depends on transmission and persistence as well as other factors. For example, high relative humidity often is essential to the production of fungal conidia, structures that grow on the external surface of an insect host (McCoy *et al.,* 1988). A greater entomopathogen population density simply increases the probability of contact between the pathogen and an uninfected host. The fourth area is activation of latent infections. It is likely that many entomopathogenic viruses produce latent infections in which the virus infects an insect but does not replicate or cause progressive disease for one or more host generations. If the host insect is subjected to some sort of stress then the latent infection can become active, resulting in viral replication and death of the host (Burand *et al.,* 1986; Tanada and Fuxa, 1987).

Research on environmental manipulation for microbial control has been limited. The primary reason is that the research emphasis for entomopathogens is mainly on microbial insecticides. Additionally, knowledge of ecology/epizootiology is required for such research, which can be risky in terms of obtaining publishable results. The research has largely focused on viruses and fungi, probably because these groups have the best ability to produce disease epizootics with a high case fatality rate (i.e., a high percentage of infected insects are killed by the disease). There also have been attempts at environmental manipulation of nematodes, perhaps because this group generally is more dependent on suitable environmental conditions than other groups. Research on environmental manipulation falls into two major categories: environmental manipulation to enhance natural epizootics and environmental manipulation in conjunction with, but not simultaneous with, entomopathogen release or application.

A.   Enhancement of Natural Epizootics

Environmental manipulation to enhance natural epizootics is preferred over manipulation in conjunction with entomopathogen release because the costs are lower. This approach also epitomizes the concept of "natural" control of an insect pest. Additionally, it adapts well to the concepts of sustainable agriculture and integrated pest management, not only in taking advantage of long-term natural control but also in that by its very nature the method integrates into normal agricultural or resource management practices. This method requires knowledge of the epizootiology of the targeted entomopathogen and some capability to predict whether epizootics can occur in a system. For example, the Soviet Union at one time had a nationwide network to predict whether entomopathogen epizootics would occur or could be induced (Klassen, 1975). Enhancement of natural epizootics has been researched with three groups of entomopathogens, the fungi, viruses, and nematodes.

1.   Fungi

Enhancement of natural epizootics of certain fungi has concentrated on agronomic practices, primarily to assist persistence and fungal population growth. Certain parameters for planting soybeans can affect epizootics of the fungus *Nomuraea rileyi* (Farlow). Samson in lepidopteran pests (Sprenkel *et al.,* 1979). Fungal epizootics reached the highest prevalence rates in early planted soybean in narrow rows at high seeding rates. These conditions resulted in early closure of the soybean canopy between plants or rows which apparently increased the relative humidity, resulting in improved fungal growth. Another fungus, *Beauveria bassiana* (Balsamo) Vuillemin, was affected by soil tillage. Prevalence of this fungus was significantly greater in *Ostrinia nubilalis*

(Hübner) infesting no-till corn than in corn in plowed or chisel regimes (Bing and Lewis, 1993). Similarly, populations of the fungi *B. bassiana, Metarhizium anisopliae* (Metschnikoff) Sorokin, and *Paecilomyces* spp. were greater in no till than in tilled plots in a soybean-wheat double cropping system (Sosa-Gomez and Moscardi, 1994). Another example of this technique was particularly interesting because it was heavily based on mathematical modeling of epizootics. This research resulted in recommendations implemented by extension specialists. Adherence to early season insecticide treatment thresholds along with early harvesting could increase profits by 20% by inducing epizootics of *Erynia sp.* in *Hypera postica* (Gyllenhal) (Brown and Nordin, 1986; Brown, 1987). The fungus probably was aided by the tendency of weevil larvae to aggregate during warm, humid conditions.

Manipulation of moisture has been another means to increase fungal prevalence. Extra irrigation increased fungal activity against *Therioaphis maculata* (Buckton) in alfalfa (Hall and Dunn, 1957) and irrigation induced an epizootic of *Entomophthora* sp. in an aphid population in lucerne (Walters and Bishop, 1978). Overhead irrigation induced epizootics of *Erynia* spp. in *Acyrthosiphon pisum* (Harris) infested legumes but drip irrigation did not (Pickering *et al.,* 1989). Similarly, water sprays to increase relative humidity in mushroom hothouses were the major factor inducing epizootics of *Erynia* sp. in fungus gnats (Huang *et al.,* 1992).

There clearly is potential to conserve fungal populations and epizootics by delaying or not applying chemical fungicides normally used on crops. Field prevalence of *N. rileyi,* a pathogen of lepidopteran pests in soybean in certain areas, has been significantly reduced by applications of benomyl (Johnson *et al.,* 1976; bin Husin, 1978), chlorothalonil, maneb, thiobendazole, fentin hydroxide (bin Husin, 1978), benlate, Du Ter, and Bravo (Horton *et al.,* 1980). The lepidopteran pests included in these studies of *N. rileyi* were *Anticarsia gemmatalis* Hübner, *Plathypena scabra* (Fabricius), *Pseudoplusia includens* (Walker), and *Heliothis* spp. Interestingly, another pathogen of lepidopteran soybean pests, *Entomophthora gammae* Weiser, was not adversely affected by benomyl, chlorothalonil, maneb, or thiobendazole (bin Husin, 1978; Livingston *et al.,* 1981). Four Entomophthoraceae were the subjects of a study of epizootics in *Aphis fabae* Scopoli infesting field beans (Wilding, 1982). Of four fungicides, captafol, mancozeb, tridemorph, and benomyl, only the latter reduced prevalence of *Erynia neoaphidis* Remaud. and Henn. *Entomophthora planchoniana* Coru, *Neozygites fresenii* Remaud. and Kell., and *Conidiobolus obscurus* Remaud. & Kell. were unaffected by the four fungicides in these field tests.

2. Viruses

Natural prevalence of entomopathogenic viruses can be enhanced by aiding their transport to a point of contact with the insect pest. For example, sprinkler irrigation enhanced transport of the nucleopolyhedrosis viruses (NPVs) of *A. gemmatalis* and *P. includens* throughout soybean plants (Young, 1990). Manipulation of viral epizootics also has been successful in pastures, a low value crop that often does not justify relatively expensive means of pest control. Oversowing, intensive rotational stocking, stocking paddocks immediately after cutting hay, and direct drilling rather than ploughing of winter feed crops were recommended to control *Wiseana* spp. with *Wiseana* NPV, *Wiseana* entomopox virus, and *Wiseana* granulosis virus (Kalmakoff and Crawford, 1976; Crawford and Kalmakoff, 1977).

Viral prevalence has been improved by two particularly novel methods of environmental manipulation. A trench mortar has been used to blow NPV contaminated forest litter up into trees, resulting in contamination of the foliage and initiation of a viral epizootic in larvae of *Lymantria dispar* (L.) (Podgwaite, 1985). Many insects infected by NPVs are known to transmit the viruses to their offspring and it is possible that stages or even generations of the insects have latent infections in which the virus is not replicating or harming its host (Tanada and Fuxa, 1987). When the insect becomes stressed, as with sublethal dose of insecticides, this supposedly can activate these latent infections resulting in epizootics. Sublethal sprays of chemical insecticides were used in the U.S.S.R. to activate viral infections in insect pests (Klassen, 1975), a method that has potential usefulness in any host-pathogen system with a high prevalence of latent infections. Application of chemical insecticide in this manner as well as any other method that stresses the insect can be considered an example of environmental manipulation in that the host insect is part of the virus' environment.

3. Nematodes

There has been only one attempt at enhancement of natural epizootics of nematodes. Tillage, weed management, and irrigation were investigated to enhance prevalence of *Heterorhabditis heliothidis* (Khan, Brooks, Hirschmann) in *Diabrotica undecimpunctata howardi* (Barber) infesting corn (Brust, 1991). No-till and the presence of weeds significantly increased the numbers of nematodes in soil bioassays but irrigation had no effect. These results were consistent with subsequent crop damage and yield. Although pesticides adversely affected a nematode in the laboratory (Ishibashi and Takii, 1993) there was no attempt to conserve nematode populations in the field by eliminating or changing the timing of pesticide applications.

B. Enhancement of Entomopathogen Application

Environmental manipulation can enhance the results of releasing or applying an entomopathogen in the field. Research in this area has indicated that all of the other three approaches to microbial control, short-term insecticide, seasonal colonization, and introduction-establishment can benefit from an "improved" environment. The review in this section will include only examples of manipulations before or after, but not simultaneous with, the release or application of entomopathogens. Manipulation simultaneous with application is best considered part of the application process itself and therefore is not consistent with the concept of environmental manipulation. Methods to enhance entomopathogen application have included enhancement of pathogen persistence and, occasionally, transmission. The application process usually delivers the entomopathogen to a site of contact with the insect and, therefore, enhancement of transport or transmission usually is not as important as enhancement of natural epizootics. Enhancement of entomopathogen application has been researched with three groups, fungi, viruses, and nematodes.

1. Fungi

Research of environmental manipulation of fungi in conjunction with application has been limited. An increase of relative humidity in a glasshouse enhanced control resulting from a release of *Entomophthora fresenii* Nowak against *A. fabae* (Dedryver, 1979). In another study, the fungicide mancozeb reduced but did not eliminate prevalence of *B. bassiana* after the fungus was applied against *Leptinotarsa decemlineata* (Say) (Clark *et al.,* 1982). Fungi usually are heavily dependent on environmental factors such as relative humidity and air movement, so environmental manipulation in conjunction with fungal application is an area ripe for further research.

2. Viruses

Enhancement of viral application has been attempted in three different types of ecosystems. No-till management of soybean increased amounts of *A. gemmatalis* NPV in soil up to 2 years after its application, at which time there was still sufficient virus to initiate epizootics (Moscardi, 1989). In contrast, tillage of soybean increased activity after application of the NPV of *A. gemmatalis* in soybean by transporting virus from the soil to leaves (Young and Yearian, 1986). In a pasture habitat, cattle were allowed to graze beginning several days after spray application of *Spodoptera frugiperda* (J. E. Smith) NPV. Viral prevalence increased in the presence of cattle, probably due to increased transport of the virus from the soil reservoir (Fuxa, 1991). Thus, enhancement in conjunction with viral application may be particularly useful for improving viral transmission.

In coconut palms, a release of the baculovirus of *Oryctes rhinoceros* (L.) can control this insect for years. Researchers have recommended that once the virus is in the pest population five dead palms be left standing per hectare. However, all the other dead trunks should be collected into piles, allowing cover crops to overgrow them (Zelazny *et al.*, 1992). This strategy apparently concentrates the beetle population, resulting in improved transmission of the virus.

3. Nematodes

Attempts to enhance nematode applications have been limited to one of two methods, focused on irrigation or effects of agrichemicals. Irrigation was necessary for establishment of *Neoaplectana carpocapsae* Weiser, *Neoaplectana glaseri* Steiner, and *H. heliothidis* after spray application to control *Popillia japonica* Newman in golf course turf (Shetlar *et al.*, 1988) (but see Chapter 13). Irrigation before and after spray application of *Heterorhabditis bacteriophora* Poinar was necessary to achieve consistent control of *P. japonica* and *Cyclocephala borealis* Arrow in Kentucky bluegrass (Downing, 1994). Agrichemicals including molinate, methyl parathion plus benomyl plus thiabenzole, and fentin hydroxide lowered parasitism of *Culex quinquefasciatus* Say by *Romanomermis culicivorax* Ross and Smith. Urea, carbofuran, and ammonium sulfate had no effects or uncertain effects on this nematode (Walker and Meek, 1987).

IV. SELECTION OF SYSTEMS FOR ENVIRONMENTAL MANIPULATION

In discussion of environmental manipulation for microbial control a question arises as to how to choose crop (resource)/pest/entomopathogen systems to target for research. Characteristics of the pest, resource, or crop and ecosystem are critical to the environmental manipulation approach. Careful selection of target systems can help researchers to avoid loss of resources on unsuitable systems and, perhaps, wasting a resource manager's time and money.

Criteria for choosing systems for environmental manipulation can be complex. These criteria differ to some extent from those underlying the other three approaches to microbial control (short-term insecticide, introduction-establishment, and seasonal colonization). The criteria for environmental manipulation can be placed into three categories: criteria that are difficult to circumvent, characteristics that can be improved with research, and additional factors to consider. This categorization has an operational advantage in that the possible use of an entomopathogen against a particular pest can be evaluated quickly by the "criteria difficult to circumvent." Negative responses in this category will weigh heavily against the development of environmental manipulation. Criteria difficult to circumvent are those that are inherent to the ecosystem or entomopathogen (and, therefore, difficult to change) and yet are important to the potential success of environmental manipulation in suppressing pest damage. For example, environmental manipulation is less likely to succeed against a direct pest than an indirect pest but an insect's function as a direct pest is not likely to be amenable to change by human intervention. Certain of these criteria and factors have been adapted from Burges (1981), Jutsum (1988), Barbosa and Segarra-Carmona (1993), and Fuxa (1995).

There are several characteristics of the pest and ecosystem that can be difficult to circumvent. Environmental manipulation is more likely to succeed against indirect pests than direct pests because most entomopathogens act relatively slowly and will not prevent direct damage, i.e., that to a part of a crop plant that is used directly by humans. Pests that chew open vegetative areas or that live in soil are more likely to become infected by many entomopathogens than insects with sucking mouthparts, insects that bore into plant structures, or aquatic insects; although fungi and nematodes provide exceptions to certain of these generalizations. The chance for success is better against single pests rather than pest complexes due to host specificity of most entomopathogens and against pests with moderate to high economic injury levels due again to the slow action of entomopathogens.

Numerous pest generations provide an opportunity for entomopathogens to increase in numbers, an essential component of this approach, which usually requires natural production of the entomopathogen in the environment. The environment must be generally favorable for the entomopathogen even though one or more environmental conditions are manipulated. The biological characteristics of the resource or crop should favor the manipulation. For example, a low growing plant increases the probability that certain cultural methods might transport a virus from soil onto vegetation. Environmental manipulation is advantageous in crops of low value due to its low cost. However, this method often requires activity in addition to the usual ones in crop production or resource management and thus it must be able to compete economically with alternative control measures.

The entomopathogen must have certain characteristics to succeed in the environmental manipulation approach. It must be able to replicate extensively in epizootics, whether enhanced or otherwise, and it must do so reliably when the enhancement manipulation is performed. Persistence at the point of contact with the insect often is a target of the manipulation but the entomopathogen also must be able to persist for relatively long time periods: up to one year or more in its reservoir, usually soil. A broad host range can be beneficial to pathogen reproduction or in pest complexes. Environmental manipulation requires a virulent pathogen, preferably one that kills the pest relatively quickly, so that the user sees some result of the manipulation. For the same reason this approach requires that the pest population be suppressed below economic injury levels. The method must be cost-effective and have one or more advantages (e.g., cost or environmental safety) over competing control methods. It also requires perhaps more than the other approaches to microbial control, i.e., more than compatibility with resource management or agricultural practices. It requires environmental safety, though it is difficult to imagine a more environmentally sound means of pest management.

In many cases, the possibility of environmental manipulation in a particular entomopathogen/pest system can be improved with research. This approach requires an extensive knowledge base about the pest, entomopathogen, ecosystem, and management practices already in place. An existing infrastructure can be essential, including an extension network, appropriate equipment already in the hands of the users (no capital outlay), and researchers already studying various aspects of the system being manipulated, among others. Eventually, it may be possible that pest population quality, primarily in terms of pest resistance to the entomopathogen, could be a concern if environmental manipulation became used extensively against a particular pest.

Finally, other factors are worth considering in the selection of a system for research and implementation of environmental manipulation. Pest population age structure can be a concern because the great majority of entomopathogens are more efficacious against younger insects. Pests of an enclosed resource, such as plants grown in a greenhouse, are potential targets of this method because of the possibility of controlling moisture levels. Entomopathogens such as certain nematodes and fungi that can actively invade their insect hosts through the cuticle or body openings and perhaps even search for hosts over short distances might provide more opportunities than pathogens that are passively ingested. This is, in part, because manipulations of transport to increase disease prevalence can result in imprecise placement of the pathogen.

Environmental manipulation has certain advantages over the approaches to microbial control that require environmental release of pathogen units because certain requirements for those approaches have caused major problems in implementation. The requirements for other approaches that are of little or no concern in environmental manipulation include market size (host specificity), cost of pathogen production, patentability, registration, persistence in storage, formulation, screening for efficacious species or strains, and habitat stability.

V. CONCLUSION

Environmental manipulation for microbial control of insects deserves more research attention than it has received in the past. Many resource (crop)/pest systems include one or more entomopathogens that occur naturally and occasionally cause epizootics. An assist at a critical point in their life cycle by a method that is environmentally sound and often inexpensive has the potential in many cases to result in season-long suppression of the pest population and damage. Many methods to enhance entomopathogens, such as host plant resistance to increase pest susceptibility, have not even begun to be explored in field research. To become a significant method of pest control environmental manipulation primarily requires research funding, administrative encouragement of research that has only a moderate success rate, and an efficient extension network. The most important area for research will be the ecology or epizootiology of target entomopathogen/pest systems with the objectives of identifying systems in which epizootics are possible and crucial points in the entomopathogen's life cycle where it might be assisted to enhance epizootics.

Acknowledgments
        This paper was approved for publication by the Director of the Louisiana Agricultural Experiment Station as manuscript number 96-17-0360.

REFERENCES

Andreadis, T. G. (1987). Transmission. *In* "Epizootiology of Insect Diseases." (J. R. Fuxa, and Y. Tanada, eds.), pp. 159-176. John Wiley Sons. New York, NY.

Barbosa, P., and Segarra-Carmona, A. (1993). Criteria for the selection of pest arthropod species as candidates for biological control. *In* "Steps in Classical Arthropod Biological Control." (R. G. Van Driesche, and T. S. Bellows, eds.), pp. 5-23. Thomas Say Publ. Entomol., Entomol. Soc. Amer. Lanham, MD.

Benz, G. (1987). Environment. *In* "Epizootiology of Insect Diseases." (J. R. Fuxa, and Y. Tanada, eds.), pp. 177-214. John Wiley and Sons. New York, NY.

Bing, L. A., and Lewis, L. C. (1993). Occurrence of the entomopathogen *Beauveria bassiana* (Balsamo) Vuillemin in different tillage regimes and in *Zea mays* L. and virulence towards *Ostrinia nubilalis* (Hübner). *Agric., Ecosys., Environ.* 45, 147-156.

bin Husin, A. R. (1978). "Effects of Foliar Fungicides on the Entomopathogenic Fungi, *Nomuraea rileyi* (Farlow) Samson and *Entomophthora gammae* Weiser, and the Abundance of Defoliating Caterpillar Populations on Soybean." M.S. Thesis, Department of Entomology, Louisiana State University, Baton Rouge, LA.

Brown, G. C. (1987). Modeling. *In* "Epizootiology of Insect Diseases." (J. R. Fuxa, and Y. Tanada, eds.), pp. 43-68. John Wiley and Sons. New York, NY.

Brown, G. C., and Nordin, G. L. (1986). Evaluation of an early harvest approach for induction of *Erynia* epizootics in alfalfa weevil populations. *J. Kans. Entomol. Soc.* 59, 446-453.

Brust, G. E. (1991). Augmentation of an endemic entomogenous nematode by agroecosystem manipulation for the control of a soil pest. *Agric., Ecosys., Environ.* 36, 175-184.

Burand, J. P., Kawanishi, C. Y., and Huang, Y.-S. (1986). Persistent baculovirus infections. *In* "The Biology of Baculoviruses. Volume 1, Biological Properties and Molecular Biology." (R. R. Granados, and B. A. Federici, eds.), pp. 159-175. CRC Press. Boca Raton, FL.

Burges, H. D. (1981). Strategy for the microbial control of pests in 1980 and beyond. *In* "Microbial Control of Pests and Plant Diseases 1970-1980." (H. D. Burges, ed.), pp. 797-836. Academic Press. London, U.K.

Clark, R. A., Casagrande, R. A., and Wallace, D. B. (1982). Influence of pesticides *on Beauveria bassiana,* a pathogen of the Colorado potato beetle. *Environ. Entomol.* 11, 67-70.

Crawford, A. M., and Kalmakoff, J. (1977). A host-virus interaction in a pasture habitat. *J. Invert. Pathol.* 29, 81-87.

Dedryver, C. A. (1979). Déclenchement en serre d'une épizootie a *Entomophthora fresenii* sur *Aphis fabae* par introduction d'inoculum et régulation de l'humidité relative. *Entomophaga* 24, 443-453.

Downing, A. S. (1994). Effect of irrigation and spray volume on efficacy of entomopathogenic nematodes (Rhabditida: Heterorhabditidae) against white grubs (Coleoptera: Scarabeidae). *J. Econ. Entomol.* 87, 643-646.

Fuxa, J. R. (1991). Release and transport of entomopathogenic microorganisms. *In* "Risk Assessment in Genetic Engineering." (M. Levin, and H. Strauss, eds.), pp. 83-113. McGraw-Hill. New York, NY.

Fuxa, J. R. (1995). Ecological factors critical to the exploitation of entomopathogens in pest control. *In* "Biorational Pest Control Agents. Formulation and Delivery." (F. R. Hall, and J. W. Barry, eds.), pp. 42-67. Amer. Chem. Soc. Washington, DC.

Fuxa, J. R. (1997). Microbial control of insects: status and prospects for IPM. *In* "IPM System in Agriculture." Vol. 2. Biocontrol in Emerging Biotechnology. (R. K. Upadhyay, K. G. Mukerji, and R. L. Rajak, eds.), pp. 57-104. Aditya Books (P) Ltd., New Delhi, India.

Fuxa, J. R., and Tanada, Y. (1987). Epidemiological concepts applied to insect epizootiology. *In* "Epizootiology of Insect Diseases." (J. R. Fuxa, and Y. Tanada, eds.), pp. 3-21. John Wiley and Sons. New York, NY.

Hall, I. M., and Dunn, P. H. (1957). Fungi on spotted alfalfa aphid. Calif. Agric. 11, 5-14.

Horton, D. L., Carner, G. R., and Turnipseed, S. G. (1980). Pesticide inhibition of the entomogenous fungus *Nomuraea rileyi* in soybeans. *Environ. Entomol.* 9, 304-308.

Huang, Y., Zhen, B., and Li, Z. (1992). Natural and induced epizootics of *Erynia ithacensis* in mushroom hothouse populations of yellow-legged fungus gnats. *J. Invert. Pathol.* 60, 254-258.

Ishibashi, N., and Takii, S. (1993). Effects of insecticides on movement, nictation, and infectivity of *Steinernema carpocapsae*. *J. Nematol.* 25, 204-213.

Johnson, D. W., Kish, L. P., and Allen, G. E. (1976). Field evaluation of selected pesticides on the natural development of the entomopathogen, *Nomuraea rileyi,* on the velvetbean caterpillar in soybean. *Environ. Entomol.* 5, 964-966.

Jutsum, A. R. (1988). Commercial application of biological control: status and prospects. *Phil. Trans. Roy. Soc. Lond. Series B* 318, 357-373.

Kalmakoff, J., and Crawford, A. M. (1976). Virus control of porina. *New Zealand J. Agr.* August, 41-42.

Klassen, W. (1975). "Impressions of Applied Insect Pathology in the U.S.S.R." U.S. Dept. of Agriculture, A. R. S. publication, Hyattsville, MD.

Livingston, J. M., Yearian, W. C., Young, S. Y., and Stacey, A. L. (1981). Effect of benomyl on an *Entomophthora* epizootic in a *Pseudoplusia includens* population. *J. Georgia Entomol. Soc.* 16, 511-514.

McCoy, C. W., Samson, R. A., and Boucias, D. G. (1988). Entomogenous fungi. *In* "CRC Handbook of Natural Pesticides. Volume V, Microbial Insecticides. Part A, Entomogenous Protozoa and Fungi." (C. M. Ignoffo, ed.), pp. 151-236. CRC Press. Boca Raton, FL.

Moscardi, F. (1989). Use of viruses for pest control in Brazil: the case of the nuclear polyhedrosis virus of the soybean caterpillar, *Anticarsia gemmatalis. Mem. Inst. Oswaldo Cruz,* 84 (suppl. III), 51-56.

Pickering, J., Dutcher, J. D., and Ekbom, B. S. (1989). An epizootic caused by *Erynia neoaphidis* and *E. radicans* (Zygomycetes, Entomophthoraceae) on *Acyrthosiphon pisum* (Hom., Aphididae) on legumes under overhead irrigation. *J. Appl. Entomol.* 107, 331-333.

Podgwaite, J. D. (1985). Strategies for field use of baculoviruses. *In* "Viral Insecticides for Biological Control." (K. Maramorosch, and K. E. Sherman, eds.), pp. 775-797. Academic Press, Orlando, FL.

Shetlar, D. J., Suleman, P. E., and Georgis, R. (1988). Irrigation and use of entomogenous nematodes, *Neoaplectana* spp. and *Heterorhabditis heliothidis* (Rhabditida: Steinernematidae), for control of Japanese beetle (Coleoptera: Scarabaeidae) grubs in turfgrass. *J. Econ. Entomol.* 81, 1318-1322.

Sosa-Gomez, D. R., and Moscardi, F. (1994). Effect of till and no-till soybean cultivation on dynamics of entomopathogenic fungi in the soil. *Fla. Entomol.* 77, 284-287.

Sprenkel, R. K., Brooks, W. M., Van Duyn, J. W., and Deitz, L. L. (1979). The effects of three cultural variables on the incidence of *Nomuraea rileyi,* phytophagous Lepidoptera, and their predators on soybeans. *Environ. Entomol.* 8, 334-339.

Steinhaus, E. A. (1975). "Disease in a Minor Chord." Ohio State University Press, Columbus, OH.

Tanada, Y., and Fuxa, J. R. (1987). The pathogen population. *In* "Epizootiology of Insect Diseases." (J. R. Fuxa, and Y. Tanada, eds.), pp. 113-157. John Wiley and Sons. New York, NY.

Walker, T. W., and Meek, C. L. (1987). Long term effects of riceland agrichemicals on postparasites and adults of *Romanomermis culicivorax* (Nematoda: Mermithidae). *J. Entomol. Sci.* 22, 302-306.

Walters, P. J., and Bishop, A. L. (1978). Effects of overhead sprinkler irrigation on numbers of and disease incidence in BGA populations. *In* "Lucerne Aphid Workshop." pp. 118-121. Agr. Res. Cent., Tamworth Dept. Agr. New South Wales, Australia.

Wilding, N. (1982). The effect of fungicides on field populations of *Aphis fabae* and on the infection of the aphids by Entomophthoraceae. *Ann. Appl. Biol.* 100, 221-228.

Young, S. Y. (1990). Influence of sprinkler irrigation on dispersal of nuclear polyhedrosis virus from host cadavers on soybean. *Environ. Entomol.* 19, 717-720.

Young, S. Y., and Yearian, W. C. (1986). Movement of a nuclear polyhedrosis virus from soil to soybean and transmission in *Anticarsia gemmatalis* (Hübner) (Lepidoptera: Noctuidae) populations on soybean. *Environ. Entomol.* 15. 573-580.

Zelazny, B., Lolong, A., and Pattang, B. (1992). *Oryctes rhinoceros* (Coleoptera: Scarabeidae) populations suppressed by a baculovirus. *J. Invert. Pathol.* 59, 61-68.

CHAPTER
15

DEPLOYMENT OF THE PREDACEOUS ANTS AND THEIR
CONSERVATION IN AGROECOSYSTEMS

Ivette Perfecto and Antonio Castiñeiras

I. INTRODUCTION

The conservation and use of ants for biological control has been underestimated by pest managers for decades. The negative reputation of ants is related to a lack of understanding of the ecological role of ants in agroecosystems and forests. Ironically, the use of ants for biological control of insect pests was the first reported case of conservation biological control in the literature. Today, there are several examples of the use and conservation of ants as biological control agents. In this chapter, we examine some of these examples and draw generalizations that may help guide future biological control programs with ants.

II. EXAMPLES OF THE DEPLOYMENT OF ANTS FOR BIOLOGICAL CONTROL

A. Ants for the Control of Cocoa *(Theobroma cacao* L.) Pests

1. Malaysia and Indonesia

In the early 1900s, cocoa farmers in Indonesia began introducing the ant *Dolichoderus thoracicus* Smith after noticing less damage due to a variety of pests in areas where the ant was abundant. Although this practice was discontinued as pesticides became widely available, in the 1980s farmers in Malaysia reinitiated the practice (Khoo and Chung, 1989). Today a substantial amount of ecological research on the ant community and its interactions with other organisms in the cocoa agroecosystem has formed a more solid basis for the use of ants as biological control of cocoa pests in Malaysia and Indonesia (Khoo and Chung, 1989; Way and Khoo, 1991, 1992; Hierbaut and Van Damme, 1992; Khoo and Ho, 1992; See and Khoo, 1996). One interesting aspect of the use of *D. thoracicus* in cocoa is that this species has overcome the negative reputation associated with ants that tend homopterans. Its deterrent activity on the mirids *Helopeltis antonii* Signoret, *H. theivora* Waterhouse, and *H. theobromae* Mill. as well as the lepidopteran *Conopomorpha cramerella* (Snellen), and even rats, seems to have outweighed its association with a number of mealybug species (Khoo and Chung, 1989; Way and Khoo, 1992; See and Khoo, 1996). Furthermore, some of the management strategies for this species consist of introducing the mealybugs along with the ants to areas from where the ant is absent, introducing the mealybugs in areas where the ant is present but not abundant and leaving the proximal ends of the harvested pods on the trees to conserve mealybugs (Khoo and Chung, 1989).

Another interesting aspect of the use of *D. thoracicus* in cocoa is the management of the ant community to enhance the biological control by this species. In Malaysian cocoa plantations the three-dimensional ant mosaic *(sensu* Leston, 1973) consists of three dominant species and a large number of subordinate species. *Dolichoderus thoracicus* is one of the dominants and *Oecophylla smaragdina* (F.) and *Anoplolepis longipes* (Jerdon) are the other two. Although *O. smaragdina* and *A. longipes* have potential as biological control agents elsewhere they are not as effective as *D. thoracicus* in controlling populations of the main cocoa pests (Way and Khoo, 1992). To increase the abundance of *D. thoracicus* in areas where the other two species are abundant *O. smaragdina* and *A. longipes* populations have to be reduced. The intercrop of coconut palms with cocoa could prove to be an excellent management tool for giving *D. thoracicus a* competitive advantage over the other dominant species. *Dolichoderus thoracicus* commonly uses the crowns and curled leaflets of fallen fronds of coconut palms as nesting sites. The rapid colonization of artificial nests in cocoa plantations by *D. thoracicus* suggests that nesting sites are a limiting resource (Way and Khoo, 1992) and that the species could benefit from (i.e., could be conserved with) additional sites provided by the palm leaves.

2. Brazil

The ant community of plantations in the Bahia region of Brazil have been intensively studied over the past ten years and some species have been identified as potential control agents of major cocoa pests. However, the augmentation and conservation of ants for biological control in this region have not yet reached the same level of sophistication as in Southeast Asia. An ant mosaic has been identified with between three (Madeiros *et al.*, 1995) to seven (Majer *et al.*, 1994) dominant species. Of these, two species have been tentatively identified as potential biological control agents against cocoa pests and worthy of conservation and perhaps augmentation: the Ponerinae *Ectatomma tuberculatum* (Olivier) and the Dolichoderinae *Azteca chartifex* Forel (Majer and Delabie, 1993).

Ectatomma tuberculatum is a common ant in cocoa plantations in Bahia as well as many forested habitats in the neotropics (Majer and Delabie, 1993). In some farms in Brazil 90% of the cocoa trees have *E. tuberculatum* nests at their base (Delabie, 1990). This species dominates large continuous traces of canopy in cocoa plantations (Majer *et al.*, 1994) and shows sufficient long-term permanence in individual cocoa trees (Madeiros *et al.*, 1995). This species is also known by the indigenous people of Guatemala as a good predator of cotton pests (Gotwald, 1986). Although it has been reported to be associated with homopterans, including *Planococcus citri* (Risso), a pest in cocoa, the associations do not appear to be strong (Madeiros *et al.*, 1995) and therefore should not be a cause of major concern for the conservation of this ant in the cocoa agroecosystem.

The second species being considered for biological control of pests is *A. chartifex*. Traditionally, some Bahia cocoa producers promoted this species by distributing nest fragments into their plantations after noticing that pods in trees with ants looked better than those in trees where the ants were absent. The Kayapo Indians used members of the *A. chartifex* group to limit leaf-cutter ants in the State of Pará (Overall and Posey, 1984). As with many other ant species the main concern regarding the conservation of this species for biological control is its association with homopterans, including *P. citri* (Majer and Delabie, 1993; Madeiros *et al.*, 1995). However, as more information is gathered and our understanding of the ecology of the system improves, it seems that the balance is turning in favor of the ant. As with the case of *D. thoracicus* in cocoa plantations in Malaysia and Indonesia, it seems that the positive impact of the ant outweighs its potential negative impacts due to association with homopterans.

Over the last few decades the tendency in cocoa plantations has been to decrease vegetational diversity and increase pesticide spraying. In addition, the surrounding natural ecosystems (mainly forests) have been disappearing. All of these changes, most likely, will have an effect on the structure of the ant community and therefore on the biological control of insect pests within the cocoa agroecosystem. In Section V we will revisit this example and place it within the context of conservation biology.

B. *Oecophylla* Species for the Control of Tree Crop Pests in Asia and Africa

Oecophylla longinoda Latr. and *0. smaragdina* have been widely used as biological control agents of pests on tree crops in Asia and Africa. Way and Khoo (1992) reviewed the literature on *Oecophylla* species and found them to be beneficial predators in coconuts, oil palm, cocoa, coffee, citrus, mango, *Eucalyptus,* and other timber trees. *Oecophylla smaragdina* has been used against pests in citrus plantations in China for over 1600 years (Yang, 1982). A Chinese publication dating from 304 AD narrates how colonies of these ants were sold in the market to be used for the control of insect pests in citrus (Needham, 1986; Huan and Yang, 1987). Still today, farmers introduce colonies of ants into their plantations and place bamboo sticks between branches of adjacent trees to spread ant patrolling to trees that lack colonies (Yang, 1982).

The use of *0. longinoda* in coconut plantations represents an excellent example of conservation biological control. This ant species occurs in Africa and is the main natural enemy of the coconut bug *Pseudotheraptus wayi* Brown, which has been reported to cause crop loss of up to 67% in Tanzania (Vanderplank, 1959, 1960). Threshold spraying for *P. wayi* has reduced pesticide application by more than half, thus reducing the potential harmful effects of pesticides on *0. longinoda* (Lohr and Oswald, (1990). This highly aggressive species forms nests and forages in 76 different species of trees and bushes (Varela, 1992). Other ant species compete with *0. longinoda* for nesting sites and food, in some cases displacing it entirely (Oswald, 1991; Zerhusen and Rashid, 1992; Rapp and Salum, 1995). *Pheidole megacephala* (F.) is considered to be the strongest and most widely distributed competitor of *0. longinoda.* In coconut plantations this species nests in the ground among the roots, at the base of the palm trunk. Both of these species prey on a variety of insects and tend homopterans. However, *P. megacephala* has a wider food range: collecting nectar and pollen as well as small seeds from the ground. They both also prey on *P. wayi* (including eggs). However trees occupied by *0. longinoda* have much better control than those occupied by *P. megacephala* (Rapp and Salum, 1995).

In order to improve biological control of the coconut bug, the use of selective baits has been incorporated into integrated control programs. In coconut plantations in Zanzibar, Tanzania, selective and long-lasting control of *P. megacephala* has been achieved with the use of AMDRO, a fire ant bait (Oswald, 1991; Zerhusen and Rashid, 1992). Treatment results in rapid colonization of the palms by *0. longinoda* and a significant reduction of the coconut bug (Oswald, 1991). Studies have reported that between 50 and 87% of the treated palms are colonized by *0. longinoda* once *P. megacephala* disappears (Zerhusen and Rashid, 1992). Another promising and more environmentally benign method for reducing competition from *P. megacephala* is the manipulation of the undergrowth vegetation.

The way in which vegetation affects the distribution of ants in coconut plantations has been studied since the 1950s (O'Connor, 1950; Way, 1953; Brown, 1959; Stapley, 1971; Rapp and Salum, 1995). Although there are a variety of opinions regarding the effect of undergrowth, particularly weeds, on the distribution of *O. longinoda* and *P. megacephala,* most evidence indicates that weeds in the immediate area of the palms tend to reduce competition between these two species (O'Connor, 1950; Way, 1953; Brown, 1959; Stapley, 1971; Rapp and Salum, 1995). Rapp and Salum (1995) reported that after the cessation of weed control the number of palms colonized by *O. longinoda* increased from 696 trees to 1776 in one year. They argued that *P. megacephala* stopped attacking *O. longinoda* and started foraging on weeds and bushes on the ground. The management of the undergrowth vegetation for altering competition interaction among ant species has already been incorporated into IPM practices in the Ivory Coast (Fateye and De Taffin, 1989). This example illustrates the complicated interactions that need to be taken into consideration in conservation biological control.

C. Conservation and Deployment of *Pheidole megacephala* in Cuba

Pheidole megacephala occurs throughout the tropics and subtropics (Crowell, 1968; Ogata, (1982). This polyphagous ant (Greenslade, 1972) has been reported as a predator of more than 20 species of arthropods (Castiñeiras, 1985a). In coffee (Da Fonseca *et al.*, 1971 ; Krantz *et al.,* 1978) and pineapple (Krantz *et al.,* 1978; Mc Ewen *et al.,* 1979; Reimer *et al.,* 1991) plantations it is considered a pest because it protects mealybugs that transmit viral diseases. Nevertheless, its pest status on pineapple has changed lately because it has been demonstrated that P. *megacephala* does not move *Dysmicoccus neobrevipes* Beardsley among pineapple plants and it does not decrease mealybug mortality by consuming honeydew (Jahn and Beardsley, 1996). In citrus groves, *P. megacephala* restricts the action of parasitoids and predators of scale insects (Steyn, 1955; Castiñeiras and Fernández, 1983; Castiñeiras and Paez, 1989) but as it is nocturnal it has less effect on other natural enemies than the diurnal ant species (Steyn, 1954). Also, as discussed above, in coconut plantations it is considered a pest because of its competitive interaction with *O. longinoda,* an efficient predator of coconut pests.

Pheidole megacephala is extensively used in Cuba for biological control of the sweet potato weevil *Cylas formicarius elegantulus* (Summ.) (Castiñeiras *et al.,* 1991a) and the banana weevil *Cosmopolites sordidus* (Germ.) (Castiñeiras *et al.*, 1991b). The ants prevent weevils from laying eggs on the subterranean parts of the sweet potato and banana plants when they nest around the roots (Castiñeiras, 1989). Due to its effectiveness as a biological control agent this species is protected by vegetable growers in Cuba. Areas with a high density of colonies are fenced and designated as "reservoirs of *P. megacephala*" and the colonies are fed every two weeks with molasses and table leftovers. Pesticide applications are prohibited in those areas. The reservoirs are then used for the collection of colonies that will be transported to the sweet potato or banana fields. For their use in biological control programs colonies of *P. megacephala* are taken from the soil by placing traps on top of the nests during the rainy season (May-November). Traps consist of packages of 10 to 20 banana leaves tied together. One trap may collect a colony of about 22 queens and 62,000 workers and immatures (Castiñeiras, 1985b). Commercial control of *C. f. elegantulus* or *C. sordidus* is achieved with at least nine colonies per hectare. The effect of *P. megacephala* on *C. sordidus* is more evident in the second harvest because weevil populations and cumulative damage increase in banana plantations after the first harvest. A recommendation for conservation biological control is that sweet potato and banana fields should not be sprayed with chemical insecticides after ant colonies are introduced (Castiñeiras *et al.*, 1985; Castiñeiras *et al.*, 1990). However, bioinsecticides such as *Bacillus thuringiensis kurstaki* (Berl.) (Castiñeiras and Calderón, 1982), *Beauveria bassiana,* or *Metarhizium anisopliae* are compatible with the use of *P. megacephala* (Castiñeiras *et al.*, 1990).

The conventional spray program for the sweet potato weevil in Cuba consisted of 6 to 12 chemical pesticide treatments that achieved 85% control of the pest. The overall IPM program for *C. f. elegantulus* now consists of (1) the selection of land cleared of sweet potato tubers infested with weevils, (2) the use of certified seeds, (3) pesticide sprays during the first four weeks after planting, if necessary, and (4) the introduction of colonies of *P. megacephala* six weeks after planting. If chewing damage caused by Lepidoptera *(Herse cingulata* (F.) and *Spodoptera* spp.) is observed after the introduction of the ants then *B. thuringiensis* may be sprayed. Control of the sweet potato weevil with the IPM program can be increased to 90% at a lower cost than the conventional spray program. A similar IPM program is used in banana plantations (Castiñeiras *et al.*, 1991b). The case of *P. megacephala* in Cuba is unique in conservation biological control in that a formal conservation biological control program has been put in place solely for the protection and augmentation of this species.

III. THE ROLE OF ANTS AS NATURAL BIOLOGICAL CONTROL AGENTS

In addition to the above-mentioned programs of biological control where particular species of ants are introduced, conserved, or augmented to control specific pests there are many reports on the action of ants in maintaining potential pest populations in check without any human intervention to ensure such control.

A. "Milpas" in Central America

For peasant farmers in Mexico, Central America, and some parts of South America who cannot afford pesticides and do not have access to technical advice for implementing sophisticated biological control programs the natural ant community living in their fields can represent insurance against pest outbreaks. In the "milpa" system (maize in combination with other crops such as beans and squash), which forms the basic production system of small farmers in these regions, ants have been reported to maintain a variety of pest populations under control. The ant community in the milpa system is dominated by fire ants *(Solenopsis geminat* (F.)), several *Pheidole* species, and in some areas *Ectatomma* species (Risch, 1980; Risch and Carroll, 1982a,b; Perfecto, 1990, 1991a; Perfecto and Sediles, 1992). Studies conducted in Nicaragua and Mexico indicated that the combination of *S. geminata* (the tropical fire ant), *Pheidole radoskowzii* Mayr and *Ectatomma ruidum* Roger is effective in controlling populations of *Spodoptera frugiperda* (J. F. Smith), *Dalbulus maidis* (De Long & Wolcott) as well as other maize pests under both rain-fed and irrigated conditions (Risch, 1980; Risch and Carroll, 1982a,b; Perfecto, 1990, 1991a; Perfecto and Sediles, 1992). *Ectatomma ruidum* is particularly effective against pupae of *S. frugiperda,* removing up to 97% of artificially placed pupae from the ground of unplowed, unsprayed fields in Nicaragua (Perfecto, 1990). In Mexico, Risch (1980) reported up to 80% predation of rootworm eggs in the soil by ants, particularly *S. geminata.* The control of insect pests by ants in the milpa system was clearly demonstrated through the experimental application of insecticide to reduce ant foraging activity (Perfecto, 1990; Perfecto and Sediles, 1992). Plots where ants had been reduced had higher levels of insect pests. In a similar experiment, Risch and Carroll (1982a) reported 98% reduction of weevils in plots with colonies of *S. geminata* as compared to plots from which *S. geminata* had been eliminated with selective poisonous baits.

B.   Ants as Natural Biological Controls of Cotton Pests

Cotton consumes the largest quantities of pesticides worldwide. Most of the pesticides goes to control insect pests such as the boll weevil *Anthonomus grandis* Boheman and the cotton bollworm *Helicoverpa zea* (Boddie). The first documented cases of the pesticide treadmill or the vicious cycle by which more and more pesticide must be applied to a crop, occurred in cotton (Bottrell and Adkisson, 1977). One of the reasons for this appears to be that under conditions of no or few pesticide applications cotton fields harbor a rich fauna of natural enemies, including ants. Furthermore, the extrafloral nectaries of the cotton plant have been demonstrated to attract a rich ant fauna which maintains constant patrol on plants (Koptur, 1992). In Brazil, ants have been reported to remove 20% of adult boll weevil (Fernández *et al.*, 1994). *Pheidole* species have also been reported to kill up to 95% of the cotton leafworm (*Alabama argillacea* (Hübner)) in the soil. This insect is a typical secondary pest which results from the application of insecticides (Vellani *et al.*, 1984). In North America *Solenopsis* species were shown to have the greatest impact against *A. argilicea*. The red imported fire ant *Solenopsis invicta* Buren consumed up to 85% of the weevil larvae in a Texas cotton field (Sterling, 1978). In addition, Agnew and Sterling (1981) provided evidence that this species can also prey on a large number of pupae and adult weevils.

IV.   DRAWING GENERALIZATIONS FROM THE EXAMPLES

A.   Perennial versus Annual Cropping Systems

For a long time it was thought that ants could not be used in annual cropping systems because of the disturbance generated by plowing. For this reason, most of the research on ants as biological control agents was undertaken on perennial crops or forests. The general idea is that after the disturbance generated by plowing every cropping season, ant nests are destroyed and subsequently many colonies either die or move to field edges. It was assumed that a long period of time was necessary for the reestablishment of ant colonies in the cultivated area. Furthermore, the lack of a permanent canopy in annual cropping systems does not allow the establishment of colonies of arboreal species.

The examples from the milpa systems in Mexico and Central America as well as sweet potato in Cuba clearly demonstrate that ants can play an important role as biological control agents in annual cropping systems as well. The problem of destruction of nests can be overcome in three main ways. First, as in the case of cotton and maize, the ant community that gets established in these systems is adapted to high levels of disturbance (Perfecto, 1991b). The species that dominate in these communities tend to have very high colonization rates, fast rate of colony growth, and high mobility. These opportunistic species, the classic example of which is *S. geminata,* form a "pioneering community" which is able to recolonize the fields and respond rapidly to population levels of other insects in the system. The second way to overcome the impact of habitat disruption is by reintroducing colonies into the fields. This augmentation strategy is used by farmers in Cuba and it appears that even with the additional labor cost implied by the reintroduction the use of *P. megacephala* is economically feasible (Castiñeiras *et al.,* 1991b). Finally, the best way to avoid the negative impacts of tillage is by not plowing the land. As conservation tillage becomes more common reports on biological control by ants in annual cropping systems are bound to become more common as well. The impact of conservation tillage on the ant community will be discussed in Section IV,E,2,a.

B.   Ant-Homopteran Mutualism

The mutualistic association between ants and homopterans is pervasive in many agricultural systems (Way, 1963). Although at first glance this association appears to be a detrimental one for agricultural production it does not need to be so. As can be concluded from the examples of *D. thoracicus, A. chartifex,* and *0. longinoda* the association of ants with homopterans can have a positive effect on crop production. In some cases homopterans may be essential for successful biological control with ants (Khoo and Chung, 1989). Homopterans provide a stable source of food for the ants allowing them to reach higher population levels which could result in a better control of other more noxious herbivores. A net benefit of the ant-homopteran mutualism can be obtained under two situations. First is when the Homoptera that are protected by the ant are not significant pests of the crop, as in the case of the mealybugs and *D. thoracicus* in cocoa (Khoo and Chung, 1989). A second circumstance is when the Homoptera are indeed pests but a more noxious pest is effectively controlled by the ants when the homopterans are present, as in the case of *P. citri* and *E. tuberculatum,* also in cocoa. The ant-homopteran association can be detrimental to the crop when these conditions are not met. The mutualism can be particularly detrimental when the homopteran is a disease vector. In these cases a very small population of homopterans causes economic damage and the benefits resulting from ant predation on other herbivores may not be enough to offset the damage. Overall, these examples point at the complex interactions between ants, homopterans, other herbivores, other predators, and the crop. When using ants as biological control agents, these interactions have to be evaluated, at least qualitatively, in order to develop effective and sustainable pest management programs.

C. Good versus Bad Ants

With the possible exception of leafcutter ants which defoliate plants in order to feed their mutualist fungi there is not an ant species that can be considered intrinsically harmful for agriculture. Frequently, species that are considered pests in a particular agroecosystem or under certain circumstances can be effective biological control agents under others. Ants can be detrimental because of their associations with homopterans because they displace other more effective predators or even because they are seed predators and remove or damage seeds of crops. But these very same species of ants can also be beneficial. The most striking example of this "yin/yang" characteristic is found in *P. megacephala.* As part of the research for this chapter, we conducted a literature search for the past ten years for *P. megacephala.* The result was that 66% of the 42 papers refer to this species as a pest. In some cases this was due to the fact that *P. megacephala* is also considered an invasive ant which displaces endemic species (Gallespie and Reimer, 1993; Reichel and Andersen, 1996) but in many cases it was because it tended homopterans which caused economic damage (De Barro, 1990; Reimer and Bearsdley, 1990). It other cases it displaced other predators (Cudjoe *et al.,* 1993; Reimer *et al.,* 1993) including other ants which were effective biological control agents (Rapp and Salum, 1995). However, this is the same species which has proved to be so successful as a biological control agent in Cuba (Castiñeiras *et al.,* 1991a). Another interesting and well known example is that of *S. invicta,* the red imported fire ant. Although the Department of Agriculture of the U.S.A. has invested several hundred million dollars since the late 1950s in an eradication program for this species (Logfren, 1986) it is reported to be the most effective predator of major cotton pests (Sterling, 1978). Finally, even seed predators can be considered beneficial under certain circumstances and detrimental under others. In Nicaragua, *S. geminata* was reported to remove almost 100% of all the tomato seed in the Sébaco Valley (A. Sediles, Universidad Agricola de Nicaragua, pers. commun.). Nevertheless, Carroll and Risch (1984) have suggested that this same species may be effective in controlling certain weeds (through seed predation) in Mexican milpas.

D. The Ant Community

When considering the deployment and conservation of ants for biological control it is important to realize that we will be dealing with the ant community and not with individual species. Interactions among ant species have been intensively studied for the past 30 years (Hölldobler and Wilson, 1990). The role of competition in structuring ant communities is well established now (Greenslade, 1971; Carroll and Jansen, 1973; Levins *et al.,* 1973; Vepsäläinen and Pisarski, 1982; Perfecto, 1994). Both interference and exploitation competition have been demonstrated to play a major role in structuring ant communities. In the words of E. 0. Wilson, "the worst enemy of an ant is another ant." The strength of these interactions makes the use of ants for biological control a challenging enterprise.

The previous examples illustrate just how complex the situation can be. In the cocoa farms in Brazil there could be up to seven dominant species and numerous other subordinates (Majer *et al.,* 1994). Most of these species are generalist predators and are most likely competing with one another for food and nesting sites. However, the coexistence of all these species in the same agroecosystem suggest that they may be partitioning their niche (Torres, 1984). In other words, some species may be preying on large items while other smaller ones may prey on smaller items, some species may be foraging in the canopy of the trees while others forage in the ground or the base of the trunk, some species may forage during the day while others are nocturnal, etc. This niche stratification seems to be what contributes to a high ant diversity in the coffee agroecosystem (Perfecto and Vandermeer, 1994). Although managing an entire community is certainly more difficult than managing a single species, which is why most programs follow the single species approach, this complexity also provides flexibility to the pest manager. Furthermore, the diversity of species can maintain natural control on a variety of endemic pest species. Finally, the role of subordinate species should not be underestimated. There is evidence that these nondominant species play an important role in pest suppression as well (Way *et al.,* 1989).

E. Habitat Manipulation for the Conservation of Ants

In many cases of biological control with ants it is apparent that the environment plays a major role in the effectiveness of the control. Manipulating the habitat is an important part of conservation biological control.

1. Insecticide reduction

Conservation biological control has historically consisted of the reduction of insecticides and/or the use of nonpersistent synthetic or microbial pesticides with the purpose of decreasing negative impacts on the natural enemies of the pests (see Chapter 11). The impact of insecticides on the ant community and therefore on the effectiveness of the control of pests, has been demonstrated in several agroecosystems (Julia and Mariau, 1978; Basedow, 1993). In Cuba, a key component of the biological control program for the sweet potato weevil with *P. megacephala* is the strict elimination of synthetic insecticides and the use, only when necessary, of microbial formulations. Likewise the integrated pest management program of *P. wayi* in coconut combines the use of insecticides and the action of ants. Insecticides are only applied to palms that are not occupied by *0. longinoda* (Julia and Mariau, 1978).

2. Alteration of cultivation practices

a. Conservation tillage.   The most frequent observation of arthropod predators regarding conservation tillage is a negative correlation between the intensity of tillage and the number of soil- and litter-inhabiting predatory arthropods (Stinner and House, 1990). Although most studies of the influence of tillage on predators are conducted in the temperate zone where carabids, and not ants, are the most abundant soil-dwelling predators, a few studies document the same pattern for ants (Tonhasca, 1993). Pest reduction in no-tillage corn has been reported for Costa Rica, where ants are the most abundant soil-dwelling predators (Shenk and Sounders, 1984). Unfortunately, in these studies no data were taken on ant abundance. Robertson *et al.* (1994) reported higher incidence of ants in no-tillage fields of grains in Australia. Other studies have shown no influence of tillage intensity on either pest or predators (Mack and Beckman, 1990).

     The most obvious impact of plowing on soil inhabiting ants is the destruction of their nests. As discussed in Section IV, A, this destruction causes some mortality but many colonies simply move their nest to the edge of the field (Perfecto, 1991b). In a study on the effect of plowing on the ant community in a field in Nicaragua, Perfecto (1991b) demonstrated that although plowing initially results in a change in the composition of the ant community after a period of several months, if no further plowing is conducted, the community returns to its original composition. Although plowing does not eliminate ants from the fields on a permanent basis the change in the composition of the community can be important for biological control. For example, highly disturbed fields in the neotropics tend to be dominated by opportunistic species like *S. geminata*. Although this species can be an effective predator, if the peak of *S. geminata* in the field does not coincide with the stage of the pest when it needs control the ant would not be effective. What this suggests is that in conventional tillage systems the use of ants for biological control has to be carefully evaluated and the dynamics of the ant community, as its relates to tillage, needs to be well understood.

b. Pruning and shade elimination. Pruning of trees or the complete elimination of shade trees may have an enormous impact on the diversity and density of ants. Although no study has been conducted to examine the effect of pruning on ants, several studies (especially in coffee) have examined the effect on ants of the reduction or elimination of shade trees from plantations (Nestel and Dickschen, 1990; Perfecto and Vandermeer, 1994; Perfecto and Snelling, 1995). Studying the ant community in a gradient of coffee plantations going from plantations with high density of shade to shadeless plantations, Perfecto and Snelling (1995) reported a significant decrease in ant diversity. Although the relationship between ant diversity and pest control is not well understood we can speculate that a diverse ant community can offer more safeguards against pest outbreaks than a community dominated by just a few species. In Colombia, preliminary reports point to lower levels of the coffee borer, the main coffee pest in the region, in shaded coffee plantations. There are some indications that a nondominant small ant species is responsible for the control (J. Monterey, Centro de Manejo Integrado de Plagas, CATIE, Nicaragua, pers. commun.). Apparently, this species does not live in unshaded plantations. Cocoa is another crop that is traditionally cultivated under shade trees. The ant species that have been so successful in controlling pests in cocoa (see Sections II,A,1 and II,A,2) are all species that flourish under shaded conditions.

One of the most obvious consequences of pruning or shade elimination, with regard to the ant community is the change in microclimatic conditions. In particular, microclimate becomes more variable with more extreme levels of humidity and temperature. A recent study documented changes in the composition of the ant community with shade and leaf litter manipulation (Perfecto and Vandermeer, 1996), similar to those that occur after plowing (Perfecto, 1991b). Many other factors can contribute to lower diversity of ants in unshaded plantations. As one may expect, most of the tree-nesting species tend to disappear when trees are eliminated. In cases of severe pruning many nesting sites (in tree branches) also can disappear, along with alternative food sources such as nectar from flowering trees. In addition to the obvious effects of pruning and shade elimination, more indirect ones have been reported. For example, it has been suggested that changes in microclimatic conditions induce changes in the interaction coefficients among ant species. These changes in competitive interaction can result in an alteration of the species composition of the ant community in coffee plantations (Perfecto, 1994). Based on this evidence an important component of conservation biological control with ants in coffee plantations appears to be the reintroduction of shade.

3. Vegetational and structural diversity

The conservation of natural enemies by the direct enhancement of vegetational diversity has been a subject of intense study for almost thirty years (Root, 1973; Andow, 1991; Chapter 9). It was hypothesized that the lower levels of herbivores in diverse agroecosystems were a result of higher levels of natural enemies: the so called "enemies hypothesis" (Root, 1973). Among the reported factors that contribute to higher levels of natural enemies in diversified agroecosystems are the availability of diverse microhabitats, greater availability of food sources (such as prey, nectar, and pollen), alternative hosts, and shelter; all of which encourage colonization and population build-up of natural enemies. The effect of vegetational diversity on ant abundance and predation efficiency is still debatable.

Several studies have reported increases in ant foraging activity associated with lower levels of intensification which include vegetational diversity, among other factors (Altieri and Schmidt, 1984; Perfecto and Snelling, 1995). One of these studies reported a decline in ant predation levels with vegetational simplification (Altieri and Schmidt, 1984). On the other hand, Letourneau (1987) reported lower levels of ants in a maize/cowpea/squash triculture as compared with a squash monoculture, and Nestel and Dickschen (1990) reported higher ant foraging activity in a shaded monoculture plantation compared with that in a polyculture plantation. Yet, other studies have reported no changes in ant levels with increased vegetational diversity (Perfecto and Sediles, 1992; Tonhasca, 1993). Given the lack of agreement of the empirical data on the effect of vegetational diversity on ant predation efficiency we should be using general principles to guide the management of system-specific programs. The examples of the control of *P. wayi* in coconut and mirids in cocoa provide us with good case studies on how vegetational diversity can enhance the efficiency of the ant predators. In the case of coconut, weeds seem to enhance the efficiency of *O. longinoda* by attracting *P. magacephala* away from the coconut palms and therefore allowing a better control by *O. longinoda*. In cocoa, intercropping with coconut may provide an alternative nesting site to *D. thoracicus,* therefore increasing its population levels. In both of these cases, it is not the intrinsic diversity that accomplished an enhanced biological control but more a specific condition that is achieved by diversifying the system.

V. THE INTERFACE BETWEEN AGROECOLOGY AND CONSERVATION BIOLOGY

The Bahian cocoa farms in Brazil provide an excellent example of the parallels between agroecology and conservation biology. The emphasis of agroecology is production whereas the emphasis of conservation biology is biodiversity and conservation of endemic and rare species, or habitats. Cocoa, native to the Amazon basin, has been traditionally cultivated in a system known as "cabruca," in which the tallest and healthiest trees of the forest are maintained and the understory is substituted by cocoa trees. This traditional and highly diverse system now covers about 400,000 hectares in Bahia. However, this system is rapidly disappearing and being substituted either by other crops or by a more intensive form of cultivating cocoa under the canopy of a limited number of planted shade trees such as *Erythrina spp.*, *Inga* spp., and *Splundias* spp. Because of their diversity and abundance, ants are the most important component of the fauna in cocoa plantations worldwide. Leston (1978) reported 130 ant species in one hectare of a secondary forest in Bahia, while J. H. C. Delabie (unpublished data) found 105 soil-surface species and 70 litter species on one hectare of an experimental cocoa area in the same region. As discussed earlier, ants offer a number of benefits to cocoa plantations, but can also have detrimental effects.

The wise management of the ant community based on knowledge of the ecology of the community as a whole as well as individual species is the key for promoting the beneficial impacts while limiting the negative ones. For example, there seems to be a consensus that the negative effect of *Wasmannia auropunctata* (Roger) outweighs its positive effects within the Bahia's cocoa agroecosystem. This species has been identified as one of the three codominant species in the agroecosystem. However, it only becomes well established after other dominants have declined, which usually happens after pesticide applications. It is also possible that the decline of other beneficial codominants is associated with the reduction of canopy cover (as it happens in coffee plantations), opening up the opportunity for *W. auropunctata* to become dominant. In other words, the traditional "cabruca" system may conserve the biodiversity of ants in a production system which at the same time benefits economically from this diversity.

Other ant species that seem to be disappearing from the region are the army ants of the genus *Eciton* and with them their associated fauna, especially birds. It is possible that the diverse cocoa plantations in the Bahia region had been acting as a refuge of biodiversity. This is indeed the case of shaded coffee plantations in the mid-elevation regions of Central America and Mexico (Perfecto *et al.,* 1996). However, the potential conservation impact of cocoa plantations is not limited to ants and their associated fauna. Just recently a new bird species (and genus) was identified from a production cocoa plantation in Bahia. Researchers speculate that with the high rate of deforestation in the region, diverse cocoa plantations became the last refuge for this rare bird. Unfortunately, conservation biologists and agroecologists have failed to recognize these points of interceptions in their goals (but see Chapter 2). Conservation for biological control is only the first step in merging production oriented goals with purely conservation oriented goals.

REFERENCES

Agnew, C. W., and Sterling, W. L. (1981). Predation of boll weevils in partially-open bolls by the red imported fire ant. *Southwest. Natur.* 6, 215-219.

Altieri, M. A., and Schmidt, L. L. (1984). Abundance patterns and foraging activity of ant communities in abandoned, organic and commercial apple orchards in northern California. *Agric., Ecosys., Environ.* 11, 341-352.

Andow, D. (1991). Vegetational diversity and arthropod population response. *Annu. Rev. Entomol.* 36, 561-586.

Basedow, T. (1993). Predatory arthropods in cabbage terraces under different conditions in the Cordillera Region of Luzon, Philippines. *Bull. Entomol. Res.* 83, 313-319.

Bottrell, D. G., and Adkisson, P. L. (1977). Cotton insect pest management. *Annu. Rev. Entomol.* 22, 451-481.

Brown, E. S. (1959). Immature nut fall of coconuts in the Solomon Islands. ll. Changes in ant populations and their relation to vegetation. *Bull. Entomol. Res.* 50, 523-558.

Carroll, C. R., and Jansen, D. H. (1973). Ecology of foraging ants. *Annu. Rev. Ecol. Syst.* 4, 231-257.

Carroll, C. R., and Risch, S. J. (1984). The dynamic of seed harvesting in early successional communities by tropical ants, *Solenopsis geminata. Oecologia* 61, 388-392.

Castiñeiras, A. (1985a). "Morphology, ecology and use of *Pheidole megacephala* (Hymenoptera: Formicidae) for the biological control of the sweet potato weevil, *Cylas formicarius elegantulus* (Coleoptera: Curculionidae)." Ph.D. Thesis. Univ. of Havana, School of Biology, Havana, Cuba.

Castiñeiras, A. (1985b). Evaluation of traps for colonies of *Pheidole megacephala* (Hymenoptera: Formicidae). *Ciencia y Técnica Agric. Protec. Plantas* (Havana, Cuba) 8, 99-106.

Castiñeiras, A. (1989). Relationships of *Pheidole megacephala* (Hymenoptera: Formicidae) with *Cylas formicarius elegantulus* (Coleoptera: Curcullonidae) in sweet potato fields. *Ciencia y Técnica Agric. Protec. Plantas* (Havana, Cuba) 4, 15-19.

Castiñeiras, A., and Calderón, A. (1982). Susceptibility of *Pheidole megacephala to* three microbial pesticides: Dipel, Bitoxibacillin and *Beauveria bassiana* under laboratory conditions. *Ciencia y Técnica Agric. Protec. Plantas* (Havana, Cuba). Supl. 1, 61-66.

Castiñeiras, A., and Fernández, X. (1983). Population increase of *Toumeyella cubensis* (Homoptera: Coccidae) in association with *Pheidole megacephala* (Hymenoptera: Formicidae). *Ciencia y Técnica Agric. Protec. Plantas* (Havana, Cuba) 6, 11-14.

Castiñeiras, A., and Paez, M. J. (1989). Relationships of *Pheidole megacephala* (Hymenoptera: Formicidae) with citrus insects. *Ciencia y Técnica Agric. Protec. Plantas* (Havana, Cuba) 12, 43-50.

Castiñeiras, A., Borges, A., and Obregón, 0. (1991a). Biological control of *Cylas formicarius elegantulus (Summ.)*. *Les Colloques* (Paris) 58, 417-422.

Castiñeiras, A., Cabrera, T., Calderón, A., López, M., and Luján, M. (1991b). Biological control of *Cosmopolites sordidus (Germ.)*. *Les Colloques* (Paris) 58, 424-428.

Castiñeiras, A., Calderón, A., López, M., and Dierksmeyer, G. (1990). Effect of entomopathogenic fungi and biocides used in banana plantations in Cuba on *Pheidole megacephala*. *Ciencia y Técnica Agric. Protec. Plantas* (Havana, Cuba) 13, 37-44.

Castiñeiras, A., Obregón O., and Borges, A. (1985). Persistence of the biological activity of seven chemical insecticides on *Pheidole megacephala,* under laboratory conditions. *Ciencia y Técnica Agric. Protec. Plantas* (Havana, Cuba) 8, 29-43.

Crowell, K. L. (1968). Rates of competitive exclusion by the Argentine ant in Bermuda. *Ecology* 49, 551-555.

Cudjoe, A. R., Neuenschwander, P., and Copland, M. J. W. (1993). Interference by ants in biological control of the cassava mealybug *Pheanocuccus manihoti* (Hemiptera: Pseudococcidae) in Ghana. *Bull. Entomol. Res. 83*, 15-22.

Da Fonseca, A., Passon, J., and Honorato, A. (1971). "Manual of Coffee Pests." Institute of Coffee of Angola, Luanda.

De Barro, P. J. (1990). Natural enemies and other species associated with *Saccharicoccus sacchari* (Cockrerell) (Hemiptera: Pseudococcidea) in the Bundaberg area, Southeast Queensland. *J. Austr. Entomol. Soc.* 29, 87-88.

Delable, J. H. C. (1990). The ant problem of cocoa farms in Brazil. *In* "Applied Myrmecology: A World Perspective." (R. K. Van Der Meer, K. Jeffe, and A. Cedeño, eds.), pp. 555-559. Westview Press. Boulder, CO.

Fateye, A., and De Taffin, G. (1989). Integrated control of *Pseudotheraptus devastans* and related species. *Oleagineaux* 44, 11.

Fernández, W. D., Oliviera, P. S., Carvalho, S. L., and Habib, M. E. M. (1994). *Pheidole* ants as potential biological control agents of the boll weevil, *Anthonomus grandis* (Col., Curculionidae), in Southeast Brazil. *J. Appl. Entomol.* 118, 437-441.

Gallepsie, R. G., and Reimer, N. (1993). The effect of alien predatory ants (Hymenoptera: Formicidae) on Hawaiian endemic spiders (Araneae: Tetragnathidae). *Pacific Sci.* 47, 22-33.

Gotwald, W. H. (1986). The beneficial economic role of ants. *In* "Economic Impact and Control of Social Insects." (S. B. Vinson, ed.), pp. 290-313. New York Praeger, New York, NY.

Greenslade, P. J. M. (1971). Interspecific competition and frequency changes among ants in Solomon Islands coconut plantations. *J. Appl. Ecol.* 8, 323-349.

Greenslade, P. J. M. (1972). Comparative ecology of 4 tropical ant species. *Insect Soc.* 19, 195-212.

Hierbaut, M., and Van Damme, P. (1992). The use of artificial nests to establish colonies of the black cocoa ant *(Dolichoderus thoracicus* Smith) used for biological control of *Helopeltis theobromae* Mill. in Malaysia. *Med. Fac. Landbouw. Univ. Gent* 57, 533-542.

Hölldobler, B., and Wilson, E. 0. (1990). "The Ants." Cambridge University Press, Cambridge, U.K.

Huang, H. T., and Yang, P. (1987). The ancient cultured citrus ant. *BioSci. 37,* 665-671.

Jahn, G. C., and Beardsley, J. W. (1996). Effects of *Pheidole megacephala* (Hymenoptera: Formicidae) on survival and dispersal of *Dysmicoccus neobrevipes* (Homoptera: Pseudococcidae). *J. Econ. Entomol.* 89, 1124-1129.

Julia, J. F., and Mariau, D. (1978). La punaise du cocotier: *Pseudotheraptus* sp. in Cote d'lvoire. 1. Etudes pre-alables a la mise au point d'un methode de lutte integree. *Oleagineaux* 33, 65-75.

Khoo, B., and Ho, C. T. (1992). The influence of *Dolichoderus thoracicus* (Hymenoptera: Formicidae) on losses due to *Helopeltis thievora* (Heteroptera: Miridae), Black pod disease, and mammalian pests in cocoa in Malaysia. *Bull. Entomol. Res.* 82, 485-491.

Khoo, K. C., and Chung, G. F. (1989). Use of the black cocoa ant to control mirid damage in cocoa. Plant. *Kuala Lampur* 65, 370-383.

Koptur, S. (1992). Extrafloral nectary-mediated interactions between insects and plants. *In* "Insect-Plant Interactions." (E. Bernays, ed.), pp. 81-129. CRC Press. Boca Raton, FL.

Krantz, J., Schmutterer, H., and Koch, W. (1978). "Diseases, pests and weeds in tropical crops." John Wiley and Sons. Chichester, U.K.

Leston, D. (1973). The ant mosaic-tropical tree crops and the limiting of pest and diseases. *PANS* 19, 311-341.

Leston, D. (1978). A neotropical ant mosaic. *Ann. Entomol. Soc. Amer.* 71, 649-653.

Letourneau, D. K. (1987). The enemies hypothesis: tritrophic interactions and vegetational diversity in tropical agroecosystems. *Ecology* 68, 1616-1622.

Levins, R., Pressick, L. M., and Heatwole, H. (1973). Coexistence patterns in insular ants. *Amer. Sci.* 61, 463-472.

Logfren, C. S. (1986). History of imported fire ants in the United States. *In* "Fire Ants and Leaf-cutting Ants." (C. S. Logfren, and R. K. Van Der Meer, eds.), pp. 36-47. Westview Press. Boulder, CO.

Lohr, B., and Oswald, S. (1990). An integrated pest management approach towards the control of the low-density coconut pest, *Pseudotheraptus wayi* Brown (Hemiptera: Coreidae). *Integr. Pest Manag. Trop. Subtrop. Crop. Sys.* 2, 395-404.

Mack, T. P., and Beckman, C. B. (1990). Effects of two planting dates and three tillage systems on the abundance of lesser cornstalk borer (Lepidoptera: Pyralidae), other selected insects, and yield in peanut fields. *J. Econ. Entomol.* 83, 1034-1041.

Madeiros, M. A., Fowler, H. G., and Bueno, O. C. (1995). Ant (Hym., Formicidae) mosaic stability in Bahian cocoa plantations: implications for management. *J. Appl. Entomol.* 119, 411-414.

Majer, J. D., and Delabie, J. H. C. (1993). An evaluation of Brasilian cocoa farm ants as potential biological control agents. *J. Plant Prot. Trop.* 10, 43-49.

Majer, J. D., Delabie, J. H. C., Smith, M. R. B. (1994). Arboreal community patterns in Brazilian cocoa farms. *Biotropica* 26, 73-83.

Mc Ewen, F. L., Hapai, W. J., and Su, T. H. (1979) Laboratory tests with candidate insecticides for control of *Pheidole megacephala (F.)*. *Proc. Haw. Entomol. Soc.* 13, 119-124.

Needham, J. (1986). "Science and civilization in China. Biology and Biological Technology." Vol. VI, Part 1: Botany. Cambridge University Press, Cambridge, U.K.

Nestel, D., and Dickschen, F. (1990). The foraging kinetics of ground ant communities in different Mexican coffee agroecosystems. *Oecologia* 84, 58-63.

O'Connor, B. A. (1950). Premature nutfall of coconuts in the British Solomon Islands Protectorate. *Fiji Agric. J.* 21, 21-42.

Ogata, K. (1982). Taxonomic study of the ant genus *Pheidole* Westwood of Japan, with a description of a new species (Hymenoptera: Formicidae). *Kontyu* 50, 189-197.

Oswald, S. (1991). Application of the selective fire ant bait AMDRO against the harmful brown house ant, *Pheidole megacephala,* for improvement of the biological control of the coconut bug, *Pseudotheraptus wayi,* by the beneficial red weaver ant, *Oecophylla longinoda. Z. Pflkrankh. Pflanzenschutz* 98, 358-363.

Overall, W. L., and Posey, P. A. (1984). Uso de formigas do genero *Azteca* por controle de saúvas entre os indios Kaiapos do Brasil. *Attini* 16, 2.

Perfecto, I. (1990). Indirect and direct effects in a tropical agroecosystem: the maize-pest-ant system in Nicaragua. *Ecology* 71, 2125-2134.

Perfecto, I. (1991a). Ants (Hymenoptera: Formicidae) as natural control agents of pests in irrigated maize in Nicaragua. *J. Econ. Entomol.* 84, 65-70.

Perfecto, I. (1991b). Dynamics of *Solenopsis geminata* in a tropical fallow field after ploughing. *Oikos* 62, 139-144.

Perfecto, I. (1994). Foraging behavior as a determinant of asymmetric competitive interactions between two ant species in a tropical agroecosystem. *Oecologia* 98, 184-192.

Perfecto, I., and Sediles, A. (1992). Vegetational diversity, the ant community and herbivorous pests in a tropical agroecosystem in Nicaragua. *Environ. Entomol.* 21, 61-67.

Perfecto, I., and Snelling, R. (1995). Biodiversity and tropical ecosystem transformation: ant diversity in the coffee agroecosystem in Costa Rica. *Ecol. Applic.* 5, 1084-1097.

Perfecto, I., and Vandermeer, J. H. (1994). Understanding biodiversity loss in agroecosystems: Reduction of ant diversity resulting from transformation of the coffee ecosystem in Costa Rica. *Entomology (Trends Agric. Sci.)* 2, 7-13.

Perfecto, I., and Vandermeer, J. H. (1996). Microclimatic changes and the indirect loss of ant diversity in a tropical agroecosystem. *Oecologia* 108, 577-582.

Perfecto, I., Rice, R., Greenberg, R., and Van der Voolt, M. (1996). Shade coffee: a disappearing refuge of biodiversity. *BioSci.* 46, 598-608.

Rapp, G., and Salum, M. S. (1995). Ant fauna, pest damage and yield in relation to the density of weeds in coconut sites in Zanzibar, Tanzania. *J. Appl. Entomol.* 119, 45-48.

Reichel, H., and Andersen, A. N. (1996). The rainforest ant fauna of Australia's northern territory. *Austr. J. Zool.* 44, 81-95.

Reimer, N. J., and Beardsley, J. W. (1990). Effectiveness of hydramethylnon and fenoxycarb for the control of the bigheaded ant (Hymenoptera: Formicidae), an ant associated with mealybug wilt of pineapple in Hawaii. *J. Econ. Entomol.* 83, 74-84.

Reimer, N. J., Cope, M., and Yasuda, G. (1993). Interference of *Pheidole megacephala* (Hymenoptera: Formicidae) with biological control of *Coccus viridis* (Homoptera: Coccidae) in coffee. *Environ. Entomol.* 22, 483-488.

Reimer, N. J., Glancey, B. M., and Beardsley, J. W. (1991). Development of *Pheidole megacephala* (Hymenoptera: Formicidae) colonies following ingestion of fenoxycarb and pyriproxyfen. *J. Econ Entomol* . 84, 56-60.

Risch, S. J. (1980). Ants as important predators of rootworm eggs in the neotropics. *J. Econ. Entomol.* 74, 88-90.

Risch, S. J., and Carroll, C. R. (1982a). Effect of a keystone predaceous ant, *Solenopsis geminate*, on arthropods in a tropical agroecosystem. *Ecology* 63, 1979-1982.

Risch, S. J., and Carroll, C. R. (1982b). The ecological role of ants in two Mexican agroecosystems. *Oecologia* 55, 114-119.

Robertson, L. N., Kettle, B. A., and Simpson, G. B. (1994). The influence of tillage practices on soil macrofauna in semi-arid agroecosystems in northeastern Australia. *Agric., Ecosys. Environ.* 48, 149-156.

Root, R. B. (1973). Organization of plant arthropod association in simple and diverse habitats: the fauna of collards (Brassica oleracea). *Ecol. Monogr.* 43, 95-124.

See, Y. A., and Khoo. (1996). Influence of *Dolichoderus thoracicus* (Hymenoptera: Formicidae) on cocoa pod damage by *Conopomorpha cramerella* (Lepidoptera: Gracillariidae) in Malaysia. *Bull. Entomol. Res.* 86, 467-474.

Shenk, M., and Sounders, J. L. (1984). Vegetation management systems and insect responses in the humid tropics of Costa Rica. *Trop. Pest Manag.* 30, 186-193.

Stapley, J. H. (1971). "The Problem of Immature Nutfall and the Ant Complex in the Solomon Islands." *Report to the Ministry of Agriculture, Solomon Islands.*

Sterling, W. L. (1978). Fortuitous biological suppression of the boll weevil by the red imported fire ant. *Environ. Entomol.* 7, 564-568.

Steyn, J. J. (1954). The effect of the cosmopolitan brown house ant *(Pheidole megacephala* (F.)) on citrus red scale *(Aonidiella aurantii* Mask.) at Letaba. *J. Entomol. Soc. So. Africa* 17, 252-264.

Steyn, J. J. (1955). The effect of mixed ant populations on red scale *(Aonidiella aurantii* Mask.) on citrus at Letaba. *J. Entomol. Soc. So. Africa* 18, 93-103.

Stinner, B. R., and House, G. J. (1990). Arthropods and other invertebrates in conservation-tillage agriculture. *Annu. Rev. Entomol.* 35, 299-318.

Tonhasca, A., Jr. (1993). Effects of agroecosystem diversification on natural enemies of soybean herbivores. *Entomol. Exp. Appl.* 69, 83-90.

Torres, J. A. (1984). Niches and coexistence in ant communities in Puerto Rico: repeated patterns. *Biotropica* 16, 284-295.

Vanderplank, F. L. (1959). Studies on the coconut pest *Pseudotheraptus wayi* Brown (Coreidae) in Zanzibar. II. Some data on the yields of coconuts in relation to damage caused by the insect. *Bull. Entomol. Res.* 50, 135-149.

Vanderplank, F. L. (1960). The bionomics and ecology of the red tree ant *Oecophylla* sp. and its relationship to the coconut bug *Pseudotheraptus wayi* Brown (Coreidae). *J. Anim. Ecol.* 29, 15-33.

Varela, A. (1992). The role of *Oecophylla longinoda* (Formicidae) in control of *Pseudotheraptus wayi* (Coreidae) on coconuts in Tanzania. Ph.D. Thesis. Imperial College, Silwood Park, London, U. K.

Vellani, H. C., Campos, A . R., Gravena, S., and Busoli, A. C. (1984). Surto de curuquere do algodoeiro *(Alabama argillacea)* com epizootia de *Nomuraea rileyi* e declinio de predadores apos tratamentos com Semivol ¿ 30. *Ecossistema* 9, 63-66.

Vepsèlèinen, K., and Pisarski, B. (1982). Assembly of island ant communities. *Ann. Zool. Fennici* 19, 327-335.

Way, M. J. (1953). The relationship between certain ant species with particular reference to biological control of the Coreid *Theraptus* sp. *Bull. Entomol. Res.* 44, 669-691.

Way, M. J. (1963). Mutualism between ants and honeydew-producing Homoptera. *Annu. Rev. Entomol.* 8, 307-344.

Way, M. J., and Khoo, K. C. (1991). Colony dispersion and nesting habits of the ant, *Dolichoderus thoracicus* and *Oecophylla smaragdina* (Hymenoptera: Formicidae), in relation to their success as biological control agents in cocoa. *Bull. Entomol. Res.* 81, 341-350.

Way, M. J., and Khoo, K. C. (1992). Role of ants in pest management. *Annu. Rev. Entomol.* 37, 479-503.

Way, M. J., Cammel, M. E., Bolton, B., and Kanagaratnam, P. (1989). Ants (Hymenoptera: Formicidae) as egg predators of coconut pests, especially in relation to biological control of the coconut caterpillar, *Opisina arenosella* Walker (Lepidoptera: Xyloryctidae), in Sri Lanka. *Bull. Entomol. Res.* 79, 219-233.

Yang, P. (1982). Biology of the yellow citrus ant, *Oecophylla smaragdina* and its utilization against citrus insect pests. *Acta Sci. Nat. Univ. Sunyantseni* 3, 102-105.

Zerhusen, D., and Rashid, M. (1992). Control of the bighearted ant *Pheidole* megacephala Mayr (Hym.: Formicidae) with the fire ant bait 'AMDRO' and its secondary effects on the population of the African weaver ant *Oecophylla longinoda* Latreille (Hym.: Formicidae). *J. Appl. Entomol.* 113, 258-264.

CHAPTER
16

CONSERVATION OF APHIDOPHAGA
IN PECAN ORCHARDS

James D. Dutcher

I. INTRODUCTION

Three foliage feeding aphid species infest pecan during the season (Tedders 1978). Aphid feeding damage causes early defoliation leading to reductions in staminate and pistillate flowers and nut production the following season (Dutcher *et al.*, 1984). Even though pecan aphid populations are regulated by aphidophaga and entomopathogenic fungi the aphids remain at a population size sufficient to produce enough honeydew to support the growth of sooty mold (Tedders, 1986; Edelson and Estes, 1987). Conservation biological control techniques increase the population levels of aphidophaga and entomopathogenic fungi in pecan orchards (Tedders, 1983; Bugg and Dutcher, 1989; Bugg *et al.*, 1990; Dutcher, 1995). These include reducing the frequency of pesticide sprays, planting legumes as intercrops in the orchard to produce alternate prey aphids for aphidophaga, and partitioning of the foraging behavior of the red imported fire ant with trunk sprays of insecticide that prevent ants from reaching aphids and mealybugs in the tree, yet allowing ants to remain on the orchard floor as predators of pecan weevil larvae. Implementation of the conservation biological control techniques is often confounded by interactions with climate, secondary predators, and predator behavior. Combining the use of intercrops and removal of ants as secondary predators from the tree crown enhances aphidophaga. Details of these approaches are provided below.

A. Pecan Culture

Pecan *Carya illinoensis* Wangenheim (K. Koch) is indigenous to North America and is grown for nut production, lumber, and shade. Nut production is of three cultural types (Sparks, 1992). Improved production was originally initiated in 1846, in Louisiana by Antione a slave, when he grafted the first cultivar (Centennial) (Sparks, 1992). *Native production* entails the use of natural stands of trees which are harvested either each year or whenever the crop is sufficiently large to justify the expense of clearing the understory and processing the crop. *Seedling production* consists of the use of trees grown from seed (often from one parent tree) that are planted singly or in an orchard. The crop has an assortment of nut sizes, kernels of variable quality, and a wide range of maturity dates. Third, *improved cultivar production* occurs when scions of selected, improved cultivars are grafted onto a seedling rootstock and the trees are planted in an orchard. Improved cultivars are selected from three sources, native stands, seedling orchards, and breeding programs. Most commercial cultivars are selected from seedling orchards. Pecan is wind pollinated and improved orchards usually have two or more cultivars selected so that pollen shed in one cultivar coincides with stigma receptivity in the other. Many orchards have seedling trees as a source of pollen. Improved orchards are intensively managed with respect to irrigation, fertilization, and pest control whereas native and seedling orchards have low-input management (Smith *et al.,* 1995). Improved orchards have higher production and production costs and are easier to harvest than native or seedling orchards due to maintenance of a flat harvesting surface and uniform maturation of the nut crop.

B.   Pecan Insect Control

Pest management of pecan ranges from low input management and a reliance on natural control to a preventive chemical control approach. Improved and seedling orchards rely on preventative chemical control of diseases (Latham, 1995) and integrated management of insect pests (Harris, 1983). Seedling and native pecan trees are typically not treated with pesticides as frequently as trees of improved cultivars. Many native producers rely totally on natural enemies for insect control (Goff, 1996). In improved cultivar orchards phytophagous insects and mites have sustained population growth from season to season due to an abundance of foliage and fruit. The pecan crop is susceptible to damage from nut and foliage feeding insects and mites from budbreak to nut maturity (Moznette *et al.*, 1940). This period ranges from 137 to 198 days for improved cultivars (Sparks, 1992). Preventative applications of foliar pesticides that typically have a residual activity of seven to 14 days are not cost-effective.

Producers use rigorous insect (Harris, 1983; Ellis, 1985) and climate (Pickering *et al.,* 1990b) monitoring systems to determine the optimum spray dates for insect and mite control. Pecan weevil *(Curculio caryae* (Horn)), hickory shuckworm (Cydia *caryana* (Fitch)), blackmargined aphid *(Monellia caryella* (Fitch)), yellow pecan aphid *(Monelliopsis pecanis* Bissell), black pecan aphid *(Melanocallis caryaefoliae (Davis)),* pecan nut casebearer *(Acrobasis nuxvorella* Neunzig), pecan phylloxera *(Phylloxera devistatrix* Pergande), kernel-feeding hemipterans (including southern green stink *bug (Nezara viridula* (L.)), brown stink bug *(Euschistus servus* (Say)), leaf-footed *bug (Leptoglossus* spp.), pecan spittlebug *(Cladstoptera achatina* Germar), pecan leaf scorch mite *(Eotetranychus hicoriae* (McGregor)), and pecan serpentine leafminers *(Stigmella juglandifoliella* (Clemens)) are insect and mite pests that often have to be controlled with insecticides to prevent damage and possible production losses (Payne *et al.,* 1979). Producers of improved cultivars typically apply one spray for pecan nut casebearer, one spray for early season hickory shuckworm, and three sprays for the late season pest complex (i.e., pecan weevil, kernel-feeding hemipterans, leafminers, aphids, mites, and late season hickory shuckworm). Chemical control of foliage-feeding insects is achieved by mixing a specific larvicide, aphidicide, or miticide with the principle insecticide for nut pests (Dutcher *et al.,* 1984). Preventive insecticide applications are still effectively used in pecan orchards. Phylloxeras are controlled with a preventive spray of chlorpyrifos soon after budbreak. Aphids are controlled with soil application of aldicarb in the spring and summer as an alternative to scouting and foliar aphidicide sprays (Dutcher and Harrison, 1984). Producers of improved orchards in the western states, where pecan aphids are the major insect pest problem, rely totally on biological control of pecan aphids (LaRock and Ellington, 1996).

II. PECAN APHID BIOLOGY AND CONTROL

The primary aphid pests of pecan are blackmargined aphid, yellow pecan aphid, and the black pecan aphid. Feeding niches are partitioned based on leaf vein size so that all three species can coexist (Tedders, 1978). Reduction or complete loss of the next seasons crop can result from depletion of the carbohydrates by the combined infestation of all three species (Dutcher, 1985; Dutcher *et al.,* 1984; Wood *et al.,* 1987). In the southern U.S.A., where pecan is an indigenous species, a large complex of beneficial insects and fungal pathogens regulate pecan aphids (Edelson and Estes, 1987; Liao *et al.,* 1984; Pickering *et al.,* 1990a). Pecan aphids have a higher reproductive rate than most aphid species (Kaakeh and Dutcher, 1992a) and can tolerate slightly higher temperatures (Kaakeh and Dutcher, 1993a). Natural control does not always regulate aphid populations below the level that produces sufficient honeydew to sustain the growth of sooty mold. High temperatures and rainfall reduce pecan aphid populations only temporarily (Kaakeh and Dutcher, 1993a,b). Consequently, producers will usually have to apply an insecticide to control the aphids.

Once insecticides are used, aphid control quickly declines in four stages. First, populations of aphidophaga are destroyed by the insecticide. Second, the aphids develop to unusually high numbers in the absence of natural control. Third, more insecticide is applied and insecticide resistant aphids become the dominant genotype in the population making the insecticide worthless as an aphid control. Finally, the producer is left without any type of aphid control, natural or chemical. Chemical control of pecan aphids with certain foliar insecticide applications has become ineffective (Dutcher and Htay, 1985; Dutcher, 1997) and costly for producers. Enhancement and conservation of natural enemies is an effective alternative for some producers (LaRock and Ellington, 1996). Problems with aphid resurgence following destruction of the natural enemy complex by the broad-spectrum insecticides in pecan aphids are quite common (Dutcher and Htay, 1985) and point to the importance of natural enemies in the pecan system and the interactions between insecticides and natural enemies.

III. NATURAL ENEMIES OF PECAN APHIDS

Aphid predators, including spiders (Bumroongsook *et al.,* 1992), ladybeetles (primarily, *Olla v-nigrum (Mulsant), Coccinella septempunctata L., Hippodamia convergens* Guerin-Mineville, *Cycloneda sanguinea (L.),* and *Harmonia axyridis* Pallas), and lacewings (primarily *Chrysoperla rufilabris (Burmeister), C. quadripunctata* Burmeister, and *Micromus posticus* (Walker)), play an important role in maintaining pecan aphid populations at low levels (Edelson and Estes, 1987; Tedders, 1986). Fungal entomopathogens can decimate a pecan aphid population in two to three days when environmental conditions favor the disease and these pathogens cause higher mortality when fungicide application is reduced (Pickering *et al.,* 1990a). Indigenous natural enemies of the pecan aphid complex cause significant reductions in the aphid populations (Liao *et al.,* 1985) and in the absence of predators blackmargined aphid populations are able to increase rapidly on pecan foliage (Alverson and English, 1990; Kaakeh and Dutcher, 1992a). Continued development of blackmargined and yellow pecan aphid populations are dependent on leaf age and previous infestation level (Bumroongsook and Harris, 1992). Generalist predators can be cultured on alternate prey aphids in cover crops on the pecan orchard floor (Bugg *et al.,* 1990). Ants interact with other insects in agricultural systems (Way and Khoo, 1992) and red imported fire ants are common secondary predators in most southeastern pecan orchards, preying on aphidophaga and removing them from the tree (Tedders *et al.*, 1990). The red imported fire ant is an important control of pecan weevil, reducing larval populations in the soil by one third (Dutcher and Sheppard, 1981). Management of red imported fire ants can be achieved by treating the mounds directly, partitioning ant foraging with trunk sprays of chlorpyrifos (Dutcher *et al.*, 1995), or culturing alternate prey aphids on hemp sesbania (Bugg and Dutcher, 1993); a plant that naturally repels red imported fire ant (Kaakeh and Dutcher, 1992b).

IV. CONSERVING NATURAL ENEMIES OF PECAN INSECTS

Natural enemies of phytophagous insects and mites associated with pecan prevent most of these pests from causing significant injury to the tree and the nut crop. Pests are reduced in pecan orchards for a longer time by biological than by chemical control. Broad-spectrum insecticides and monocultures create two sets of adverse conditions for natural enemies of insect pests. However, direct toxicity of the insecticides to natural enemies can be minimized by (1) using selective insecticides, (2) identifying biological windows in which sprays can be applied (i.e., when primary pests are abundant and natural enemies of secondary pests are rare), (3) melding existing information into control decisions with expert systems, and (4) determining the economic injury level of a pest so that sprays are not applied against a pest population that is too low to cause significant production losses.

In addition, plant diversity in a monoculture can be increased by the use of intercrops, cover crops, refugia, and weedy culture (Dutcher, 1993). Native trees rarely have significant insect pest problems due, in part, to low input management (Goff, 1996). These trees are too tall to spray and understory vegetation often is not removed until harvest time, thus the two main adverse conditions for the success of natural control do not exist. Pecan producers with improved and seedling orchards, practice many conservation biological control techniques. Understory plant diversity in seedling and improved orchards is artificially increased with cover crops, intercrops, trap crops, and banker plants (Tedders, 1983; Mizell and Schiffhauer, 1987; Bugg and Waddington, 1994; Ree, 1995). Banker plants are open rearing units for beneficial insects. The plants are infested with alternate prey species that produce predators and parasitoids that also feed the pest. The alternate prey is not a pest of the main crop (e.g., Stacey, 1977). Cover crops can be sown over the entire orchard floor in pecan orchards (Bugg *et al.,* 1991b) as well as in almond, walnut, apple, pear, cherry, peach, and citrus (Bugg and Waddington, 1994).

Increasing the plant diversity on the orchard floor is often associated with an increase in insect diversity; some insects are beneficial to production and other insects are noxious pests. Aphidophaga associated with pecan (Edelson and Estes, 1987) are enhanced by planting either crape myrtle as a banker plant (Mizell and Schiffhauer, 1987) or a series of annual plants as intercrops (Bugg *et al.,* 1991a). Crape myrtle plants are perennial shrubs that can be planted directly in the orchard without disrupting the harvesting surface. Crape myrtle plants can also be grown in the greenhouse and brought into the pecan orchard when the predators and parasitoids are needed for pecan aphid control.

Intercrops have been planted experimentally in pecan orchards of the southern U.S.A. (Bugg *et al.,* 1991a) during the cool seasons (Bugg *et al.,* 1991b; Bugg and Dutcher, 1993) and warm seasons (Bugg and Dutcher, 1989). In these experiments the intercrops provided alternative aphid prey for predators and parasitoids of the pecan aphid during periods when the pecan aphid populations were low in the tree. Movement of aphidophaga between the intercrop and the tree is difficult to quantify in pecan due to a tree height of 10 to 30 meters and tree crowns that can exceed 7200 cubic meters. However, pecan aphids may attract ladybeetles into the tree from the cover crop, if there are no aphids in the cover crop. Aphids in the intercrop may attract aphidophaga out of the trees. Cutting the intercrop when aphidophaga are needed in the tree may be important to pecan aphid control. Intercrops have to be removed before harvest as the nuts are mechanically harvested from the ground. Intercrops that harbor kernel-feeding hemipterans also need to be harvested before pecans produce seed. Trap crops reduce kernel-feeding hemipteran damage by attracting the hemipterans away from the primary crop to an alternative host plant. The alternate host plant, sown adjacent to the orchard, is treated with insecticide to kill the hemipterans before they damage the pecan (Ree, 1995). The trap crop leads to a more effective method of insecticide use since the insecticide is not applied to beneficial insects in the tree crown.

Pecan nut casebearer control is improved in pecan orchards by (1) the use of prediction models to determine the best time for the implementation of management tactics, (2) the replacement of broad-spectrum insecticides with formulations of *Bacillus thuringiensis* Berliner toxins for control of populations, and (3) improved pest monitoring with pheromone baited traps (Harris *et al.,* 1995). Hickory shuckworm chemical controls are applied over a four week period after the shell hardening stage of nut development. Formulations of *B. thuringiensis* toxins effectively control hickory shuckworm populations and are less harmful to aphidophaga than contact insecticides.

Many producers physically remove infestations by clearing pecan shucks from the orchard after harvest and reducing the overwintering population. A pheromone trap has been developed to improve monitoring of hickory shuckworm (Collins *et al.,* 1995).

Producers cut back on pecan weevil sprays with improved monitoring of adult emergence (Tedders and Wood, 1995), edaphic factors (Alverson, 1985), and pecan nut development (Harris, 1985) and by early harvest before the weevils reenter the soil from infested pecans (Dutcher and Payne, 1981). Potential biological control of pecan weevil larvae and adults includes entomopathogens (such as bacteria, fungi, and rickettsia), and entomopathogenic nematodes (Sikorowski, 1985), predaceous domestic fowl and swine, wild birds, small mammals (Tedders, 1985), red imported fire ants (Dutcher and Sheppard, 1981), and parasitic flies (Harrison and Gardner, 1997). Pecan weevil trapping is very effective in describing the adult emergence patterns that can extend to 80 days in the field (Tedders and Wood, 1995). Multiple applications of carbaryl needed for pecan weevil control are very effective in controlling pecan weevil but are also destructive to aphidophaga. Many producers initially use the broad-spectrum insecticides to reduce high pecan weevil populations and then alternate to insecticides that are less toxic to aphidophaga in subsequent seasons if low pecan weevil populations are detected in traps (Dutcher and Payne, 1981). Insecticides have variable toxicity to different insect species. Certain compounds are toxic to pests but relatively nontoxic to beneficial insects (Dutcher and Payne, 1985). Certain insecticides used against late season insect pests have a differential toxicity favorable to lacewing larvae (Hurej and Dutcher, 1994a) but none have been tested that are favorable to ladybeetle adults and larvae (Hurej and Dutcher, 1994b; Mizell and Schiffhauer, 1990).

Hymenopterous parasitoids have been reported for pecan nut casebearer and hickory shuckworm (Gunasena and Harris, 1988), pecan leafminers (Heyerdahl and Dutcher, 1985a), walnut caterpillar (Harris and Pravia, 1977), and pecan aphids (Tedders, 1978; Edelson and Estes, 1987). Few attempts have been made to manage parasitoids in the pecan system. Tedders (1977) introduced parasitoids for aphid control. Heyerdahl and Dutcher (1985a) found 36 species associated with the four leafminer species. A sampling scheme was developed for serpentine leafminers (Heyerdahl and Dutcher, 1985b) to identify the optimum insecticide application time period during moth emergence when parasitoids were still within the mines. Watterson and Stone (1982) found that one primary parasitoid and five secondary parasitoids were associated with blackmargined aphid in far west Texas.

V. ENHANCEMENT TECHNIQUES IN PECAN ORCHARDS

Crimson clover *(Trifolium incarnatum* L.) and hairy vetch *(Vicia villosa* (Roth)) are planted as winter cover crops in commercial pecan orchards to improve soil nitrogen, promote soil structure, suppress harmful weeds, and provide refuge for natural enemies of pecan aphids. These cover crops sustain populations of pea aphid *Acyrthosiphon pisum* Harris (Bugg *et al.,* 1991b). Hemp sesbania *(Sesbania exaltata* [Rafinesque-Schmaltz] Cory) and hairy indigo *Indigofera hirsute* L. are planted as summer intercrops between the tree rows to enhance aphidophaga (Bugg and Dutcher, 1993). Both crops sustain a high population density of cowpea aphid, *Aphis craccivora* Koch, an alternate prey for generalist predators of the pecan aphid. The selection of the intercrop plant depends on the season and the alternate prey species. A few of the many interactions between intercrop plant species, alternate prey, and predators have been observed in controlled experiments. The cowpea aphid develops faster on cowpea, hemp sesbania, and hairy indigo than on crimson clover or hairy vetch (Kaakeh and Dutcher, 1993c). The pea aphid develops faster on crimson clover and hairy vetch than on hemp sesbania or hairy indigo (Kaakeh and Dutcher, 1993d). Cowpea aphids develop on intercrops for the entire summer and aphidophagous insect abundance differs between plant species.

American jointvetch *(Aeschynomene americana* L.), cowpea *(Vigna unguiculata* L), and alyceclover *(Alysicarpus vaginalis* [L.] De Candolle) maintain cowpea aphids, aphidophaga, and red imported fire ants. Red imported fire ants prefer foraging on certain cover crops. Generally, intercrop plants with higher red imported fire ant abundance had lower abundance of aphidophaga (Bugg and Dutcher, 1989). Among the crops tested as summer intercrops (Bugg and Dutcher, 1993), hemp sesbania was the only ant repellent plant and it had the longest sustained abundance of aphidophaga and alternate prey aphids. Hemp sesbania extract was repellent and caused mortality to red imported fire ants (Kaakeh and Dutcher, 1992b).

Red imported fire ants are opportunists in agricultural habitats where they are introduced; feeding on honeydew as well as pest and beneficial insect populations (Way and Khoo, 1992). In a comprehensive set of experimental and empirical studies on red imported fire ant activity in pecan orchards, Tedders *et al.* (1990) found that mowing of orchard ground cover caused a change in the ratio of red imported fire ant on the ground to those in pecan trees. A higher proportion of the ants forage in the tree in mowed plots. Ants forage to a height of 9 m in a tree, and can form nests in crotch angles and among the nut clusters of the tree. The ants collect the honeydew from leaf surfaces rather than directly from the pecan aphids. Apparently, the ants rarely prey on pecan aphids, but commonly prey on lacewings, ladybeetles, and syrphids.

Dutcher (1995) and Dutcher *et al. (*1995) conducted field experiments in an improved pecan orchard in Georgia to determine the effects of warm season intercrops and the imposed restriction of foraging by red imported fire ants on the abundance of blackmargined aphids and aphidophagous insects in pecan. Aphid abundance was measured in trees in four intercrop treatments: hairy indigo, hemp sesbania, hairy indigo and hemp sesbania mixture, and mowed sod. Within each plot one-half of the trees were treated with a trunk spray of chlorpyrifos to prevent ants from foraging in the tree crown and the other one-half of the trees were not treated. Season-long monitoring of the insects in the trees indicated that coccinellid abundance was generally the same in the trees of all intercrop treatments. Though the treatments did not have an effect on blackmargined aphid abundance on all sample dates (Fig.1), the late summer peak of aphid abundance was significantly lower in trees with the hemp sesbania intercrops than in trees of the other three intercrop treatments. The exclusion of red imported significantly decreased blackmargined aphid abundance in the trees with mowed sod and trees with hairy indigo. Ant exclusion significantly increased blackmargined aphid abundance in the trees with sesbania and hairy indigo plus sesbania.

VI. DISCUSSION

Long-term orchard management practices are effective in pecan orchards to stabilize the abundance of imported and indigenous natural enemies. The producer has a long-term investment in the orchard and can implement conservation techniques over several years. Conservation techniques of intercropping with sesbania alone, the combination of mowed sod and ant exclusion, and the combination of intercropping with hairy indigo and ant exclusion reduced pecan aphid populations in Georgia. Whether these reductions will result in an increase in pecan production is not known. Pecan production is highly variable from season to season and even pecan nutrition experiments do not indicate consistent increases in production following the application of fertilizer (Worley, 1995). Thus, the documentation of increased production as a result of the use of conservation tactics may be difficult. The development and implementation of an alternative to chemical control of pecan aphids will require greater monitoring and another level of decision making for producers. Fortunately, communication networks (Pickering *et al.,* 1990b), identification of aphid resistant pecan cultivars (Kaakeh and Dutcher, 1994), and further research on host plant-aphid interactions (Smith and Kaakeh, 1997) will develop more technological support for the pecan production system.

REFERENCES

Alverson, D. R. (1985). Influence of edaphic factors on emergence of the pecan weevil, *Curculio caryae* (Horn) (Coleoptera: Curculionidae). *In* "Pecan Weevil: Research Perspective." (W. W. Neel, ed.), pp. 11-26. Quail Ridge Press, Brandon, MS.

Alverson, D. R., and English, W. R. (1990). Dynamics of pecan aphids, *Monelliopsis pecanis* and *Monellia caryella* on field-isolated single leaves of pecan. *J. Agric. Entomol.* 7, 29-38.

Mean Number of Aphids/Terminal

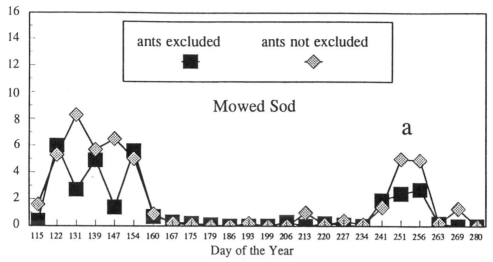

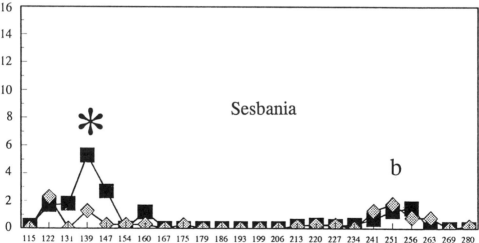

Figure 1. Blackmargined aphid abundance in pecan tree plots sown with different intercrops. One-half of the trees in each plot were treated (ants excluded) with a trunk spray of insecticide to prevent ants from foraging in the tree crown. The other one-half of the trees were not treated (ants not excluded). An asterisk above a population peak indicates that the aphid densities are significantly different within the intercrop treatment (P<.05, LSD Test) between trees with and without the trunk spray. Letters above the late season population peaks indicate the comparison of intercrop means on the day 251. Intercrop means with the same letter are not significantly different (P< 0.05, LSD Test) Intercrops were planted on Day 130 (May 10, 1994) and harvested on Day 280 (Oct. 6, 1994).

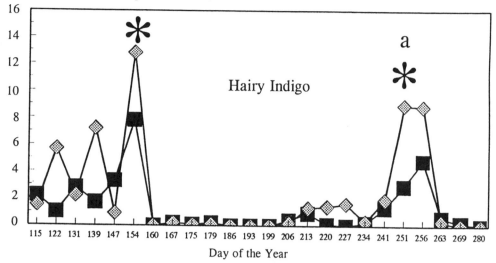

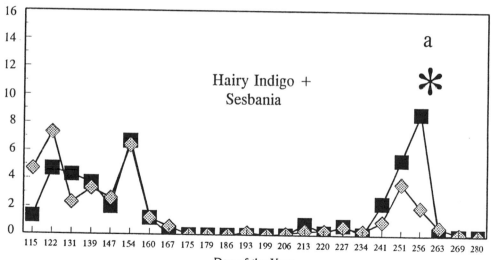

Bugg, R. L., and Dutcher, J. D. (1989). Warm-season cover crops for pecan orchards: horticultural and entomological implications. *Biol. Agric. Hort.* 6, 123-148.

Bugg, R. L., and Dutcher, J. D. (1993). *Sesbania exaltata* (Rafinesque-Schmaltz) Cory (Fabaceae) as a warm-season cover crop in pecan orchards: effects on aphidophagous Coccinellidae and pecan aphids. *Biol. Agric. Hort.* 9, 215-229.

Bugg, R. L., and Waddington, C. (1994). Using cover crops to manage arthropod pests of orchards: a review. *Agric., Ecosys. Environ.* 50, 11-28.

Bugg, R. L., Phatak, S. C., and Dutcher, J. D. (1990). Insects associated with cool-season cover crops in southern Georgia: Implications for biological control in truck crops and pecan. *Biol. Agric. Hort.* 7, 71-45.

Bugg, R. L., Sarrantonio M., Dutcher, J. D., and Phatak, S. C. (1991a). Understory cover crops in pecan orchards: possible management systems. *Amer. J. Altern. Agric.* 6, 50-62.

Bugg, R. L., Dutcher, J. D., and McNeill, P. J. (1991b). Cool-season cover crops in the pecan orchard understory: effects on coccinellidae (Coleoptera) and pecan aphids (Homoptera: Aphididae). *Biol. Cont.* 1, 8-15.

Bumroongsook, S., and Harris, M. K. (1992). Distribution, conditioning, and interspecific effects of blackmargined aphids and yellow pecan aphids (Homoptera: Aphidiae) on pecan. *J. Econ. Entomol.* 85, 187-191.

Bumroongsook, S., Harris, M. K., and Dean, D. A. (1992). Predation of blackmargined aphids (Homoptera: Aphididae) by spiders on pecan. *Biol. Cont.* 2, 15-18.

Collins, J. K., Hedger, G. H., and Eikenbary, R. D. (1995). An update on hickory shuckworm research. *In* "Sustaining Pecan Productivity Into the 21st Century." (M. W. Smith, W. Reid, and B. W. Wood, eds.), pp.5-9. Second National Pecan Workshop Proceedings, Wagoner, Oklahoma, July 23-26, 1994. *US Dept. Agric., Agric. Res. Serv.* ARS 1995-3. National Technical Information Service. Springfield, VA.

Dutcher, J. D. (1985). Impact of late season aphid control on pecan tree vigor parameters. *J. Entomol. Sci.* 20, 55-61.

Dutcher, J. D. (1993). Recent examples of conservation of arthropod natural enemies in agriculture. *In* "Pest Management: Biologically Based Technologies." (R. D. Lumsden, and J. L. Vaughn, eds.), pp. 101-108. Amer. Chem. Soc. Washington, DC.

Dutcher, J. D. (1995). Intercropping pecan orchards with legumes: entomological implications. *In* "Sustaining Pecan Productivity Into the 21st Century." (M. W. Smith, W. Reid, and B. W. Wood, eds.), pp. 10-24. Second National Pecan Workshop Proceedings, Wagoner, Oklahoma, July 23-26, 1994. *U.S. Dept. Agric., Agric., Res Serv. ARS.*1995-3. National Technical Information Service. Springfield, VA.

Dutcher, J. D. (1997). Loss of pyrethroid efficacy to blackmargined aphid (Homoptera: Aphididae) populations on pecan. *In* "New Developments in Entomology." (K. Bondari, ed.), pp. 195-204. Research Signpost. Trivandrum, India.

Dutcher, J. D., and Harrison, K. F. (1984). Applications of reduced rates of systemic insecticides for control of foliar pecan arthropods. *J. Econ. Entomol.* 77, 1037-1040.

Dutcher, J. D., and Htay, U. T. (1985). Resurgence and insecticide resistance problems in pecan aphid management. *In* "Aphids and Phylloxeras of Pecan." (W. W. Neel, W. L. Tedders, and J. D. Dutcher, eds.), pp. 17-29. *Univ. Ga. Agric. Exp. Stn. Spec. Publ.* 38, Ga. Agric. Exp. Stn. Athens, GA.

Dutcher, J. D., and Payne, J. A. (1981). Pecan weevil *(Curculio caryae,* Coleoptera: Curulionidae) bionomics: a regional research problem. *Misc. Publ. Entomol. Soc. Amer.* 12, 45-68.

Dutcher, J. D., and Payne, J. A. (1985). The impact of pecan weevil control strategies on non-target arthropods. *In* "Pecan Weevil: Research Perspective." (W. W. Neel, ed.), pp. 39-50. Quail Ridge Press. Brandon MS.

Dutcher, J. D., and Sheppard, D.C. (1981). Predation of pecan weevil larvae by red imported fire ants. *J. Ga. Entomol. Soc.* 16, 210-213.

Dutcher, J. D., Estes, P. M., and Worley, R. E. (1995). "Enhancement of Aphidophaga with Summer Cover Crops in Pecan - A SRIPM Project." Final Report to U. S. Dept. Agric. Coop. State Res. Serv. Southern Region IPM Program. Project No. GEO-RF330-206. 116 p.

Dutcher, J. D., Worley, R. E., Daniell, J. W., Moss, R. B., and Harrison, K. F. (1984). Impact of six insecticide-based arthropod pest management strategies on pecan yield and quality and return bloom under four irrigation-soil fertility regimes. *Environ. Entomol.* 13, 1644-1653.

Edelson, J. V., and Estes, P. M. (1987). Seasonal distribution of predators and parasites associated with *Monelliopsis pecanis* Bissell and *Monellia caryella* (Fitch) (Homoptera: Aphidae). *J. Entomol. Sci.* 22, 336-347.

Ellis, H. C. (1985). Extension problems in pecan entomology. *In* "Aphids and Phylloxeras of Pecan." (W. W. Neel, W. L. Tedders, and J. D. Dutcher, eds.), pp. 1-4. *Univ. Ga. Agric. Exp. Stn. Spec. Publ.* 38. Ga. Agric. Exp. Stn. Athens, GA.

Goff, B. (1996). Growing pecans in Pointe Coupe parish - natives flourish in this high-rainfall, minimum-input setting. *Pecan South* 29, 26-27.

Gunasena, G. H., and Harris, M. K. (1988). Parasites of hickory shuckworm and pecan nut casebearer with five new host-parasite records. *Southwest Entomol.* 13, 107-111.

Harris, M. K. (1983). Integrated pest management of pecans. *Annu. Rev. Entomol.* 28, 291-338.

Harris, M. K. (1985). Pecan phenology and pecan weevil biology and management. *In* "Pecan Weevil: Research Perspective." (W.W. Neel, ed.), pp. 51-58. Quail Ridge Press, Brandon MS.

Harris, M. K., and Pravia, G. A. (1977). Egg parasites of the walnut caterpillar in Texas. *Southwest. Entomol.* 2, 170-178.

Harris, M. K., Jackman, J. A., Ree, B., and Knutson, A. (1995). Pecan nut casebearer update. *In* "Sustaining Pecan Productivity Into the 21st Century." (M. W. Smith, W. Reid, and B. W. Wood, eds.), pp. 28-30. Second National Pecan Workshop Proceedings, Wagoner, Oklahoma, July 23-26, 1994. *U.S. Dep. Agric., Agric. Res Serv.* ARS1995-3. National Technical Information Service. Springfield, VA.

Harrison, R. D., and Gardner. W. A. (1997). Biological control of nut weevils. *In* "New Developments in Entomology." (K. Bondari, ed.), pp. 205-212. Research Signpost. Trivandrum, India.

Heyerdahl, R., and Dutcher, J. D. (1985a). Hymenopterous parasitoids of pecan leafminers. *J. Entomol. Sci.* 20, 411-421

Heyerdahl, R., and J. D. Dutcher. (1985b). Management of the serpentine leafminer (Lepidoptera: Nepticulidae). *J. Econ. Entomol.* 78, 1121-1124.

Hurej, M., and Dutcher, J. D. (1994a). Indirect effect of insecticides used in pecan orchards to (sic) larvae of *Chrysoperla rufilabris* (Neuroptera: Chrysopidae). *J. Entomol. Sci.* 29, 450-456.

Hurej, M., and Dutcher, J. D. (1994b). Indirect effect of insecticides on convergent lady beetle (Coleoptera: Coccinellidae) in pecan orchards. *J. Econ. Entomol.* 88, 1632-1635.

Kaakeh, W., and Dutcher, J. D. (1992a). Estimation of life parameters of *Monelliopsis pecanis, Monellia caryella* and *Melanocallis caryaefoliae* (Homoptera: Aphididae) on single pecan leaflets. *Environ. Entomol.* 21, 632-639.

Kaakeh, W., and Dutcher, J. D. (1992b). Foraging preference of red imported fire ants, *Solenopsis invicta* Buren, among three species of summer cover crops and their extracts. *J. Econ. Entomol.* 85, 389-394.

Kaakeh, W., and Dutcher, J. D. (1993a). Survival of yellow pecan aphids and black pecan aphids (Homoptera: Aphididae) at different temperature regimes. *Environ. Entomol.* 22, 810-817.

Kaakeh, W., and Dutcher, J. D. (1993b). Effect of rainfall on population abundance of aphids (Homoptera: Aphididae) on pecan. *J. Entomol. Sci.* 28, 283-286.

Kaakeh, W,, and Dutcher, J. D. (1993c). Population parameters and probing behavior of cowpea aphid (Homoptera: Aphididae) on preferred and non-preferred host cover crops. *J. Entomol. Sci.* 28, 145-155.

Kaakeh, W., and Dutcher, J. D. (1993d). Rates of increase and probing behavior of *Acyrthosiphon pisum* (Homoptera: Aphididae), on preferred and non-preferred host cover crops. *Environ. Entomol.* 22, 1016-1021.

Kaakeh, W., and Dutcher, J. D. (1994). Probing behavior and density of *Monelliopsis pecanis, Monellia caryella,* and *Melanocallis caryaefoliae* (Homoptera: Aphididae) on pecan cultivars. *J. Econ. Entomol.* 87, 951-956.

LaRock, D. R., and Ellington, J. J. (1996). An integrated pest management approach, emphasizing biological control, for pecan aphids. *Southwest. Entomol.* 21, 153-166.

Latham, A. J. (1995). Pecan scab management in humid regions. *In* "Sustaining Pecan Productivity Into the 21st Century." (M. W. Smith, W. Reid, and B. W. Wood, eds.), pp. 41-44. Second National Pecan Workshop Proceedings, Wagoner, Oklahoma, July 23-26, 1994. *U.S. Dept. Agric., Agric. Res. Serv.* ARS1995-3. National Technical Information Service. Springfield, VA.

Liao, H. T., Harris, M. K., Gilstrap, F. E., Dean, D. A., Agnew, C. W., Michels, G. J., and Mansour, F. (1984). Natural enemies and other factors affecting seasonal abundance of the blackmargined aphid on pecan. *Southwest. Entomol.* 9, 404-420.

Liao, H. T., Harris, M. K., Gilstrap, F. E., and Mansour, F. (1985). Impact of natural enemies on the blackmargined pecan aphid, *Monellia caryella* (Homoptera: Aphidae). *Environ. Entomol.* 14, 122-126.

Mizell, R. F., and Schiffhauer, D. E. (1987). Seasonal abundance of the crapemyrtle aphid *Sarucallis kahawaluokalani,* in relation to the pecan aphids, *Monellia caryella* and *Monelliopsis pecanis* and their common predators. *Entomophaga* 32, 511-520.

Mizell, R. F., and Schiffhauer, D. E. (1990). Effects of pesticides on pecan aphid predators, *Chrysoperla rufilabris* (Neuroptera: Chrysopidae), *Hippodamia convergens, Cycloneda sanguinea (L.), Olla v-nigrum* (Coleoptera: Coccinellidae), and *Aphelinus perpallidus* (Hymenoptera: Encyrtidae). *J. Econ. Entomol.* 83, 1806-1812.

Moznette, G. F., Nickels, C. B., Pierce, W. C., Bissell, T. L., Demaree, J. B., and Cole, J. R. (1940). Insects and diseases of the pecan and their control. *U. S. Dept. Agric. Farmers' Bull.* 1829, 60 p.

Payne, J. A., Malstrom, H. L., and KenKnight, G. E. (1979). Insect pests and diseases of pecan. *U. S. Dept. Agric. Agric. Ref. Man.,* S-5, 43 p.

Pickering, J., Dutcher, J. D., and Ekbom, B. S. (1990a). The effect of a fungicide on fungal-induced mortality of pecan aphids (Homoptera: Aphididae) in the field. *J. Econ. Entomol.* 83, 1801-1805.

Pickering, J., Hargrove, W. W, Dutcher, J. D., and Ellis, H. C. (1990b). RAIN: A novel approach to computer-aided decision making in agriculture and forestry. *Comput. Electron. Agric.* 4, 275-285.

Ree, B. (1995). Overview of pest management in the west. *In* "Sustaining Pecan Productivity Into the 21st Century." (M. W. Smith, W. Reid, and B. W. Wood, eds.), pp. 25-27. Second National Pecan Workshop Proceedings, Wagoner, Oklahoma, July 23-26, 1994. *U.S. Dept. Agric., Agric. Res. Serv.* ARS1995-3. National Technical Information Service. Springfield, VA.

Sikorowski, P. (1985). Pecan weevil pathology. *In* "Pecan Weevil: Research Perspective." (W. W. Neel, ed.), pp. 87-102. Quail Ridge Press. Brandon MS.

Smith, M. W., Reid, W., and Wood, B. W. (1995). Sustaining Pecan Productivity Into the 21st Century: Second National Pecan Workshop Proceedings, Wagoner, Oklahoma, July 23-26, 1994. *U. S. Dept. Agric., Agric. Res. Serv.* ARS1995-3. National Technical Information Service. Springfield, VA.

Smith, M. T., and Kaakeh, W. (1997). Aphid-host plant interactions in pecan. *In* "New Developments in Entomology." (K. Bondari, ed.), pp. 213-224. Research Signpost. Trivandrum, India.

Sparks, D. (1992). "Pecan Cultivars - The Orchard's Foundation." Pecan Production Innovations. Watkinsville, GA.

Stacey, D. L. (1977). "Banker" plant production of *Encarsia formosa* and its use in the control of glasshouse whitefly on tomatoes. *Plant Pathol.* 26, 63-66.

Tedders, W. L. (1977). *Trioxys pallidus* and *Trioxys complanatus* as parasites of *Monellia costalis, Monelliopsis nigropunctata* and *Tinocallis caryaefoliae. Ann. Entomol. Soc. Amer.* 70, 687-690.

Tedders, W. L. (1978). Important biological and morphological characteristics of the foliar-feeding aphids of pecans. *U.S. Dept. Agric. Tech. Bull.* No. 1579, 29 p.

Tedders, W. L. (1983). Insect management in deciduous orchard ecosystems: habitat management. *Environ. Manag.* 7, 29-34.

Tedders, W. L. (1985). Potential for the control of pecan weevil with selective biological agents. *In* "Pecan Weevil: Research Perspective." (W.W. Neel, ed.), pp. 117-127. Quail Ridge Press. Brandon, MS.

Tedders, W. L. (1986). Biological control of pecan aphids. *Proc. Southeast. Pecan Growers Assoc.* 79, 149-152.

Tedders, W. L., and Wood, B. W. (1995). Update on pecan weevil. *In* "Sustaining Pecan Productivity Into the 21st Century." (M. W. Smith, W. Reid, and B. W. Wood, eds.), pp. 31-37. Second National Pecan Workshop Proceedings, Wagoner, Oklahoma, July 23-26, 1994. *U.S. Dept. Agric., Agric. Res. Serv.* ARS1995-3. National Technical Information Service. Springfield, VA.

Tedders, W. L., Reilly, C. C., Wood, B. W., Morrison, R. K., and Lofgran, C. S. (1990). Behavior of *Solenopsis invicta* (Hymenoptera: Formicidae) in pecan orchards. *Environ. Entomol.* 19, 44-53.

Watterson, G. P., and Stone, J. D. (1982). Parasites of blackmargined aphids and their effect on aphid populations in far-west Texas. *Environ. Entomol.* 11, 667-669.

Way, M. J., and Khoo, K. C. (1992). Role of ants in pest management. *Annu. Rev. Entomol.* 37, 479-503.

Wood, B. W., Tedders, W. L., and Dutcher, J. D. (1987). Energy drained by three pecan aphid species (Homoptera: Aphididae) and their influence on in-shell pecan production. *Environ. Entomol.* 5, 1045-1056.

Worley, R. E. (1995). Pecan nutrition. *In* "Sustaining Pecan Productivity Into the 21st Century." (M. W. Smith, W. Reid, and B. W. Wood, eds.), pp. 119-132. Second National Pecan Workshop Proceedings, Wagoner, Oklahoma, July 23-26, 1994. *U. S. Dept. Agric., Agric. Res. Serv.* ARS1995-3. National Technical Information Service, Springfield, VA.

CHAPTER
17

CONSERVATION BIOLOGICAL CONTROL OF SPIDER
MITES IN PERENNIAL CROPPING SYSTEMS

J. Nyrop, G. English-Loeb, and A. Roda

I. INTRODUCTION

Spider mites in the family Tetranychidae can be common and economically significant pests in perennial crop production systems. These mites usually reach damaging levels only when natural enemies are scarce or nonexistent. A particularly important group of natural enemies of spider mites are predaceous mites in the family Phytoseiidae. When certain phytoseiid species are conserved, spider mite numbers often remain below economically damaging levels. In fact, in some systems these pests may be driven to such low numbers that they are difficult to find. Both the absence of phytoseiid mite predators and subsequent high densities of spider mites can often be traced to the use of chemical pesticides that are toxic to the predators. Therefore, a first step in conserving phytoseiid predators and realizing biological control of spider mites is avoiding or minimizing use of pesticides toxic to these natural enemies. Beyond this obvious (though often times difficult to achieve) management strategy, what else might be done and are there special considerations for perennial systems?

Answering these questions is the goal of this chapter. Unfortunately, we can not simply lay out an easy to follow recipe since such a set of instructions does not exist and perhaps never will. Instead, we approach this goal by first addressing two related questions, (1) what types of phytoseiid predators are usually credited with successful conservation biological control of spider mites in perennial systems? and (2) what dynamical patterns do these predators and their prey follow? We use the answer to these questions to posit design and management actions that might enhance conservation of phytoseiids in perennial systems as well as to suggest areas of inquiry that might provide useful information for this purpose.

For heuristic purposes, two dynamical patterns can be used to describe biological control of spider mites. Each of these patterns is, in turn, generally characterized by predators being either specialists that feed almost exclusively on tetranychid mites or generalist predators that feed on a wide range of substances. We recognize that we have setup a strawman with this distinct dichotomy when, in fact, phytoseiid predators are often neither purely specialists or generalists but a mix of these extremes (McMurtry and Croft, 1997). We also recognize that complexes of predators may occur which also makes this dichotomy less realistic. Nonetheless, for the purpose of trying to identify patterns we still find this dichotomy useful.

In the first of these models, specialist predators readily respond to increases in spider mite numbers through dispersal and reproductive processes so that there is a rapid increase in predator numbers as prey become numerous. Predators then consume all or nearly all the prey in a region of habitat. Following this overexploitation of the prey, predators either die or disperse away from the place where prey populations are very low. Successful biological control in this script depends on a rapid response by predators to increases in prey numbers and on high rates of prey consumption. When biological control is realized under this scenario it may occur rather quickly. Prey and predator may persist in some assemblage of smaller populations (e.g., in a metapopulation context); however, prey and predator usually become locally extinct.

In the second model, predators may respond numerically to increases in prey number, but this need not be the case. Instead, predators are often found in moderate to high numbers independent of tetranychid prey. Predators that contribute to these dynamical patterns often have a wide diet breadth and may feed on pollen, fungi, other mites , and even the host plant.. These predators persist in the pest's habitat even when spider mites are scarce or nonexistent. Because these predators do not quickly increase in number in response to prey abundance they are often incapable of exercising control over prey in a short period of time. More frequently, these predators must be allowed to increase in numbers, even over years, before biological control is realized.

In perennial cropping systems we contend that generalist phytoseiid predators often play a primary role in conservation biological control. Furthermore, when more specialized predators are key players in the biological control drama, the dynamical pattern between these predators and their prey is similar to that displayed by generalist predators and their prey. The essence of this pattern is that predators persist even when target pests are scarce in proximity to places where pest mites may be abundant. Discovering what allows the population dynamics of more specialized predators and their prey to match those of generalist predators and their prey is often the key to making conservation biological control work.

In what follows, we first review the evidence from perennial crop systems in North America with respect to characteristics of successful and unsuccessful phytoseiid predators. We then present results from experiments in New York apple agroecosystems that examine this in greater detail. Finally, we elucidate factors that contribute to, or constrain, persistence of generalist phytoseiid predators in perennial systems.

II.   SUCCESSFUL MITE BIOLOGICAL CONTROL IN PERENNIAL CROPPING SYSTEMS

We conducted a survey to determine if conservation mite biological control was successful in several perennial cropping systems and, if so, what species of predator(s) was responsible for suppression of the pest mite. Then we explored what attribute(s) of the predators appeared to influence their success. The survey was a compilation of findings reported in the literature, conversations with experts across the continent who work with mite pests in perennial systems, and our understanding of these systems. The edited volume by Helle and Sabelis (1985) provided much of the natural history information concerning phytoseiid biology. We viewed mite biological control as successful when no acaricides were needed to prevent mite damage for more than two growing seasons. We used this criterion because in perennial systems a conserved biological control agent should limit pest abundance for more than a single year. The survey was not exhaustive but was designed to provide a broad perspective on the conservation of predatory mites in perennial systems in the United States and Canada.

The survey revealed distinct regionalization of successful and persistent control of tetranychid mites by a few species of predatory mites (Table 1). In the northeast (Massachusetts, Michigan, New York, and Ontario Canada) and mid-Atlantic region (North Carolina, Virginia) the focus of mite biological control in perennial systems has been in apples where two phytoseiid species predominate, *Amblyseius fallacis* (Garman) and *Typhlodromus pyri* (Scheuten). In a later section of this chapter, we explore in detail the dynamics between these two predators and European red mite (*Panonychus ulmi* (Koch)). Here we present a more general description of these dynamics that was forged by our and other workers' observations and experiences. *Amblyseius fallacis* provides sporadic and unpredictable biological control of European red mite in apples in the northeast even when chemical pesticides toxic to the predator are not used. *A. fallacis* is most often found in apple trees late in the growing season thereby frequently necessitating the use of miticides. In particular, during hot summers,

A. fallacis seems to provide little control of European red mite. This may be because pest mite numbers increase rapidly before predators are numerous in the trees and because low humidity conditions that may be associated with high temperatures are unfavorable to *A. fallacis*. During cooler summers, however, predators are better able to colonize and increase in numbers in the arboreal habitat before *P. ulmi* become numerous.

Table 1. Predators commonly found in perennial production systems in North America that offer some level of control of phytophagous mites.

Predator	Pest(s)[a]	System(s)	Region[b]	Level of Control	Predator Diet
Amblyseius fallacis (Garman)	ERM	Apple	Northeast[c], Southeast[d], Northwest[e]	Inconsistent Inconsistent Inconsistent	mites, perhaps pollen
	TSSM	Mint Strawberry	Northwest[e] Northwest[f]	Consistent Consistent	
Euseius hibisci (Chant)	ERM	Almonds	Southwest	?	fungi, plant juices, honey dew, pollen, mites, citrus thrips
	ABM	Avocado	Southwest	Consistent	
	CRM	Citrus	Southwest[g]	Consistent	
Euseius tularensis Congdon	CRM	Citrus	Southwest[h]	Consistent	fungi, plant juices, honey dew, pollen, mites, citrus thrips
Metaseiulus occidentalis (Nesbitt)	ERM CRM, PSM, SM, TSSM	Almonds	Northwest	Consistent	Mites
	ERM ABM PSM	Apple- ERM Avocado- ABM Grape- PSM	Northwest[e,i] Northwest Southwest	Consistent	

Continues

Typhlodromus	ERM	Apple	Northeast	Consistent	fungi, plant
pyri			and		juices, pollen,
Scheuten			Northwest[e,i]		mites
	TSSM	Caneberry	Northwest[e,i]	Consistent	
	ERM	Grape-	Northeast	Consistent	
		ERM	and		
			Northwest	?	

[a]ABM = Avocado brown mite, *Oligonychus punicae* (Hirst), BAM = Brown almond mite, *Bryobia rubrioculus* (Scheuten), CRM = Citrus red mite, *Panonychus citri* (McGregor), ERM = European red *mite, Panonychus ulmi* (Koch), PSM = Pacific spider mite, *Tetranychus pacificus* McGregor, SM = Strawberry mite, *Tetranychus turkestani (Ugarov* & Nikolski), TSSM = Two-spotted spider mite, *Tetranychus urticae Koch*
[b]Northeast = Michigan, New England States, Ontario; Southeast = North Carolina; Northwest = British Columbia, Northern California, Oregon, Washington; Southwest = Southern California
[c]B. D. Solymar, Brock University, St. Catherines, Ontario
[d]J. F. Walgenbach, Mountain Horticultural Crops Res. Stn. Fletcher, NC 28732
[e]B. A. Croft, Dept. of Entomology, Oregon State University, Corvallis, OR 97331
[f]D. A. Raworth, Agric. Canada Res. Stn. Vancouver, British Columbia
[g]J. G. Morse, Dept. Of Entomology, University of California, Riverside, CA 92521
[h]E. Grafton-Cardwell, Kearney Agric. Cent. University of California, Parlier, CA 93648
[i]J. E. Dunley, Tree Fruit Res. Cent., Washington State University, Wenatchee, WA 98801

Where it is found, *Typhlodromus pyri* is an effective and persistent biological control agent albeit one that requires one or more years before its full impact can be seen. The success of this predator can be attributed to two factors. First, it remains in the tree year round, showing little tendency to disperse even when tetranychid prey are scarce. Second, but related to the first, *T. pyri* can feed and survive on a number of food items, including pollen, fungi, and other mites, such as the apple rust mite *Aculus schlechtendali* (Nalepa). Because of this general food habit, *T. pyri* can maintain itself at relatively high densities when European red mite prey are scarce (Walde *et al.,* 1992; McMurtry and Croft, 1997). Although *T. pyri* does not respond rapidly to increases in European red mite numbers once the predator is established in apple trees it persists in the tree at densities that prevent European red mite from becoming abundant.

In the northwest (British Columbia, Oregon, Washington), predatory mites control phytophagous mites in a variety of perennial crops. In mint, *A. fallacis* provides consistent biological control of *T. urticae*. Persistence of this predator in mint has been attributed to two factors; the humid environment provided by the crop which promotes survival of the predator and minimal use of pesticides because relatively few pests require control. In hops, *A. fallacis* also persists from one year to the next; however, modification of cultural practices is necessary to ensure biological control of *T. urticae*. Predators remain in the plant crown below the soil surface after harvest and overwinter there. The current practices of removing foliage at the base and lower portions of the plant and piling dirt around the base, however, reduces the number of overwintering predators that are available to colonize the plant in the Spring (Croft *et al.,* 1993; B. Croft, pers. commun.). In perennial strawberries, *A. fallacis* remains active all winter, tracking spider mite populations and providing consistent, effective biological control.

In tree fruit systems in the northwest three phytoseiids predominate; *A. fallacis, Metaseiulus occidentalis* (Nesbitt), and *T. pyri. Amblyseius fallacis* controls mite pests in apple and cherry root stock plantings but not in commercial orchards. Its presence in roots stock is probably due to the more humid environment (Croft et *al.,* 1993; Nyrop and Roda, unpublished). Apple and cherry seedlings are regularly watered providing a humid environment for *A. fallacis* whereas in early Spring, it is hypothesized that the apple orchard canopy does not provide enough humidity for this predator. *Amblyseius fallacis* are not sufficiently abundant in commercial orchards early enough in the growing season to offer effective mite control, a situation analogous to the northeast. In more humid regions, *T. pyri* does well in orchards that are not treated with pesticides toxic to the predator. This predator persists on alternative food sources (e.g., pollen) and maintains numbers sufficient to prevent pest mites (primarily European red mite) from reaching damaging levels. The dynamics between the predator and prey are similar to those reported in New York (see Section III).

In drier climates, *M. occidentalis* provides successful biological control of European red mite and two spotted spider mite. The success of this predator is based on three life history traits: it persists in an arid (20 to 30% humidity) environment, it readily disperses into orchards, and it has sufficiently high reproductive and prey consumption rates so that it can suppress moderate to high densities of prey. However, even with these biological attributes, successful use of *T. occidentalis* in a conservation biological control program hinges on maintaining prey in the crop so that a cycle of boom and bust dynamics between prey and predator does not occur. In this regard, *M. occidentalis* is quite different from *T. pyri* whose numbers can remain quite high even when other mite prey are absent. However, by maintaining alternate prey in the system, dynamics of *M. occidentalis* becomes more like that of *T. pyri.*

In the southwest (primarily California) conservation biological control of spider mites has been successful in four perennial crops; citrus, almonds, avocados, and grapes. In almonds and grapes, *M. occidentalis* can provide consistent biological control of spider mites. However, as in apples, prey must be maintained in the crop to prevent the predator from dispersing away or from starving. In grapes, the primary pest is the Pacific mite *T. pacificus* (McGregor) and alternative prey are tydeiids and the Willamette mite *Eotetranychus willamettei* Ewing. In almonds, several species of spider mites may occur as pests: *P. citri, P. ulmi, Bryobia rubrioculus* (Scheuten), *T. urticae, T. pacificus,* and *T. turkestani* (Ugarov and Nikolski). It is perhaps the number of potential pest mites and their different temporal patterns that provide the key to maintaining *M. occidentalis* in this system.

In avocado and citrus, generalist predators are credited with providing successful biological control of pest mites. These mite predators belong to the genus *Euseius* which are mostly found in arboreal habitats and are polyphagous: some species achieve their highest reproduction when feeding on pollen. In citrus, *Euseius tularensis* (Congdon) effectively controls citrus red mite *(P. citri* (McGregor)) although this may be, in part, because the trees can tolerate high densities of mites. This predator can persist in the system even when citrus mite numbers are low due to its ability to feed on other food items such as pollen and the tree itself. Predators actively feed on growing tissue abundant in spring and fall and pruning trees stimulates a growth flush which has been correlated with increased numbers of predators. While *E. tularensis* is an effective predator, citrus mite outbreaks can occur during hot and dry weather, possibly due to its negative effects on the predator.

In avocado orchards, spider mite management of *Oligonychus punicae* (Hirst) mainly occurs through predation by *Euseius hibisci* (Chant) and along coastal California by *T. limonicus* (Garrard and McGregor). Avocados tolerate moderate infestations of *0. punicae* and damaging populations occur only sporadically unless orchards are treated with broad-spectrum pesticides. *Euseius hibisci* offers natural control, but usually is not able to suppress outbreak populations due to its relatively low reproductive rate. In addition, it does not aggregate and oviposit on leaves infested with *0. punicae* and it is inhibited from attacking prey protected under webbing. *Euseius hibisci* is also believed to control *Eotetranychus sexmaculatus* (Riley), a more severe pest. *Euseius hibisci* is also found in almonds where it can control pest mites.

Several tentative conclusions can be made based on our survey of conservation biological control in perennial systems in North America. First, persistent biological control of spider mites in perennial cropping systems is certainly possible. Second, effective, persistent biological control depends on maintaining predaceous mites in the cropping system and in close proximity to the pest mites. For generalist predators this may only require that pesticides be managed so that predaceous mites are conserved. For more specialized predators alternate prey or low levels of pest mites must be maintained. Third, specific predator species seem to be most effective on specific types of plants. For example, *A. fallacis* seems to be most effective in low herbaceous plants such as strawberries, mint, and caneberries (raspberries and blackberries) whereas its effectiveness in apples is more sporadic. Fourth, the association with plant type may also be correlated with performance in different climates and microclimates. Referring again to *A. fallacis,* it has been suggested that this predator is better adapted to more moist conditions and this may reflect its success as a biological agent in low growing plants. Fifth, it appears that there is but one specialist predator and one generalist predator that predominates in each system. If the generalist is not circumscribed by pesticides, it predominates over the specialist once spider mites are reduced to low levels. Finally, the number of predator species that have been successfully used is very low. Why this is so can only be guessed at, but the ability to tolerate pesticides is probably a key factor.

III. PATTERNS OF MITE PREDATOR-PREY DYNAMICS IN NEW YORK APPLES

In New York apple orchards two phytoseiid mite predators can be found: *T. pyri,* a polyphagous predator and *A. fallacis,* an oligophagous and more specialized predator of tetranychids. *Typhlodromus pyri* has been shown to provide effective and persistent biological control of the primary mite pest of apples, the European red mite (Hardman *et al.,* 1991; Walde *et al.,* 1992; Blommers, 1994). This predator is effective primarily because it is able to persist in moderate to high numbers in trees even when prey are scarce. Until recently it was thought that *A. fallacis* overwintered almost exclusively beneath apple trees in ground cover and moved into apple trees following increases in tetranychid mite densities therein.

Because movement from the ground cover to the trees caused a lag in the response by the predators to increases in pest mite abundance biological control was often not fully effective. However, Nyrop *et al.* (1994) showed that *A. fallacis* also overwintered in trees and often in high numbers. Thus, it might be possible to manage the apple system to better conserve *A. fallacis* in the tree habitat and thereby improve biological control. This prompted experiments to investigate the dynamics of *A. fallacis* and T. *pyri* in consort with their prey *P. ulmi.* We briefly describe the results of these experiments in order to illustrate how attributes of phytoseiids related to persistence can greatly influence predator/prey dynamics and the effectiveness of biological control. Methodology and other details can be found in Nyrop *et al.* (1994).

Dynamics of the two phytoseiids and their European red mite prey were compared over four years (1992-1996) in a planting of apple trees arranged into four blocks each approximately 1.5 ha in size. Prior to 1991, insecticides toxic to *T. pyri* and *A. fallacis* were used in this orchard. In 1991 trees in blocks 2 and 4 were inoculated with *T. pyri*. No predators were released into blocks 1 and 3 because we anticipated that *A. fallacis* would naturally invade these sites. Also at this time, the entire orchard was placed under a pesticide regime benign to both phytoseiid species. In addition to following predator/prey dynamics in these blocks during the growing season, beginning in 1992 we assessed numbers of live phytoseiids on twigs during the winter and early spring.

In 1991, as many as 70 motile European red mites per leaf were found throughout the orchard planting despite two miticide applications. At the end of the 1991 growing season predators were present in all four blocks. However, because densities were low (< 0.1/leaf), we could not estimate with precision the proportion that were *T. pyri* or *A. fallacis*. *Typhlodromus pyri* was detected in blocks 2 and 4 but not blocks 1 and 3. Before bloom in 1992, the entire orchard was treated with petroleum oil to control European red mite; *T. pyri* apparently suppressed this pest for the remainder of the growing season in blocks 2 and 4. In blocks 1 and 3, where *A. fallacis* was found exclusively, two miticide applications were required to keep pest mite densities below threshold levels (Fig. 1). This is a somewhat surprising because densities of *T. pyri* were quite low and appeared to not be abundant enough to regulate European red mite numbers. *Typhlodromus pyri* was at times more than twice as abundant as *A. fallacis,* however.

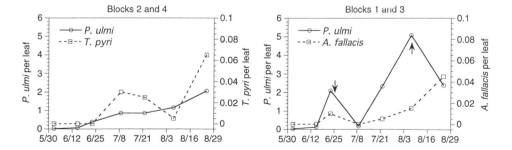

Figure 1.   Dynamics between *Typhlodromus pyri* and European red mite (Blocks 2 and 4) and *Amblyseius fallacis* and European red mite (Blocks 1 and 3) at Geneva, NY during 1992. Arrows indicate timing of miticides.

In order to persist in a temperate climate like New York phytoseiids must be able to successfully overwinter. During the winter of 1992 and 1993, numbers of live phytoseiids overwintering on tree branches were estimated eight times and the temporal patterns are shown in Fig. 2. Most striking is that numbers of *T. pyri* were constant while densities of live *A. fallacis* declined dramatically. Based on this result, it is not surprising that the numbers of *A. fallacis* are often quite low at the beginning of the growing season and that it takes time to recolonize trees from other habitats. Some *T. pyri* were recovered in blocks 1 and 3, a result either of dispersal from the blocks into which they were inoculated (2 and 4) or of increases of *T. pyri* resident in the orchard block before the start of the experiment.

The pattern of predator-prey dynamics observed in 1992 was repeated in 1993 (Fig. 3). In blocks 2 and 4, *T. pyri* completely suppressed European mite population growth whereas in blocks 1 and 3 two miticide applications were again needed to control pest mite numbers. By the end of the growing season *A. fallacis* was again numerous in blocks 1 and 3. During winter to spring of 1993-1994, phytoseiid numbers were only measured three times but patterns of survival were similar to those obtained in 1992; with *A. fallacis* disappearing after the first sample and *T. pyri* numbers remaining relatively constant for the first two samples. Numbers of *T. pyri* increased greatly on the last sample date (22.5 mites per funnel) probably

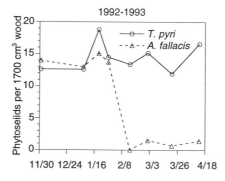

Figure 2. Survival of *Typhlodromus pyri* and *Amblyseius fallacis* during winter 1992-1993 at Geneva, NY.

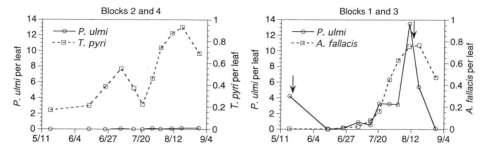

Figure 3. Dynamics between *Typhlodromus pyri* and European red mite (Blocks 2 an 4) and *Amblyseius fallacis* and European red mite (Blocks 1 and 3) at Geneva, NY during 1993.

because these predators congregate in developing flowers to feed on pollen (Nyrop, unpubl. data).

During the 1994 and 1995 growing seasons *T. pyri* and European red mite dynamics in blocks 2 and 4 were very similar to those observed the previous two years. *T. pyri* suppressed European red mite population growth (Fig. 4). In blocks 1 and 3 during 1994, *A. fallacis* disappeared from trees; all phytoseiids identified (285) were *T. pyri*. Furthermore, European red mite numbers were kept below 8 motile mites per leaf (data not shown). Numbers of phytoseiids on the trees during the winter of 1994 and 1995 were not estimated because *A. fallacis* could no longer be found in the orchard. In 1995, portions of blocks 1 and 3 were treated with a pyrethroid insecticide to induce increased numbers of European red mite and thereby encourage *A. fallacis* to colonize trees. The pyrethroid application resulted in high numbers of European red mite and by early fall, both *A. fallacis* and *T. pyri* were present in moderate numbers. The number of live phytoseiids during the winter and early spring of 1995 and 1996 was estimated in each blocks 1 and 3, 17 times (Fig. 5). Survival of *T. pyri* and *A. fallacis* was very similar, although numbers of *T. pyri* were higher. However, beginning 3 June and then for three additional weekly samples no *A. fallacis* were found in blocks 1 and 3. All phytoseiids recovered and identified

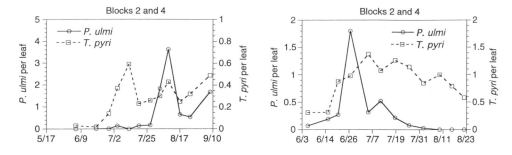

Figure 4. Dynamics between *Typhlodromus pyri* and European red mite at Geneva, NY during 1994 (left) and 1995 (right).

(85) were *T. pyri*. Dynamics of *T. pyri* and European red mite in blocks 2 and 4 were similar to those in previous years (Fig. 5).

We hypothesized that the absence of *A. fallacis* in trees in the Spring following successful overwintering may result from dispersal out of the trees after diapause termination. An experiment was conducted in the laboratory to assess the dispersal tendencies of the two predators. Predators that had been starved for 24 hours were placed on apple leaves in wind tunnels. A fine mesh screen was placed downwind from the leaves and the number of phytoseiids that dispersed over a 30 minute period was determined by enumerating predators collected from the screens. Both predators dispersed when starved, however, significantly more (p < 0.01) *A. fallacis* dispersed (59%) than *T. pyri* (34%).

Typhlodromus pyri provided consistent, effective biological control of European red mite once the predator became established and initially suppressed the pest mite. Causes for these patterns are threefold. First, *T. pyri* remained at relatively high densities by feeding on alternative foods when European red mite were scarce. During 1992 and the latter part of 1995, European red mite numbers were very low yet *T. pyri* remained abundant. Second, *T. pyri* had less propensity to disperse out of trees when tetranychid prey were scarce. Finally, *T. pyri* was better able to survive winter conditions in trees. In contrast, *A. fallacis* did not persist in trees because it tends to disperse from trees when tetranychid prey are scarce and predators in trees

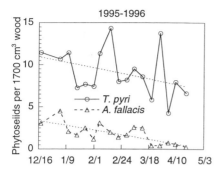
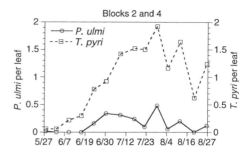

Figure 5. Survival of *Typhlodromus pyri* and *Amblyseius fallacis* during winter 1995-1996 at Geneva, NY and dynamics between *Typhlodromus pyri* and European red mite at Geneva, NY during 1996.

do not survive winters well. Even when *A. fallacis* survived in trees from fall to the following early spring this predator was generally absent from trees between just before bloom until mid-summer. The probable reason for this is dispersal out of trees after breaking diapause. Predators are active in the spring before European red mite eggs hatch so the predators have no prey to feed on. These patterns are similar to those reported for *M. occidentalis* and *T. pyri* in the northwest (Croft and McRae 1992). In this system, *M. occidentalis* persists in trees if apple rust mite are common and if other competitors (e.g., *Zetsellia mafi* (Ewing)) are eliminated.

In the *T. pyri*/European red mite system persistence of the predator in the prey's habitat is the key to effective biological control. Spatial aggregation, functional, and numerical responses have little if any bearing on the long-term outcome of these interactions.

IV. GENERALIZATIONS AND FUTURE RESEARCH

In the introduction we proposed two alternative models for explaining successful conservation biological control of spider mites in perennial systems.

Evidence from the literature, experience from field practitioners in North America, and our own research supports the notion that the ability of phytoseiid predators to persist in the cropping system is of critical importance. We find less support for the colonization model which requires rapid colonization of the crop plant from outside sources when spider mite problems develop. This model may be of more use in describing biological control when colonization can be aided by augmentative releases. Augmentative release has its place in biological control, but it violates the objective of conservation biological control to maintain pest populations below damaging levels without regular intervention. Given this objective, we believe a focus on persistence within the crop is warranted.

In this section, we will attempt to elucidate factors likely to contribute to, or constrain, persistence of phytoseiid predators in perennial systems. We consider the role of pesticides, diet specialization, tolerance to temperature and humidity conditions, overwintering ability, habitat preference, and host plant traits. During this discussion, we will also ask the question, can the cropping system and/or mites be manipulated in order to enhance phytoseiid persistence? Finally, we sketch out a number of areas of research that may be profitable to pursue in the future.

A.   Characteristics That May Enhance Persistence Traits of Phytoseiids

1.   Ability to tolerate pesticides

As mentioned in the introduction to this chapter, an ability to tolerate pesticides commonly used in most perennial crops is of overriding importance. Of the hundreds of species of phytoseiid predators that have the ability to persist in perennial systems, perhaps only a handful are able to do so in the face of some pesticide use. Evidence of this comes from comparing species of phytoseiid predators found in abandoned vines or trees with species found in commercial plantings as well as from manipulative experiments (McMurtry *et al.,* 1970; Amano and Chant, 1990; Croft, 1990).

Therefore, the ability to tolerate some pesticide use is a serious constraint on successful conservation of phytoseiid predators. A number of approaches have been attempted or suggested to alleviate this constraint. The continued development and use of more selective insecticides such as Bt-type bacteria and insect growth regulators would be of enormous benefit. We know that some fungicides, such as ethylene bisdithiocarbamates (e.g., mancozeb), can also be very deleterious to phytoseiid predators (Hagley and Bings, 1989; Zacharda and Hluchy, 1991; Gyorffy-Molnar and Polgar, 1994). There is increasing coordination of disease and insect control programs, taking into consideration the potential of some fungicides to seriously disrupt biological control of spider mites.

A second approach to reduce the negative impact of pesticides on phytoseiid predators is to breed for increased tolerance and/or resistance. Resistance to some pesticides has developed naturally in some systems although this is a slow process and not predictable (Croft, 1990). Natural variation in pesticide resistance can be enhanced through artificial selection in the laboratory (reviewed by Hoy, 1985; Croft, 1990). For example, Hoy and her colleagues have selected *M. occidentalis* for increased resistance to sulfur, organophosphates, carbamates and/or pyrethroids and then released these mites back into the field. Success in establishing and maintaining artificially selected pesticide resistant phytoseiid mites in the field has been mixed, however. The ability to genetically engineer resistance to some pesticides may expand the potential for biological control (Presnail and Hoy, 1994).

2. Ability to use alternative food items

Next to tolerance of pesticides, degree of diet specialization may be the single most important factor contributing to persistence of phytoseiids in perennial systems. A spider mite specialist such as *P. persimilis* Athias Henriot, with its high reproductive capacity when prey are available, is frequently able to drive prey populations to very low levels (Sabelis, 1985a). These low population levels lead to either death of the phytoseiid or dispersal out of the system (Burnett, 1979; Strenseth, 1985; but see Gough, 1991). Having the ability to feed on other food sources, such as pollen and fungi or even plant exudates, however, allows predators to persist in the absence of spider mite prey, sometimes at surprisingly high densities (McMurtry and Rodriguez, 1987). The distribution and abundance of such species as *T. pyri, T. victorensis* Womersley, *E. hibisci*, and *T. caudiglans* Schuster, that can feed and reproduce on pollen and perhaps fungi, is frequently found to be independent of the distribution and abundance of spider mite prey (Smith and Papcek, 1991; McMurtry, 1992; Karban *et al.*, 1995). The impact of this on predator/prey dynamics may be very important. There is good laboratory evidence that these generalist species can survive and reproduce on these alternative food sources, sometimes at rates greater than when fed spider mites (McMurtry and Rodriguez, 1987; McMurtry, 1992). The role of fungi and pollen under field conditions, however, has not been very well studied experimentally (but see Flaherty and Hoy, 1971; Kennett *et al.*, 1979; Osakabe *et al.*, 1987). The observation that *T. pyri* populations in apples will increase in the spring without any mite prey strongly argues, however, for an important role of pollen or fungi. To our knowledge, no attempt has been made to augment fungal spores in the field for the purposes of enhancing phytoseiid populations.

An ability to survive and reproduce on non-mite food resources has distinct advantages. Pollen and fungal spores are available during most of the growing season, as people allergic to some of these airborne agents can readily attest (Gregory, 1974). Although there are likely to be peaks in availability of these food items (e.g., Kennett *et al.,* 1979) as well as variation in quality for particular species of phytoseiids (Ouyang *et al.,* 1992) sufficient amounts are probably present to maintain phytoseiid populations at modest levels. Two important questions that need to be addressed, however, are, (1) for a particular crop system, what constitutes an effective minimum density to allow the predator to adequately respond to spider mite hot spots? and (2) how does the distribution pattern of the phytoseiid within a plant or field interact with predator density in influencing effectiveness? These questions seem particularly relevant given that, as a general rule, generalist phytoseiids do not show as strong a numerical or functional response to spider mite prey as specialists (Sabelis, 1985b; McMurtry, 1992; McMurtry and Croft, 1997).

There are circumstances when mite specialists can persist in a system in the absence of the primary spider mite pest. This requires availability of alternative mite prey. Work in both grapes and apples involving *M. occidentalis* supports this view. In California vineyards, the Pacific spider mite *T. pacificus* is considered the primary pest spider mite (Flaherty and Huffaker, 1970; Flaherty *et al.,* 1992). Another tetranychid mite, the Willamette mite *E. willametti,* is also frequently present and recently, probably is more common than the Pacific mite. There is good evidence from laboratory studies that *M. occidentalis* has a preference for Pacific mites although it will readily feed on Willamette mites (Hoy and Smilanick, 1981; Hanna and Wilson, 1991). Flaherty and Huffaker (1970) hypothesized that Willamette mites, or possibly thudded mites, serve as important alternative prey early in the season allowing *M. occidentalis* to persist in the system in the absence of Pacific mites. When Pacific mite populations begin to appear later in the season, the predator is present in the system and able to switch over to it, thereby exerting control. Field studies have confirmed this shared predator effect (English-Loeb *et al.,* 1993; Karban *et al.,* 1994). It is likely that these types of indirect interactions are important in other predator/prey systems as well (Holt and Lawton, 1993).

A similar pattern apparently occurs in apples on the west coast. The apple rust mite has been suggested as an important alternative prey for *M. occidentalis,* allowing this predator to persist in apple trees (Hoyt, 1969; Croft and Hoying, 1977). Given its strong numerical and functional response to European red mite and two spotted spider mite (the key mite pests), even low numbers of *M. occidentalis* are able to respond to hot spots of these mites. Without these alternative prey, however, we would hypothesize that *M. occidentalis* would become very rare or disappear from the system. We predict the predator/prey relationship would be destabilized resulting in outbreaks.

Persistence of phytoseiids in these systems may not be sufficient when the goal is to keep spider mite pests below economic thresholds. It may also be necessary, in some cases, that the predator or predators show some level of preference for spider mite prey when available. However, as long as they feed on tetranychid prey to some extent, pollen in the spring may allow phytoseiid populations to grow to sufficient densities that they are able to suppress pest mites in the absence of a strong preference for them (e.g., McMurtry and Scriven, 1966). At this point, it is difficult to predict to what extent a predator needs to show a preference for the pest species for it to still be effective. This will undoubtedly depend on a number of factors including intrinsic growth rate of the pest species, distributions and densities of both prey and predator, and economic threshold.

3. Life-history constraints

The ability to tolerate the abiotic environment, particularly temperature and humidity, plays a major role in determining geographic distribution patterns of phytoseiid predators as well as many other ectothermic organisms (Andrewartha and Birch 1954; McMurtry and Scriven 1965; Croft *et al.*, 1993). Based on mostly laboratory work, a considerable amount of within and among species variation exists among phytoseiid predators with regard to temperature and humidity tolerances (Sabelis, 1985c; van Dinh *et al.*, 1988, Croft *et al.*, 1993). Presumably these differences constrain the regions and habitats where a species can be an effective control agent although this has rarely been explicitly examined.

Given constraints on temperature and humidity tolerances, is there anything that can be done about it? Two approaches come to mind. First, we may be able to take advantage of naturally occurring intraspecific genetic variation in tolerances. There is observational and some experimental evidence suggesting this variation exists for some predator species (McMurtry, 1980). Similar to breeding for pesticide resistance, it may be possible to select for greater tolerance to temperatures or humidity in specific phytoseiid species. It might also be possible to move races of phytoseiids that have adapted to certain abiotic conditions to locations where these races may better provide biological control. For example, Ehler and Frankie (1979) reported finding *T. pyri* from oak trees in Austin, Texas. Phytoseiids from this location certainly experience drier and warmer conditions than occur in western New York and these predators may be more tolerant to heat and moisture stress.

In order to persist in a perennial system, a phytoseiid not only must be able to tolerate the abiotic conditions during the growing season but it must also be able to survive the winter. This is especially true in more temperate climates. As summarized earlier, *T. pyri* is better able to survive winters in upstate New York than *A. fallacis* and this may be a contributing factor for why it is a more effective biological control agent. Whether a species enters reproductive diapause will significantly influence its ability to overwinter in more temperate locations. Considerable within and among species variation exits in the propensity to enter diapause and the conditions under which this occurs (Overmeer, 1985; Veerman, 1992). Such variation may help explain why "effective" control agents cannot be sustained over several seasons. Food availability during the winter may play a role in the ability to overwinter. Diapausing females tend to be much more sedentary than nondiapausing females. However, they will become active on warm days. Although diapausing females are better able to survive starvation than nondiapausing females, lack of food can greatly increase mortality (e.g., James, 1989). The tendency for phytoseiids to become active earlier in the Spring than their spider mite prey indicates that the availability of alternative food may be critical at this time as well (Overmeer, 1985).

Species of phytoseiids show distinct habitat preferences (Chant, 1959; Schuster and Pritchard, 1963; McMurtry *et al.,* 1970; Hadam *et al.,* 1986; Rothman, 1988). For example, *M. occidentalis* and *T. pyri* are found primarily in trees and vines rather than herbaceous plants. Conversely, *A. fallacis,* although found in trees, appears better adapted to low growing shrubs (Croft *et al.,* 1993). We know little about the mechanisms by which these habitat preferences are expressed nor all their ecological ramifications. It is safe to say, however, that a predator with a preference for low shrubs is less likely to persist in tree environments and visa versa. This may help explain why *A. fallacis* is considered a successful biological control agent for such crops as strawberry (when grown as a perennial), mint, and cane fruit while it is less successful in tree crops such as apples. This also may help explain why low growing cover crops may not help maintain *A. fallacis* populations in fruit trees (Nyrop *et al.,* 1994).

B.   Characteristics That May Enhance Persistence Traits of the Host Plant

Being tiny, ectothermic animals phytoseiids are very dependent on microsite conditions. Hence, variation in the structure of the leaf, in particular degree of leaf pubescence, is likely to have an important influence on persistence. Although we find exceptions, there often is a positive association between leaf hairiness and phytoseiid abundance (Downing and Moilliet, 1967; Overmeer and van Zon, 1984; Duso, 1992; Karban *et al.*, 1995; Walter and O'Dowd, 1995; Walter, 1996). The reasons for this association have not been well studied. Presence of leaf hairs may help ameliorate high temperature and low humidity conditions, provide protection from predators, or help capture food items such as pollen and fungal spores.

In a survey of 20 species of grapes planted into a common garden Karban *et al.* (1995) found that the presence of hairs along veins and in vein axes were positively associated with abundance of *T. caudiglans*. Moreover, they found that these leaf characteristics were more important than spider mite prey (also see Tuovinen and Rokx, 1991). The hairs within the vein axes deserve special mention. These structures have been called acrodomatia and have been recognized by botanists for many years (Lindstrom, 1887). Leaf domatia are specialized structures situated in vein axes on the underside of leaves. These structures have been described for a wide variety of plant species across a large number of plant families from many geographic regions of the world (O'Dowd and Willson, 1989; Pemberton and Turner, 1989; Brouwer and Clifford, 1990; Willson, 1991; O'Dowd and Willson, 1991; O'Dowd and Pemberton, 1994; Rozario, 1995; Walter and O'Dowd, 1995; Walter, 1996). This survey work clearly shows that domatia are frequently associated with fungaceous and predaceous mites.

Manipulative experiments have also been conducted which support the contention that the presence of domatia favors some species of phytoseiids. Walter and O'Dowd (1992a, 1992b) showed that when domatia were experimentally "removed" from a rain forest tree and a garden shrub, respectively, that densities of phytoseiids declined relative to densities on control leaves. In a laboratory study, Rozario (as cited by Walter, 1996) found that when she added artificial domatia to leaves of cultivated grape varieties that normally lack domatia densities of *M. occidentalis* increased. A similar response was found for *T. caudiglans* when artificial shelters were added to leaf disks (Putman and Herne, 1964).

These experimental and correlative studies suggest that structures such as leaf domatia can be important factors in determining persistence of phytoseiid mites (see Chapter 5). What benefits are they providing the mites? This has not been well studied but modification of leaf microclimate may be particularly important. Phytoseiid mites appear to be sensitive to low humidity conditions (Strenseth, 1979; Swift and Blaustein, 1980; van Dinh *et al.*, 1988; Bakker *et al.*, 1993; Croft *et al.*, 1993; Grostal and O'Dowd, 1994). Leaf domatia may help ameliorate extreme environmental conditions like low humidity or low temperatures. Evidence for this comes from work by Grostal and O'Dowd (1994). In the laboratory, they found that as they lowered relative humidity in growth chambers that female *M. occidentalis* laid an increasing proportion of eggs within domatia rather than on other leaf surfaces.

Leaf domatia are very common in woody perennial plants and potentially serve an important function in plant defense against spider mites and other leaf parasites by providing a safe refuge for beneficial mites. Interestingly, domatia tend to be poorly developed in many but not all perennial crop plants. By incorporating leaf traits such as domatia or leaf pubescence into our cultivated crops we may be able to improve persistence of beneficial mites. This area is ripe for future research. Host plant characteristics may also influence ability to overwinter. Phytoseiid species vary in their site selection for overwintering and it is possible that different sites vary in protection from extreme conditions (Overmeer, 1985; Veerman, 1992). Host plant characteristics such as rugosity of bark and size and shape of bud scales may be important, although little data is available on this subject. Cultural manipulations may also help increase winter survival of phytoseiids. For example, Fischer-Colbrie and El-Borolossy (1988) (as cited by Veerman, 1992) found phytoseiid winter survival was much higher on trees which had the bark treated with a lime mixture to protect against frost damage compared to survival on trees not treated. In another example, Morris *et al.* (1996) report that the addition of debris to mint fields increases survival of *A. fallacis,* presumably by providing protection against extremes in temperature and humidity.

C. Establishing and Fostering Phytoseiid Mites in Perennial Systems

One of the somewhat surprising findings from our survey of experts was that only a handful of phytoseiid species have been found to provide reliable and sustainable control of spider mites in perennial systems in North America. What accounts for this paucity of species given the large number of potential agents found in surrounding native habitat? Given that we are fairly limited in our choices of phytoseiid predators that can persist and exert control we would like to know how to use them in more systems. Three related questions need to be addressed, What limits the distribution of these species? Can we expand their distributions? and What actions can we take to foster this expansion?

A better understanding of what limits the distribution of these successful control agents will provide us with a list of factors that we may be able to manipulate in order to foster their expanded use. Although there is a paucity of data pertaining to this issue we can suggest a number of factors to consider. First, the host plant itself may be very important and leaf characteristics may be particularly influential (see

Chapter 5). For example, bark characteristics and the shape and size of bud scales may be critical to successful overwintering of adult females. Second, some of these phytoseiid predators may require alternative food at key times during the season. For example, availability of pollen or fungal spores may be critical in the spring to allow overwintering populations to increase. Finally, overall plant architecture may have important influences on canopy microclimate. There may be ways to adjust pruning techniques to improve microclimate within the canopy for phytoseiids without being a detriment to other horticultural goals.

V. SUMMARY

It is clear that spider mite populations in perennial crops can be successfully maintained below economic thresholds by phytoseiid predators. However, in many commercial perennial cropping systems regular applications of acaricides are still made to control pest mites. We believe the discrepancy lies, in part, in the difference in pest complexes that occur in these systems and the pesticides available to economically control them. Controlling an insect, fungus, or disease often takes precedence over selective use of a pesticide to maintain predator populations. In some areas the presence of pesticide-resistant pests necessitates the use of insecticides known to decimate predatory mites populations. Given a choice, growers often select pesticides that have the lowest negative impact on natural enemies. However, the benefits of certain pesticides often outweigh the benefits of preserving the natural enemy.

Although economic and market factors may dictate what pesticides are used we believe that conservation biological control using phytoseiid predators is possible and will be substantially aided by understanding and enhancing characteristics that allow the predator to persist in the system. From our survey of the literature, communications with spider mite experts in North America, and our own experience in apples and grapes this can occur in one of two ways, (1) the predator is a generalist and can survive in the crop without spider mite prey by feeding on alternative food sources such as pollen or fungi or (2) the predator is oligophagous and persists in the crop by feeding on alternative mite prey when spider mites are driven to extinction. We find little empirical evidence, however, that spider mite specialist phytoseiids can sustain control over several seasons in perennial systems. This would require recolonization of habitats where spider mites were previously driven to extinction from outside areas. Although specialists such as *P. persimilis* are capable of rapid colonization of spider mite hot spots we speculate that time lags between the inception of these outbreaks and colonization by predators will frequently lead to biological control failures.

That persistence is an important attribute of successful conservation biological control of spider mites is clear but our understanding of the factors that favor persistence of phytoseiid predators in perennial systems is limited. We suggest that increased research activity focused on this question would pay dividends. Areas to consider include (1) among and within species tolerance to abiotic factors such as humidity and temperature during the field season and over the winter, (2) influence of crop plant leaf and bark structure on survival and reproductive success, (3) potential for augmenting alternative food sources such as pollen, and 4) among and within species variation in propensity to switch between spider mite and non-spider mite food sources. For perennial systems, we argue that persistence is likely to be a much more important variable in maintaining conservation biological control of spider mites than exactly how many prey are consumed per individual or the search efficiency.

REFERENCES

Amano, H., and Chant, D. A. (1990). Species diversity and seasonal dynamics of Acari on abandoned apple trees in southern Ontario, Canada. *Exp. Appl. Acarol.* 8, 71-96.

Andrewartha, H. G., and Birch, L. C. (1954). "The Distribution and Abundance of Animals." The University of Chicago Press, Chicago, IL.

Bakker, F. M., Klein, M. E., Mesa, N. C., and Braun, A. R. (1993). Saturation deficit tolerance spectra of phytophagous mites and their phytoseiid predators on cassava. *Exp. Appl. Acarol.* 17, 97-113.

Blommers, L. M. H. (1994). Integrated pest management in European apple orchards. *Annu. Rev. Entomol.* 39, 213-242.

Brouwer, Y., and Clifford, H. (1990). An annotated list of domatia-bearing species. *Notes from the Jodrell Laboratory* 12, 1-33.

Burnett, T. (1979). An acarine predator-prey population infesting roses. *Res. Popul. Ecol.* 26, 227-234.

Chant, D. A. (1959). Phytoseiid mites (Acarina: Phytoseiidae). Part I. Bionomics of seven species in southeastern England. Part II. A taxonomic review of the family Phytoseiidae, with descriptions of 38 new species. *Can. Entomol. Suppl.* 12.

Croft, B. A. (1990). "Arthropod Biological Control Agents and Pesticides." Wiley lnterscience. New York, NY.

Croft, B. A., and Hoying, S. A. (1977). Competitive displacement of *Panonychus ulmi* (Acarine: Tetranychidae) by *Aculus schlechtendafi* (Acarina: Eriophyidae) in apple orchards. *Can. Entomol.* 109, 1025-1034.

Croft B. A., and MacRae, I. V. (1992). Persistence of *Typhlodromus pyri* and *Metaseiulus occidentalis* (Acari: Phytoseiidae) on apple after inoculative release and competition with *Zetsellia mali* (Acari: Stigmaeidae). *Environ. Entomol.* 21, 1168-1177.

Croft, B. A., Messing, R. H., Dunley, J. E., and Strong, W. B. (1993). Effects of humidity on eggs and immatures of *Neoseiulus fallacis, Amblysieus andersoni, Metaseiulus occidentalis* and *Typhlodromus pyri* (Phytoseiidae): implications for biological control on apple, caneberry, strawberry and hop. *Exp. Appl. Acarol.* 17, 451-459.

Downing, R. S., and Moilliet T. K. (1967). Relative densities of predacious and phytophagous mites on three varieties of apple trees. *Can. Entomol.* 99, 738-741.

Duso, C. (1992). Role of *Amblyseius aberrans (Oud.), Typhlodromus pyri* Scheuten and *Amblyseius andersoni* (Chant) (Acari, Phtyoseiidae) in vineyards. III. Influence of variety characteristics on the success of *A. abberans* and *T. pyri* releases. *J. Appl. Ecol.* 114, 455-462.

Ehler, L. E., and Frankie, G. W. (1979). Arthropod fauna of live oak in urban and natural stands in Texas. II. Characteristics of the mite fauna (Acari). *J. Kans. Entomol. Soc.* 52, 86-92.

English-Loeb, G., Karban, R., and Hougen-Eitzman, D. (1993). Direct and indirect competition between spider mites feeding on grapes. *Ecol. Appl.* 3, 699-707.

Flaherty, D. L., and Huffaker, C. B. (1970). Biological control of pacific mites and Willamette mites in San Joaquin Valley vineyards. I. Role of *Metaseiulus occidentalis.* II. Influence of dispersion patterns of *Metaseiulus occidentalis. Hilgardia* 40, 267-330.

Flaherty, D. L., and Hoy, M. A. (1971). Biological control of pacific mites and Willamette mites in San Joaquin vineyards: Part III. Role of thudded mites. *Res. Popul. Ecol.* 13, 80-96.

Flaherty, D. L., Christensen, L. P., Lanini, W. T., Marois, J. J., Phillips, P. A., and Wilson, L. T. (1992). "Grape Pest Management." 2nd ed. Univ. Calif., Div. Agric. Nat. Res. Oakland, CA.

Gough, N. (1991). Long-term stability in the interaction between *Tetranychus urticae* and *Phytoseiulus persimilis* producing successful integrated control on roses in southeast Queensland. *Exp. Appl. Acarol.* 12, 83-101.

Gregory, P. H. (1974). "The Microbiology of the Atmosphere." 2nd ed. John Wiley and Sons. New York, NY.

Grostal, P., and O'Dowd, D. J. (1994). Plants, mites and mutualism: leaf domatia and the abundance and reproduction of mites on *Viburnum tinus* (Caprifoliaceae). *Oecologia* 97, 308-315.

Gyorffy-Molnar, J., and Polgar, L. A. (1994). Effects of pesticides on the predatory mite *Typhlodromus pyri* Scheuten - a comparison of field and laboratory results. *Bull. OILB/SROP.* 17, 21-26.

Hadam, J. J., AliNiazee, M. T., and Croft, B. A. (1986). Phytoseiid mites (Parasitiformes: Phytoseiidae) of major crops in Willamette Valley, Oregon, and pesticide resistance in *Typhlodromus pyri* Scheuten. *Environ. Entomol.* 15, 1255-1263.

Hagley, E. A. C., and Bings, A. R. (1989). Effects of three fungicides on populations of a phytophagous and several predacious mites (Acarina) on apple. *Exp. Appl. Acarol.* 6, 253-256.

Hanna, R., and Wilson, L. T. (1991). Prey preference by *Metaseiulus occidentalis* (Acari: Phytoseiidae) and the role of prey aggregation. *Biol. Cont.* 1, 51-58.

Hardman, J. M., Rogers, R. E. L., Nyrop, J. P., and Frisk, T. (1991). Effect of pesticide applications on abundance of European red mite (Acari: Tetranychidae) and *Typhlodromus pyri* (Acari: Phytoseiidae) in Nova Scotian apple orchards. *J. Econ. Entomol.* 84, 570-580.

Helle, W., and Sabelis M. W. (1985). "Spider Mites Their Biology, Natural Enemies and Control." Elsevier. Amsterdam, The Netherlands.

Holt, R. D., and Lawton, J. H. (1993). Apparent competition and enemy-free space in insect host-parasitoid communities. *Amer. Nat.* 142, 623-645.

Hoy, M. A. (1985). Recent advances in genetics and genetic improvement of the phytoseiidae. *Annu. Rev. Entomol.* 30, 345-370.

Hoy, M. A., and Smilanick, J. M. (1981). Non-random prey location by the phytoseiid predator *Metaseiulus occidentalis:* differential responses to several spider mite species. *Entomol. Exp. Appl.* 29, 241-253.

Hoyt, S. C. (1969). Integrated chemical control of insects and biological control of mites on apple in Washington. *J. Econ. Entomol.* 62, 74-86.

James, D. G. (1989). Overwintering of *Amblyseius victoriensis* Womersley (Acarina: Phytoseiidae) in southern New South Wales. *Gen. Appl. Entomol.* 21, 51-55.

Karban, R., Hougen-Eitzman, D., and English-Loeb, G. (1994). Predator-mediated apparent competition between two herbivores that feed on grapevines. *Oecologia* 97, 508-511.

Karban, R., English-Loeb, G., Walker, M. A., and Thaler, J. (1995). Abundance of phytoseiid mites on *Vitis* species: effects of leaf hairs, domatia, prey abundance and plant phylogeny. *Exp. Appl. Acarol.* 19, 189-197.

Kennett, C. E., Flaherty, D. L., and Hoffmann, R. W. (1979). Effect of wind-borne pollens on the population dynamics of *Amblyseius hibisci* (Acarina: Phytoseiidae). *Entomophaga* 24, 83-98.

Lundstroem, A. N. (1887). Von domatien. Pflanzenbiologische studien. II. Die anpassung der pflanzen and thiere. *Nova Acta Regiae Soc. Sci. Ups. Ser.* 313, 1-87.

McMurtry, J. A. (1980). Biosystematics of three taxa in the *Amblyseius finlandicus* group from South Africa with comparative life history studies (Acarina: Phytoseiidae). *Intern. J. Acarol.* 6, 147-156.

McMurtry, J. A. (1992). Dynamics and potential impact of 'generalist' phytoseiids in agroecosystems and possibilities for establishment of exotic species. *Exp. Appl. Acarol.* 14, 371-382.

McMurtry, J. A., and Croft, B. A. (1997). Life-styles of phytoseiid mites and their roles in biological control. *Annu. Rev. Entomol.* 42, 291-321.

McMurtry, J. A., and Rodriguez, J. G. (1987). Nutritional ecology of phytoseiid mites. *In* "Nutritional Ecology of Insects, Mites, Spiders, and Related Invertebrates." (F. J. Slansky, Jr., and J. G. Rodriguez, eds.), Wiley-Interscience. New York, NY.

McMurtry, J. A., and Scriven, G. T. (1965). Life history studies of *Amblyseius limonicus* with comparative observations of *Amblyseius hibisci* (Acarina: Phytoseiidae). *Ann. Entomol. Soc. Amer.* 59, 147-149.

McMurtry, J. A., and Scriven, G. T. (1966). Studies on predator-prey interactions between *Amblyseius hibisci* and *Oligonychus punicae* (Acarina: Phytoseiidae, Tetranychidae) under greenhouse conditions. *Ann. Entomol. Soc. Amer.* 59, 793-799.

McMurtry, J. A., Huffaker, C. B., and van de Vrie, M. (1970). Tetranychid enemies: their biological characteristics and the impact of spray practices. *Hilgardia* 40, 331-389.

Morris, M. A., Croft, B. A., and Berry, R. E. (1996). Overwintering and effects of autumn habitat manipulation and carbofuran on *Neoseiulus fallacis* and *Tetranychus urticae* in peppermint. *Exp. Appl. Acarol.* 20, 249-258.

Nyrop, J. P., Minns, J. C., and Herring, C. P. (1994). Influence of ground cover on dynamics of *Amblyseius fallacis* Garman (Acarina: Phytoseiidae) in New York apple orchards. *Agric., Ecosys., Environ.* 50, 61-72.

O'Dowd, D. J., and Pemberton, R. W. (1994). Leaf domatia in Korean plants: floristics, frequency, and biogeography. *Vegetatio* 114, 137-148.

O'Dowd, D. J., and Willson, M. F. (1989). Leaf domatia and mites on Australasian plants: ecological and evolutionary implications. *Biol. J. Linn. Soc.* 37, 191-236.

O'Dowd, D. J., and Willson, M. F. (1991). Associations between mites and leaf domatia. *Trends Ecol. Evol.* 6, 179-182.

Osakabe, M., Inouye, K., and Ashihara, W. (1987). Effect of *Amblyseius sojaensis* Ehara (Acarina: Phytoseiidae) as a predator of *Panonychus citri* Mcgregor and *Tetranychus kanzawai* Kishida (Acarina: Tetranychiae). *Appl. Entomol. Zool.* 22, 594-599.

Ouyang, Y., Grafton-Cardwell, E. E., and Bugg, R. L. (1992). Effects of various pollens on development, survivorship, and reproduction of *Euseius tularensis* (Acari: Phytoseiidae). *Environ. Entomol.* 231, 1371-1376.

Overmeer, W. P. J. (1985). Diapause. *In* "Spider Mites: Their Biology, Natural Enemies and Control." (W. Helle, and M. W. Sabelis, eds.), pp. 95-102. Vol. 1B. Elsevier. Amsterdam, The Netherlands.

Overmeer, W. P. J., and van Zon, A. Q. (1984). The preferences of *Amblyseius potentillae* (Acarina: Phytoseiidae) for certain plant substrates. *In* "Acarology IV." (D.A. Griffiths, and C. E. Bowman, eds.), pp. 591-596. Vol. 1. Horwood. Chichester, U.K.

Pemberton, R. W., and Turner, C. E. (1989). Occurrence of predatory and fungivorous mites in leaf domatia. *Amer. J. Bot.* 76, 105-112.

Presnail, J. K., and Hoy, M. A. (1994). Transmission of injected DNA sequences to multiple eggs of *Metaseiulus occidentalis* and *Amblyseius finlandicus* (Acari: Phytoseiidae) following maternal microinjection. *Exp. Appl. Acarol.* 18, 319-330.

Putman, W. L., and Herne, D. H. C. (1964). Relations between *Typhlodromus caudiglans* Schuster (Acarina: Phytoseiidae) and phytophagous mites in Ontario peach orchards. *Can. Entomol.* 96, 925-943.

Rothman, L. D. (1988). Habitat associations, patterns of abundance, and species richness of phytoseiid mites (Acari: Phytoseiidae) on a recent landfill site in Lake Ontario. *Proc. Entomol. Soc. Ontario* 119, 1-7.

Rozario, S. A. (1995). Association between mites and leaf domatia: evidence from Bangladesh, South Asia. *J. Trop. Ecol.* 11, 99-108.

Sabelis, M. W. (1985a). Life History. *In* "Spider Mites: Their Biology, Natural Enemies and Control." (W. Helle, and M. W. Sabelis, eds.), pp. 35-41. Vol. 1B. Elsevier. Amsterdam, The Netherlands.

Sabelis, M. W. (1985b). Predator-Prey Interaction. *In* "Spider Mites: Their Biology, Natural Enemies and Control." (W. Helle, and M. W. Sabelis, eds.), pp. 103-129. Vol. 1B. Elsevier. Amsterdam, The Netherlands.

Sabelis, M. W. (1985c). Development. *In* "Spider Mites: Their Biology, Natural Enemies and Control" (W. Helle, and M. W. Sabelis, eds.), pp. 45-53. Vol. 1B. Elsevier. Amsterdam, The Netherlands.

Schuster, R. O., and Pritchard, A. E. (1963). Phytoseiid mites of California. *Hilgardia* 34, 191-285.

Smith, D., and Papcek, D. F. (1991). Studies of the predatory mite *Amblyseius victoriensis* (Acarina: Phytoseiidae) in citrus orchards in south-east Queensland: control of *Tegolphus australis* and *Phyllocoptruta oleivora* (Acarina: Eriophyidae), effect of pesticides, alternative host plants and augmentative release. *Exp. Appl. Acarol.* 12, 195-217.

Strenseth, C. (1979). Effect of temperature and humidity on the development of *Phytoseiulus persimilis* and its ability to regulate populations of *Tetranychus urticae* (Acarina: Phytoseiidae, Tetranychidae). *Entomophaga* 24, 311-317.

Strenseth, C. (1985). Red spider mite control by *Phytoseiulus* in northern Europe. *In* "Biological Pest Control: The Glasshouse Experience." (N. W. Hussey, and N. E. Scopes, eds.), pp. 119-124. Cornell University Press. Ithaca, NY.

Swift, F. C., and Blaustein, L. (1980). Humidity tolerances of 3 species of phytoseiid mites (Acarina: Phytoseiidae). *Ann. NY Entomol. Soc.* 88, 77.

Tuovinen, T. A. R, and Rokx, J. A. H. (1991). Phytoseiid mites (Acari: Phytoseiidae) on apple trees and in surrounding vegetation in southern Finland. Densities and species composition. *Exp. Appl. Acarol.* 12, 35-46.

Van Dinh, N. V., Sabelis, M. W., and Janssen, A. (1988). Influence of humidity and water availability on the survival of *Amblyseius idaeus* and *A. anonymus* (Acarina: Phytoseiidae). *Exp. Appl. Acarol.* 4, 27-40.

Veerman, A. (1992). Diapause in phytoseiid mites: a review. *Exp. Appl. Acarol.* 14, 1-60.

Walde, S. J., Nyrop, J. P., and Hardman, J. M. (1992). Dynamics of *Panonychus ulmi* and *Typhlodromus pyri:* factors contributing to persistence. *Exp. Appl. Acarol.* 14, 261-291.

Walter, D. E. (1996). Living on leaves: mites, tomenta, and leaf domatia. Annu. *Rev. Entomol.* 41, 101-114.

Walter, D. E., and O'Dowd, D. J. (1992a). Leaf morphology and predators: effect of leaf domatia on the abundance of predatory mites (Acari: Phytoseiidae). *Environ. Entomol.* 21, 478-484.

Walter, D. E., and O'Dowd, D. J. (1992b). Leaves with domatia have more mites. *Ecology* 73, 1514-1518.

Walter, D. E., and O'Dowd, D. J. (1995). Life on the forest phylloplane: hairs, little houses, and myriad mites. *In* "The Forest Canopy: Aspects of Research on this Biological Frontier." (M. E. Lowman, and N. Nadkarni, eds.), pp. 325-351. Academic Press. New York, NY.

Willson, M. F. (1991). Foliar shelters for mites in the eastern deciduous forest. *Amer. Midl. Nat.* 126, 111-117.

Zacharda, M., and Hluchy, M. (1991). Long term residual efficacy of commercial formulations of 16 pesticides to *Typhlodromus pyri* Scheuten (Acari: Phytoseiidae) inhabiting commercial vineyards. *Exp. Appl. Acarol.* 13, 27-40.

CHAPTER
18

CONSERVING EPIPHYTIC MICROORGANISMS ON
FRUITS AND VEGETABLES FOR
BIOLOGICAL CONTROL

Charles L. Wilson

I. INTRODUCTION

Plant pathologists were late to recognize that microorganisms other than plant pathogens were intimately associated with the aboveground surfaces of plants and that they could affect disease development (Leben, 1965; Blakeman, 1985). Recent studies have focused on epiphytic microbial populations on the surfaces of fruits and vegetables that may serve as biological control agents against decay-inducing fungi (Droby *et al.*, 1996; Janisiewicz, 1991; Korsten *et al.*, 1994; Wisniewski and Wilson, 1992; El Ghaouth and Wilson, 1995). Research in this area has increased worldwide, resulting in the discovery of a number of antagonistic yeasts and bacteria which can be used for the biological control of postharvest decay of fruits and vegetables (Wilson and Wisniewski, 1994). Some of these antagonists have been made into commercial products and are on the market as biological control alternatives to synthetic fungicides for the control of postharvest decay (Brower, 1996).

Antagonists that have been developed as biological control agents for postharvest diseases of fruits and vegetables have been "fished out" of microecosystems on the surfaces of fruits and vegetables (Janisiewicz, 1991 ; Wilson *et al.*, 1993; Cheah *et al.*, 1996) and the soil (Pusey and Wilson, 1984) and used as augmentative biological control agents. We have only a rudimentary understanding of how these organisms are operating in their natural ecosystems. The success of some of these organisms in controlling postharvest decay would suggest that they may be active in the natural biological control of postharvest decay of fruits and vegetables and warrant conservation. Research is underway to understand the microecology of epiphytic microorganisms on fruit and vegetable surfaces. Through such studies we

335

will be able to more intelligently conserve and manage these epiphytes to enhance resistance of harvested commodities to postharvest decay.

The readiness with which antagonistic microorganisms can be isolated from fruit and vegetable surfaces (Wilson *et al.,* 1993) indicates that a natural "suppressive population" of epiphytic microorganisms (consisting of antagonists and saprophytes) occurs naturally and could be conserved for the biological control of plant diseases. Suppressive soils that inhibit disease development have long been recognized (Cook and Baker, 1983). Evidence is accumulating that similar populations of suppressive microorganisms occur on aboveground plant surfaces which we need to identify and conserve.

A great deal of empirical evidence also points to such "suppressive populations." For instance, washing fruits and vegetables prior to storage commonly accelerates their decay. It has been suggested that natural antagonists to decay organisms may be removed or reduced in the washing process. Chalutz and Wilson (1990) found that when microorganisms were plated out from the surface of unwashed citrus fruit a high concentration of bacteria and yeasts were present. It was only when these populations were diluted that decay pathogens appeared in the agar plates. This was interpreted as an indication that a population of yeasts and bacteria on the surface of unwashed citrus may be suppressing the development of decay organisms.

The application of pesticides has been shown to significantly affect nontarget epiphytic populations on foliage and in some instances to promote disease development (Andrews and Kenerley, 1978; Fokkema and De Nooij, 1981; Lim and Teik, 1982; de Jager *et al.*, 1994). This supports the hypothesis that a naturally occurring population of yeasts and bacteria on aboveground plant surfaces (the phylloplane) may suppress pathogen development (Wilson, 1989). Thus, pesticides used against pests (pathogens and insects) could conceivably be selected or developed which would favor naturally occurring suppressive microbial populations on the surfaces of plants, while being detrimental to pest species. It is imperative that we start determining the impact of our present pesticide applications on these extant antagonistic and saprophytic populations. We are probably overlooking what could be a powerful partner in our efforts to develop IPM programs to reduce synthetic pesticide applications.

A broad definition of biological control is used, in this chapter, to discuss biological control systems on the surfaces of fruits and vegetables. This definition involves not only "one-on-one" relationships between antagonists and pathogens but also host-mediated resistance responses of the host and naturally occurring and induced antimicrobial chemicals. In order to provide possible insights into how naturally occurring biological control systems may be operating and might be conserved the success realized in the biological control of postharvest diseases of fruits and vegetables with introduced antagonists is discussed. Further, the ecological milieu in which these biological control systems are operating will be examined with an eye toward how they may be manipulated and conserved.

In this chapter, I explore those instances where natural suppressive microbial populations may occur on the surface of fruits and vegetables which affect disease development. I suggest that an understanding of these interactions can serve as the basis for the development of tactics for the conservation of biological control of aerial and postharvest plant diseases.

II. DEFINING BIOLOGICAL CONTROL SYSTEMS

Initially, plant pathologists adopted the entomologist's classical definition of biological control (De Bach, 1964) which involves "the actions of parasites, predators, and pathogens in maintaining another organism's density at a lower average that would occur in their absence." This narrow "one-on-one" definition of biological control limits our thinking of "biological control systems" which may be operating in nature. Barbosa and Braxton (1993) have expanded the entomological definition of biological control to include "Parabiological Control" as a manipulation of the pest or the pest's resources to favor control of a pest.

Gabriel and Cook (1990) proposed that the many methods of pest and disease control be divided simply into biological, physical, and chemical. They include "the use of natural or modified organisms, genes, or gene products (delivered by organisms) in their definition." A distinction is made between chemicals produced and "delivered" by living organisms and chemicals "extracted" from living organisms. The former being biological control and the latter being chemical control according to their view. A fundamental difference exists between the objects to be controlled by entomologists and plant pathologists. Entomologists are targeting primarily an *organism* (the insect), whereas plant pathologists are targeting a *process* (the disease) as well as an organism (the pathogen). Strategies for controlling the disease process (therapy) can differ from those used to control the pathogen.

Where plant pathologists have gained some insight, it appears that biological control of plant diseases is much more complex than just "one-on-one" relationships between antagonists and plant pathogens. In our studies of antagonistic yeasts which control postharvest diseases of fruit and vegetables we have found a complex mode of action. Biological control of plant diseases does not occur in a vacuum. It is a dynamic process involving the microecology of the pathogen and antagonists as well as resistance responses of the host. For example, yeast antagonists may (1) compete directly with the pathogen at wound sites for nutrients and space (Droby and Chalutz,

1994), (2) attach to the walls of pathogens and produce wall degrading enzymes (Wisniewski *et al.* 1991), and/or (3) elicit defensive enzymes in the host (El Ghaouth *et al.,* in press). Therefore, this multifaceted biological control involves direct control of the pathogen and an elicitation of host defenses. Such multifaceted biological control might be more stable than direct control of a pathogen with an antagonist (Wilson and El Ghaouth, 1993).

There is growing evidence that some epiphytic populations on the surfaces of fruits and vegetables are not there incidentally, but may be under the genetic control of the host as part of its defense (Neal, *et al.,* 1973; Bird *et al.,* 1979; Bird, 1982; Gough et *al.,* 1986). If such is the case, conservation biological control could involve the preservation and augmentation of genes which would promote higher levels of antagonists on plant surfaces.

Taking the complexity of extant biological control systems into account, I propose the following definition for biological control of plant diseases: "The control of plant diseases by a natural biological process or the product of a natural biological process." Ehler (Chapter 1) has defined conservation biological control of insects "as a form of applied biological control in which natural enemies are preserved rather than increased through augmentation." He recognizes that augmentative and conservation biological control are part of a continuum and many techniques designed to enhance natural enemies have elements of both.

Although biological control of plant diseases and insects is viewed differently by plant pathologists and entomologists, "environmental modification" and "augmentation" are still the main methods of conserving biological control systems in both disciplines. The genetic control of antagonist behavior adds an additional element to "conservation biological control" in plant pathology which also may be applicable in entomology. Certain epiphytic fungal insect pathogens could be under the genetic control of the host (Jaques and Patterson, 1962). Also, it would appear that volatiles which attract to plants "body guards" such as predators/parasitoids of insect pests are under the genetic control of the host plant (Turlings, 1996; Chapter5).

In order to intelligently conserve extant biological control systems they first have to be identified or the conditions that maintain them delineated. This is a daunting task since plant diseases involve complex interactions among microorganisms, the host plant, and the environment. Once identified, biological control systems for plant diseases are also difficult to conserve. Climatic conditions substantially affect biological control systems operating in the environment and humans have few opportunities to intercede. Part of the dynamic process between pathogen and antagonist is a constantly changing environment. The success realized in the biological control of postharvest diseases of fruits and vegetables has been partially attributed to a greater control of environmental conditions such as temperature and humidity (Wilson and Wisniewski, 1989).

However, other components of the ecosystem are partially manageable such as genetically controlled host defenses, antagonists, hyperparasites, and cultural management practices. The greatest progress in plant pathology toward conserving biological control systems has been made in understanding, conserving, and augmenting genetic resistance in plants to diseases. Much of this effort has been necessitated because of human mismanagement through the use of monocultures and poorly adapted plants (Browning, 1974).

Our best understanding of biological control microecosystems which affect plant diseases is in the soil. A substantial body of knowledge now exists showing the significant effect that saprophytic and antagonistic microorganisms have in the soil on soil pathogens and disease development (Cook and Baker, 1983). We have been able to conserve and manipulate some of these biological control systems through cultural practices and the enhancement of naturally occurring antagonists and hyperparasites. Although biological control of epiphytic plant diseases has lagged behind control of soil-borne diseases considerable progress has been made recently in the biological control of postharvest diseases of fruits and vegetables (Wilson and Wisniewski, 1994).

III.   THE POSTHARVEST ENVIRONMENT: OPPORTUNITIES FOR BIOLOGICAL CONTROL

Over 25% of our harvested fruits and vegetables are lost to postharvest decay in the United States (USDA, 1965). Because of poor sanitation and the lack of refrigeration in developing countries these losses often exceed 50% (Coursey and Booth, 1972). Synthetic fungicides have been a major means of controlling postharvest decay of fruits and vegetables. However, because of health and environmental concerns (National Academy of Sciences, 1987) most of the major fungicides previously used to control postharvest decay have been removed from the market creating an urgent need for safe alternatives. As a result of these developments researchers worldwide have been pursuing biologically based alternatives to synthetic fungicides for the control of postharvest diseases (Wilson and Wisniewski, 1994).

Products are now emerging on the market as alternatives to synthetic fungicides for the treatment of postharvest diseases of fruits and vegetables. Most notable is a product called Aspire (containing the yeast antagonist *Candida oleophila* Lizuka) produced by Ecogen (Wilson *et al.,* 1996) and one called Biosave (containing the bacterial antagonist *Pseudomonas syringae* van Hall) produced by Ecoscience

(Janisiewicz and Marchi, 1992). The antagonists used in Aspire and Biosave were isolated from the phylloplane of tomato and apple, respectively. In applying antagonists artificially to fruit and vegetables for biological control it has been found that mixtures of antagonists (Falconi and Mendgen, 1994) and nutritional additives (Pusey, 1994) may result in additive or synergistic biological control. Numerous other antagonist have been found in the washes from fruits and vegetables (Cheah *et al.,* 1996). These developments indicate the potential importance of conserving extant biological control systems on the surfaces of fruits and vegetables.

The postharvest environment provides a rare opportunity to develop and conserve biological control systems which can be used to control decay of harvested commodities and extend their shelf life. In addition, postharvest disease biological control has other advantages over biological control in the field. These include the advantage that (1) the site of activity (wounds) of the biological control agent is more accessible to antagonists, (2) the plant parts to be treated are more concentrated allowing a more effective and efficient treatment with biological control agents, and (3) the economic value of harvested commodities is great enough to warrant elaborate management practices which may favor biological control.

The ease with which antagonists can be isolated from the fruit phylloplane would indicate that numerous "suppressive antagonists" are operative under natural conditions. The antagonistic epiphytic microflora on fruits and vegetables remain a rich reservoir for antagonistic microorganisms awaiting conservation. An added argument can be made for the conservation of diverse plant populations since such populations may harbor epiphytic microorganisms that could eventually be commercialized. Creating new biopesticides provides new tools for augmentative biological control. Creating conditions which conserve applied biopesticides, or better yet which conserve "yet to be commercialized" antagonists also can be effective conservation biological control.

IV. MULTIFACETED BIOLOGICAL CONTROL

It can be argued that naturally occurring biological control systems on fruits and vegetables are complex and multifaceted like those that have been artificially created and investigated (Wilson and El Ghaouth, 1993). These systems may involve (1) competition for nutrients and space, (2) antibiotics, (3) direct parasitism, and (4) induced resistance. Some of these biological interactions are mediated by the antagonist (e.g., antibiosis or direct parasitism). Others are mediated by the host (e.g., induced resistance) and still other by both the antagonist and host (e.g., competition for nutrients and space).

Mixtures of "suppressive antagonists" are probably involved in most biological control systems operating naturally on fruit and vegetable surfaces rather than in one on one interactions between an antagonist and pathogen. This was apparent when Stirling (1995) looked at the component organisms of the natural suppressive population of antagonists on avocado. Individual antagonists were not capable of exerting the level of control provided by mixed microbial populations.

Complex multifaceted biological control systems would be expected to be more stable than simple one-on-one pathogen/antagonist interactions. Pathogens would seemingly have a more difficult time developing resistance or tolerance to a complex of antagonists with multiple modes of action. Also, such a complex would seem to be more buffered against environmental influences. Because of such advantages, selection may have taken place for such complex multifaceted biological control systems in nature and their conservation serves the purpose of biological control.

V. CONSERVING AND PROMOTING NATURALLY OCCURRING EPIPHYTIC ANTAGONISTS: AN IPM PERSPECTIVE

A. Pesticides

Integrated Pest Management (IPM) programs are being implemented worldwide to reduce pesticide usage and promote natural enemies of pests. A paucity of information on antagonists acting naturally on the surfaces of plants as biological control agents prevents us presently from making their conservation part of IPM programs. However, where investigators have looked for "suppressive populations" on the surfaces of fruits and vegetables they have found that such biological control systems were widespread on vegetation (Korsten et al., 1994). If so, a powerful biological control force exists that can be brought into play to control fruit and vegetable diseases and reduce the use of synthetic pesticides.

The management of pesticide applications is an important area where we should look to conserve putative "suppressive populations" of microorganisms. Sometimes the effect of pesticides is differential, inhibiting some organisms and apparently not affecting others. Andrews (1992) found that the natural epiphytic microflora of apple was altered both quantitatively and qualitatively when standard pesticide applications were made. Fluorescent pseudomonads and lactic acid-type bacteria were among those microbial populations that were depressed by pesticides. Andrews and Kenerley (1978) suggest that we may be suppressing a natural antagonistic population through pesticide applications.

We can select more intelligently pesticides for IPM programs which will conserve existing biological control systems on plant surfaces. Pesticides have a profound effect on the microecology of nontargeted as well as targeted epiphytic microorganism on the surfaces of plants (Gibbs, 1972; Andrews and Kenerley, 1978; Fokkema and de Nooij, 1981; Stirling, 1995). Our most convincing "window" into the natural biological control systems operating on fruit surfaces of plants occurs when so called "iatrogenic diseases" occur (Griffiths, 1981). These are diseases resulting from human activity such as the application of fungicides. Most notable among these diseases have been those resulting from the application of copper- (Futado, 1969; Gibbs, 1972; Stirling, 1995) and benzimidazole-containing fungicides (Carter and Price, 1974). Increased disease by the use of benzimidazole fungicides has been explained by a reduction in competing saprophytes along with the development of resistance to the fungicide by the pathogen (Griffiths, 1981).

A recent study in Australia by Stirling (1995) on the induction of iatrogenic disease in avocado by the application of copper is a good documentation of an extant biological control system of microorganisms in an orchard sprayed regularly with copper from November to May compared to an unsprayed avocado orchard. Where copper fungicide was sprayed all categories of microorganisms on the fruit and leaves (bacteria, yeasts, and filamentous fungi) were 10 to 100-fold less numerous than in the adjacent unsprayed orchard. When fruit from the two orchards were harvested, ripened, and assessed for the presence of anthracnose and stem-end rot, there was significantly less diseased fruit from the unsprayed orchard over two successive years. Stirling (1995) suggested that anthracnose and stem-end rot may have been under natural biological control in the unsprayed avocado orchard. Since copper was detrimental to microorganisms on the leaf and fruit surface these microorganisms may have been responsible for disease suppression.

There are also secondary but significant consequences to conserving antagonists by the elimination, selective use, or substitution of harsh fungicides. Entomologists have also found natural epiphytic fungal parasites which appear to keep certain insect populations under control. The shift from wettable sulfur to synthetic fungicides (i.e., dithiocarbamates) to control apple scab in Nova Scotian apple orchards apparently resulted in the disappearance of epizootics of *Entomophthora sphaerosperma* Fresenius which held the apple sucker *Psylla mali* Schmidberger under natural control (Jaques and Patterson, 1962). The same phenomenon was observed in apple and pear orchards in Italy where fungicides used to control apple scab were implicated in pest resurgence due to the loss of a key insect pathogen *Entomophthora* sp. (Picco, 1978).

Fokkema and de Nooij (1981) discuss the differential effects of fungicides on constituents of the phylloplane microflora. The growth of saprophytic fungi is inhibited by broad-spectrum fungicides but these have little effect on bacteria. *Sporobolomyces spp., Cladosporium* spp., and *Aureobasidium pullulans* (de Bary) are inhibited by benzimidazoles but the yeasts *Cryptococcus* spp. and *Candida spp.,* which occur commonly on leaves are much less affected. After biological control systems are identified on plant surfaces it behooves us to determine how pesticides affect them. More effective control could be realized if pesticides were designed to target the pathogens and conserve or promote antagonists.

B. Cultural Practices

Cultural practices such as fertilization, irrigation, chemigation, and perhaps weed control may affect naturally occurring suppressive populations of antagonists on plant surfaces. Turner *et al.,* (1985) found that N and P fertilization enhanced the ratio of fungi to bacteria in the rhizosphere of rye grass. Nutrients applied onto plant surfaces as well as the soil may affect epiphytic microbial populations.

Attempts have been made at enhancing biological disease control by artificially changing the nutritional milieu on leaf surfaces and wounds. Morris and Rouse (1985) found that the application of simple organic compounds such as glutamine and alanine to bean leaf surfaces can alter the epiphytic bacterial populations. They were able to alter the population size of fluorescent pseudomonads and reduce disease severity caused by *Pseudomonas syringae.* The survival and efficacy of the chitinolytic producing antagonist *Bacillus cereus* Frankland and Frankland was enhanced by chitin applications to peanut leaves (Kohalis-Burelle *et al.,* 1992). Janisiewicz *et al.,* (1992) found that the amino acids L-asparagine and L-proline greatly enhance efficacy in artificially inoculated apple wounds. Through gaining a better understanding of the influence of nutrients on the surfaces of plants on pathogen and antagonist dynamics we can perhaps promote cultural practices which create a nutritional milieu on plant surfaces that favors antagonist colonization and development.

VI. ENVIRONMENTAL EFFECTS ON EPIPHYTIC MICROORGANISMS

Attempts have been made to characterize the ecology of epiphytic microorganisms on leaf surfaces (phylloplane) (Blakeman, 1985; Andrews, 1992). Our understanding of the ecology of fruit and vegetable surfaces is more rudimentary. The aboveground portions of a plant presents a more hostile environment for microorganisms than the soil (Andrew, 1992). Nevertheless, an ecological succession of microorganisms on the surfaces of plant leaves (and presumably on fruit surfaces) has been characterized (Blakeman, 1985; Andrew, 1992). In general, bacteria are the dominant flora in the early colonization of plant surfaces followed by a sharp increase in yeasts and eventually a rise in the populations of filamentous fungi.

Most infections of fruits and vegetables by decay organisms (usually filamentous fungi) occur through wounds. Therefore, the ecological succession of organisms at the wound site becomes important in understanding the biological control of these pathogens. Mercier and Wilson (1994) studied the ecological succession of microorganisms in apple wounds and found that yeasts and bacteria were the first colonizers. *Aureobasidium pullulans* and yeasts (mainly *Sporobolomyces roseus* Kluyver and van Neil) were dominant fungal wound invaders, while species of *Erwinia, Glucobacter, and Pseudomonas* were the most common bacteria isolated. We know less about latent infections of fruit and how epiphytic microorganisms may affect them.

The ecological succession of microorganisms on the phylloplane has been correlated with nutritional changes on leaf surfaces. Generally, carbohydrates become limiting to microbial growth first, then nitrogen sources. Bacteria appear to be more proficient in utilizing nitrogen sources than other microorganisms, whereas yeast utilize carbohydrates most efficiently (Andrew, 1992). This has been used to explain the succession of organisms from bacteria to yeasts which occurs commonly.

It has been discovered that pollen deposited on fruits or leaves of plants can stimulate fungal infection (Chu-Chou and Preece, 1968; Warren, 1972). Fokkema (1973) found that the effect from added nutrients provided by pollen was reduced by antagonistic microorganisms on the surfaces of rye leaves. He concluded that naturally occurring antagonists such as *Aureobasidium pullulans, Sporobolomyces* spp., and *Cryptococcus* spp. were able to compete with the pathogen *Helminthosporium sativum* Pammel *et al.* for the added nutrients provided by pollen and reduce its stimulatory effect.

It appears that nutrient competition is an important means whereby epiphytic antagonists suppress plant pathogens. Wilson and Lindow (1991) have characterized niche differentiation on plant surfaces resulting from differential utilization of carbon sources by epiphytic microorganisms. Janisiewicz (1996) examined the utilization of 35 carbon and 33 nitrogen sources by yeast antagonists in apple wounds and used this information to select mixtures of antagonists which would provide more effective biological control of blue mold. Combining yeast antagonists with different nutritional profiles resulted in increased control of blue mold, as compared with treatments containing the isolates alone.

As fruits and vegetables mature they become more "leaky" providing epiphytic microorganisms the nutrient base that they need to survive and multiply. What role "suppressive antagonists" play in deterring or possibly accelerating the normal senescence of fruits and vegetables under natural conditions is not know. Ecological studies in this area could yield information which would allow us to develop management practices that would conserve extant biological control systems and extend the shelf life and nutrient value of harvested fruits and vegetables.

VII. GENETIC CONTROL OF EPIPHYTIC ANTAGONISTS' ENVIRONMENT

Some evidence exists that epiphytic antagonists on the surfaces of plants are under the genetic control of the host plant. In fact, some plant-breeding programs for disease and insect resistance may have selected epiphytic antagonists which contribute to resistance (Bird, 1982). These findings have profound implications for the conservation of biological control of plant diseases. If this thesis is valid, genes which promote colonization and/or the development of epiphytic antagonists may be identifiable and manipulatable either through classical breeding or genetic engineering of crops.

Bird (1982) developed cotton breeding lines at Texas A&M University that are resistant to a variety of pathogens and insects as well as adverse environmental conditions. This resistance has been termed Multi-Adversity-Resistance (MAR). Bird et al. (1979) and Bird (1982) also have evidence that microorganisms on tissues (both below- and aboveground) play a role in MAR in cotton. The microbial populations isolated from the surfaces of MAR-cotton varieties contained more antagonists than those from susceptible varieties. Gough et al. (1986) found that winter wheat leaves sprayed with streptomycin become more susceptible to *Septoria tritici* Roberge and Desmazieres. They speculate that bacterial antagonists under the genetic control of the host may be responsible for resistance to leaf spot of wheat. This argument was strengthened when they were able to restore resistance by "replenishing" the antagonists eliminated by the streptomycin.

How plants might "control" specific epiphytic microbial populations on their surfaces is intriguing. It has been established that plant leaves, stems, and fruits are "leaky," resulting in nutrients being deposited on plant surfaces (Tukey, 1970). Given that inhibition of pathogens enhances fitness it is possible that a nutritional milieu could have evolved on certain plants which would favor the growth of specific antagonistic microorganisms. The ability of plants to "select" associated microbial populations is clearly shown with the crown gall bacterium (Wilson, 1978) that causes the plant to excrete nutrients which differentially favor its growth over other microorganisms.

Plants also have the potential capability to "control" microbial populations though the excretion of constitutive and induced antimicrobial compounds. Such compounds could differentially favor "suppressive populations" of antagonists over other saprophytes or pathogens. The ability of plants to signal and attract specific predaceous "bodyguards" against insect attack demonstrates how plants can influence external biological control agents in response to specific pests (Sabelis and De Jong, 1988). It has been demonstrated that plant volatiles can affect plant pathogens and perhaps antagonists (Afifi, 1975).

VIII. THE ROAD NOT TRAVELED: AN EPILOGUE

As we attempt to discover and conserve natural biological control systems on the surfaces of plants we need to keep a broad prospective on how such systems are structured and operate in nature. In studying "suppressive populations" of microorganisms complete populations of antagonists and saprophytes should be examined and their performance determined individually and in combination. Nutrients and other environmental parameters also should be investigated as to how they influence putative biological control systems. An understanding of the structure and mode of action of epiphytic biological control systems will also allow us to manage them more intelligently.

When naturally occurring biological control is discovered on the surfaces of fruits and vegetables a number of opportunities may present themselves to conserve such systems as the harvested product passes from the farmer's field to the consumer. Pre- and post-harvest practices such as cultivation, harvesting, fertilization, storage, packaging, transportation, along with pest control may influence "suppressive populations" on fruit and vegetable surfaces which favor biological control. We should attempt to determine how cultural and processing practices influence naturally occurring antagonists and saprophytes on fruit and vegetable surfaces and devise means to conserve such microbial populations so as to promote biological control.

A search should be made for genes which may control environments that favor epiphytic antagonistic microorganisms on the surfaces of plants. It is reasonable to expect to find genes which promote epiphytic "bodyguards" against plant pests. Such genes may promote physical and nutritional environments on plant surfaces that favor antagonist colonization and development. Also, we should not rule out the possibility that presently identified genes for resistance to pests may express themselves by promoting antagonistic epiphytic populations of microorganisms.

The potentially rich reservoir of microorganisms that exist on the surfaces of fruits and vegetables which contribute to biological control systems are worthy of exploration and conservation. It can be anticipated that increased research in this area will yield large dividends in conservation biological control of plant diseases and insects.

REFERENCES

Afifi, A. F. (1975). Effect of volatile substance from species of Labiatae on rhizopheric and phyllospheric fungi of *Phaseolus vulgarism. Phytopath. Z.* 83, 296-302.

Andrews, J. H. (1992). Biological control in the phyllosphere. *Annu. Rev. Phytopathol.* 30, 603-635.

Andrews, J. H., and Kenerley, C. M. (1978). The effects of a pesticide program on non-target epiphytic microbial populations of apple leaves. *Can. J. Microbiol.* 24, 1058-1072.

Barbosa, P., and Braxton, S. (1993). A proposed definition of biological control and its relationship to related control approaches. *In* "Pest Management: Biologically Based Technologies." (R. D. Lumsden, and J. L. Vaughn, eds.), pp. 21-27. Amer. Chem. Soc. Washington, DC.

Bird, L. S. (1982). The MAR (multi-adversity resistance) system for genetic improvement of cotton. *Plant Dis.* 66, 172-178.

Bird, I. S., Liberman, C., Percy, R. G., and Bush, D. L. (1979). The mechanism of multi-adversity resistance in cotton: theory and results. *Proc. Beltwide Cot.* Prod. Res. Conf., Cot. Dis. Coun. 39, 226-228.

Blakeman, J. P. (1985). Ecological succession of leaf surface microorganisms in relation to biological control. *In* "Biological Control on the Phylloplane." (C. E. Windels, and S. E. Lindow, eds.), pp. 6-30. Amer Phytopath. Soc. St. Paul, MN.

Brower, V. (1996). Ecogen Israel gets second approval. *Nature Biotechnol.* 14, 1212.

Browning, J. A. (1974). Relevance of knowledge about natural ecosystems to development of pest management programs for agroecosystems. Proc. Amer. Phytopathol. Soc. 1, 191-199.

Carter, M. V., and Price. T. V. (1974). Biological control of *Eutypa armeniacae.* II. Studies of the interaction between *E. armeniacae and Fusarium latertium,* and their relative sensitivities to benzimidazole chemicals. *Aust. J. Agric. Res.* 25, 105-119.

Chalutz, E., and Wilson, C. L. (1990). Postharvest biocontrol of green and blue mold and sour rot of citrus fruit by *Debaryomyces hansenii. Plant Dis.* 74, 134-137.

Cheah, L. H., Wilson, C. L., and Marshall, A. P. (1996). Rapid screening for antagonists against *Botrytis* storage rots using leaf and fruit tissue. *Postharvest Biol. Technol.* 8, 223-228.

Chu-Chou, M., and Preece, T. F. (1968). The effect of pollen grains on infections caused by *Botrytis cinerea. Ann. Appl. Biol.* 62, 11-22.

Cook, R. J., and Baker, K. F. (1983). "The Nature and Practice of Biological Control of Plant Pathogens." APS Press. Amer. Phytopathol. Soc. St. Paul, MN.

Coursey, D. G., and Booth, R. H. (1972). The postharvest phytopathology of perishable tropical produce. *Rev. Plant Pathol.* 51, 751-765.

DeBach, P. (1964). The scope of biological control. *In* "Biological Control of Insect Pests and Weeds." (P. DeBach, ed.), pp. 3-20. Reinhold Publ. New York, NY.

Droby, S., and Chalutz, E. (1994). Mode of action of biocontrol agents of postharvest diseases. *In* "Biological Control of Postharvest Diseases - Theory and Practice." (C. L. Wilson, and M. E. Wisniewski, eds.), pp. 63-75. CRC Press. Boca Raton, FL.

Droby, S., Chalutz, E., Wisniewski, M., and Wilson, C. L. (1996). Host response to introduction of antagonistic yeasts used for control of postharvest decay. *In* "Microbiology of Aerial Plant Surfaces." (Morris, C. E., Nicot, P. C., and Nguyen-The, C., eds.), pp. 73-90 Plenum Press. New York, NY.

El Ghaouth, A., and Wilson, C. L. (1995). Biologically-based technologies for the control of postharvest diseases. *Postharvest News and Inform.* 6, 5N-11N.

El Ghaouth, A., Wilson, C. L., and Wisniewski, M. E. Ultrastructural and cytochemical aspects of the biological control of *Botrytis cinerea by Candida saitoana* in apple fruit. *Phytopath.* (in press).

Falconi, C. J., and Mendgen, K. (1994). Epiphytic fungi on apples leaves and their value for control of the postharvest pathogens *Botrytis cinerea, Monilinia fructigena,* and *Penicillium expansum. J. Plant Dis. Prot.* 101, 38-47.

Fokkema, N. J. (1973). The role of saprophytic fungi in antagonism against *Drechslera sorokiniana (Helinthosporium sativum)* on agar plates and on rye leaves with pollen. *Physiol. Plant Pathol.* 3, 195-205.

Fokkema, N. J., and De Nooij, M. P. (1981). The effect of fungicides on the microbial balance in the phyllosphere. *EPPO* Bull. 11, 303-310.

Gabriel, C. J., and Cook, R. J. (1990). Biological control - the need for a new scientific framework. *BioScience* 40, 204-206.

Gibbs, J. N. (1972). Effects of fungicides on the populations of *Colletrichum and* other fungi in bark of coffee. *Ann. Appl. Biol.* 70, 35-47.

Gough, F. T., Mehdizadegan, E., and Krenzer, E. G. (1986). Effect of streptomycin on development of *Septoria tritici* blotch. (Abstr.) *Phytopathol.* 76, 1103.

Griffiths, E. (1981). Iatrogenic plant diseases. *Annu. Rev. Phytopathol.* 19, 69-82.

Janisiewicz, W. J. (1991). Control of postharvest diseases of fruits with biocontrol agents. *Food and Fertilizer Tech. Ctr. Tech. Bull.* 125.

Janisiewicz, W. J. (1996). Ecological diversity, niche overlap, and coexistence of antagonists used in developing mixtures for biocontrol of postharvest diseases of apples. *Phytopathol.* 86, 473-479.

Janisiewicz, W. J., and Marchi, A. (1992). Control of storage rots on various pear cultivars with a saprophytic strain of *Pseudomonas syringe. Plant Dis.* 76, 555-560.

Janisiewicz, W. J., Usall, J., and Bors, B. (1992). Nutritional enhancement of biocontrol of blue mold on apples. *Phytopathol.* 82, 1364-1370.

Jaques, R. P., and Patterson, N. A. (1962). Control of the apple sucker, *Psylla mali* Schmidt., by the fungus *Entomophthora spaerospera* (Fresenius). *Can. Entomol.* 94, 818-825.

Kohalis-Burelle, N., Backman, P. A., Rodriquez-Kabana, R., and Ploper, L. D. (1992). Potential for biological control of early leafspot of peanut using *Bacillus cereus* and chitin as foliar amendments. *Biol. Cont.* 2, 321-326.

Korsten, L., De Villiers, E. E., Rowell, A., and Kotze, J. M. (1994). A review of biological control of postharvest diseases of subtropical fruits. *In* "Postharvest Handling of Tropical Fruits." (B. R. Champ, E. Highley, and G. L. Johnson, eds.), pp. 172-185 ACIAR Proc.

Leben, C. (1965). Epiphytic microorganisms in relation to plant disease. *Annu. Rev. Phytopathol.* 3, 209-230.

Mercier, J., and Wilson, C. L. (1994). Colonization of apple wounds by naturally occurring microflora and introduced *Candida oleophila* and their effect on infection by *Botrytis cinerea* during storage. *Biol. Cont.* 4, 138-144.

Morris, C. E., and Rouse, D. I. (1985). Role of nutrients in regulating epiphytic bacterial populations. *In* "Biological Control on the Phylloplane." (C. E. Windels, and S. E. Lindow, eds.), pp. 63-82. Amer. Phytopathol Soc. St. Paul, MN.

National Academy of Sciences (1987). "Pesticides in Foods-The Delaney Paradox." National Academy Press. Washington, DC.

Neal, J. L., Larson, R. I., and Atkinson, T. G. (1973). Changes in rhizophere populations of selected physiological groups of bacteria related to substitution of specific pairs of chromosomes in spring wheat. *Plant Soil* 39, 209-219.

Picco, D. (1978). *Psylla pyri* defeated by complementary control. XX Intern. Symp. Crop Prot. *Medelingen van de Faculteit Landbouwwetenschappen Rijksuniversiteit Gent.* 43, 527-539.

Pusey, P. L. (1994). Enhancement of biocontrol agents for postharvest diseases and their integration with other control strategies. *In* "Biological Control of Postharvest Diseases -Theory and Practice." (C. L. Wilson, and M. E. Wisniewski, eds.), pp. 77-88. CRC Press. Boca Raton, FL.

Pusey, P. L., and Wilson, C. L. (1984). Postharvest biological control of stone fruit brown rot by *Bacillus subtilis. Plant Dis.* 68, 753-756.

Sabelis, M. W. and De Jong, C. M. (1988). Should all plants recruit bodyguards? Conditions for a polymorphic ESS of synomone production in plants. *Oikos* 53, 247-252.

Stirling, M. (1995). The role of epiphytic microorganisms in the suppression of *Colletotrichum gloeosporioides* on avocado. Ph.D. Dissertation, Queensland University, Queensland, Australia.

Tukey, H. B. (1970). The leaching of substances from plants. *Annu. Rev. Plant Physiol.* 21, 305-324.

Turlings, C. J. (1996). Herbivore-induced emissions of parasitoid attractants by plants. *Proc. XX Intern. Congr. Entom.* pp. 621.

Turner, S. M., Newmank, F. I., and Campbell, R. (1985). Microbial population of ryegrass root surfaces: Influence of nitrogen and phosphorus supply. *Soil Biol. Biochem.* 17, 711-715.

U.S. Department of Agriculture, Agricultural Research Service. (1965). "Losses in Agriculture." U.S. Department Agric. Handb. 291.

Warren, R. C. (1972). The effect of pollen on the fungal leaf microflora of *Beta vulgaris* L. and on infection of leaves by *Phoma betae*. *Neth. J. Plant Pathol.* 78, 89-98.

Wilson, C. L. (1978). Plant teratomas-Who's in control of them? *In* "Plant Disease, An Advanced Treatise." (J. G. Horsfall, and E. Cowling, eds.), pp. 215-223. Academic Press. New York, NY.

Wilson, C. L., and El Ghaouth, A. (1993). Multifaceted biological control of postharvest diseases of fruits and vegetables. *In* "Pest Management: Biologically Based Technologies." (R. D. Lumsden, and J. L. Vaughn, eds.), pp. 181-185. Amer. Chem. Soc. Washington, DC.

Wilson, M., and Lindow, S. E. (1991). Resource partitioning among bacterial epiphytes in the phylosphere. *Phytopathol.* 81, 1170.

Wilson, C. L., and Wisniewski, M. E. (1989). Biological control of postharvest diseases of fruits and vegetables: An emerging technology. *Annu. Rev. Phytopathol.* 27, 425-441.

Wilson, C. L., and Wisniewski, M. E. (1994). "Biological Control of Postharvest Diseases - Theory and Practice." CRC Press. Boca Raton, FL.

Wilson, C. L., Wisniewski, M. E., Droby, S., and Droby, E., (1993). A selection strategy for microbial antagonists to control postharvest diseases of fruits and vegetables. *Sci Hort.* 53, 183-189.

Wilson, C. L., Wisniewski, M. E., El Ghaouth, A., Droby, S., and Chalutz, E. (1996). Commercialization of antagonistic yeasts for the biological control of postharvest diseases of fruits and vegetables. *SIM News* 46, 237-242.

Wisniewski, M. E., and Wilson, C. L. (1992). Biological control of postharvest diseases of fruits and vegetables; Recent advances. *Hort Sci.* 27, 94-98.

Wisniewski, M. E., Biles, C., Droby, S., McLaughlin, R., Wilson, C., and Chalutz, E. (1991). Mode of action of the postharvest biocontrol yeast, *Pichia guilliermondii*. I Characterization of attachment to *Botrytis cinerea*. *Physiol. Mol. Plant Pathol* 39, 145-249.

BIOLOGICAL CONTROL OF SOIL-BORNE PATHOGENS WITH RESIDENT VERSUS INTRODUCED ANTAGONISTS: SHOULD DIVERGING APPROACHES BECOME STRATEGIC CONVERGENCE?

Philippe Lucas and Alain Sarniguet

I. INTRODUCTION

New trends in agricultural production and public concern about the use of pesticides have led to renewed interest in durable and environmentally friendly methods for controlling diseases. The most investigated alternative to fungicide use has been breeding plants for resistance and developing biological methods of control. To date, plant breeding has had more practical success than the development of biological controls. Biological control agents have been successful under some conditions but their widespread use in different ecosystems has revealed limitations suggesting that the environment has a great influence on the survival and activity of these microorganisms. Nevertheless, candidate disease antagonists are the subject of attempts to enhance their efficacy. These include genetic engineering to improve antibiotic production and the exploration of mechanisms that are important for their establishment in the courts or potential courts of infections by pathogens (Cook, 1993). The latter may be referred to as augmentation (as defined in Chapter 1).

Another approach is to conserve and thus to take advantage of naturally occurring biological controls. In both these approaches, the environment has generally been considered to be a "black box," e.g., in studies which evaluate the impact of agricultural practices on the reduction of diseases. Investigations of phenomena occurring in this "black box" have shown how the effects of beneficial microorganisms are influenced by nutrient status and other physico-chemical characteristics in soil (e.g., fluorescent pseudomonads interacting with the form of nitrogen fertilization, soil pH, or soil manganese content to control take-all of wheat caused by

Gaeumannomyces graminis (Sacc.) von Arx and Olivier var *tritici* Walker) (Smiley, 1978; Lucas and Sarniguet, 1990; Sarniguet *et al.,* 1992a,b; Huber and McCay-Buis, 1993).

Both approaches provide knowledge that should be applicable in the not too distant future. Managing the environment by stimulating natural occurring microorganisms first and then enhancing efficacy (if necessary and economically acceptable) by introducing specific biological control agents (into a more receptive environment) should be an effective complementary strategy. In this chapter, we first examine the problems inherent in studying soil-borne pathogens and controlling the diseases caused by them. We then present different hypotheses to explain soil suppression of take-all of wheat. Initial studies to develop marketable biological methods of control through augmentation of biological control agents are presented while noting limitations in their effectiveness. In addition, we discuss the epidemiology of take-all disease, efforts to model the impact of cultural practices on the different phases of the disease, and a hypothesis on their role in the conservation of native biological control agents. We then focus on recent studies on fluorescent pseudomonads (an important group of bacteria responsible for take-all suppression) , conducted at both the individual (augmentation) and population level (conservation). Finally we review the advantages and the present limits of such studies, the points on which they should interact, and those that should be considered as a continuum (see Chapter 1).

II. THE BASES OF BIOLOGICAL CONTROL OF
SOIL-BORNE PATHOGENS

Soil-borne plant pathogens affect crops throughout the world and have been the subject of considerable attention from the scientific community. This is justified not only by the economic impact *per se* of these diseases on crop production but also by the specific difficulties associated with controlling soil-borne diseases (Table 1). These difficulties are due mainly to the complexity of the soil environment (Table 2) compared to air-borne diseases. Diseases caused by soil-borne pathogens are difficult to control by applying fungicides to plants because the active ingredient is not transported to and through the root system. In addition, the soil, which is a closed environment, shields the target pathogen from fungicides applied to the soil. As a consequence high doses are required and increase risks of soil and ground water pollution and undesirable biological side effects.

The most efficient way of controlling soil-borne pathogens is still to disinfect the soil but this has the disadvantage of being nonspecific and it also has a harmful

Table 1. Characteristics of soil-borne and air-borne pathogens or diseases

Soil-borne pathogens /diseases	Air-borne pathogens/diseases	Control difficulties
Root and plant base pathogens	Stem and foliar pathogens	Upward systemic action of fungicides
Enclosed environment	Open environment	Soil = shield for target, high doses needed
High biological activity	Low biological activity	Acceptable soil treatments (side effects)

impact on the environment, especially when done by fumigation. However the latter technique can only be recommended in small-scale crop production (truck farming, greenhouse cultivation, etc.). A total of 22 different active ingredients were available for use against soil-borne pathogens in France in 1995, as 25 registered formulations (some of them combinations of active ingredients); while 225 formulations (from 80 active ingredients) were registered for use against foliar diseases (Anon., 1995).

Table 2. Environment-linked difficulties encountered in studying soil-borne and air-borne pathogens.

Soil-borne pathogen environment (soil)	Air-borne pathogen environment (air)	→ Problems
Complex (physical, chemical) Non-uniform (space scale)	Simple Uniform	Constituent analysis Identification, knowledge of niches
Buffered (time scale)	Changing	Delayed effects of experimental "actions"
Enclosed	Open	Direct observation impossible, Destructive samplings
Nutrient rich High microbial populations	Nutrient poor Low microbial populations	Great competition Great interaction between organisms

Among the 22 active ingredients for treating soil, four were nonspecific pesticides that also have some activity against nematodes, such as methyl-bromide, chloropicrin, dazomet, and metam-sodium. There is therefore a market for new, alternative methods of controlling soil-borne diseases.

Breeding plants for disease resistance is, of course, a major possibility although it is often regarded as being more difficult to do against soil-borne than against other diseases. As far as cereals are concerned breeding for resistance has been successful against eyespot *(Pseudocercosporella herpotrichoides* (Fron) Deighton) of wheat (Doussinault *et al.,* 1983) but still remains limited because of lack of known sources of resistance against other important diseases like take-all of wheat (Scott *et al.,* 1989). Most of the work done so far on the control of soil-borne diseases has focused on the use of biological control agents. Although some of the diseases that have been studied for a long time to identify potential biological control agents cannot yet be effectively controlled with biological agents there have been some successes of "microbial pesticides" registered in the United States (Cook, 1993).

A.   The Nature of Current Practices in the Biological Control of
     Soil-Borne Pathogens

There are several ways of approaching the biological control of soil-borne pathogens. Cook (1990), in a review on progress toward biological control, provides examples of control with resident antagonists and control with introduced antagonists. The first method, mainly based on crop rotation and the addition of organic amendments to soil went (according to Cook) in the wrong direction; modern agriculture tending to move toward less use of organic amendments, tillage, and crop rotations. It seems also that scientists moved toward studies of quite simplistic one antagonist/one pathogen relationships, removed from their natural plant and soil environment, without considering any interactions with other microorganisms.

A review of contributions to the sixth International Congress of Plant Pathology, held in Montreal in 1993 shows that 50 of the 93 posters or oral presentations in the three sessions on biological control of soil-borne pathogens were dedicated to introduced microorganisms, seven of which dealt with the importance of the environment on introduced biological control agents. Only 11 focused on enhancing resident antagonists through cultural practices. Sixteen investigated the mode of action of biological control agents, nine examined screening for candidates, seven reported on methods for controlling soil-borne pathogens with no evidence that they involved the biological activity of resident antagonists (Biological Control of Soilborne Pathogens, Sessions 16-2, 16-3, and 16-7; 6th International Congress of Plant Pathology, Montreal, July 28-August 6, 1993).

Cook (1990) concluded that the emphasis on introduced microorganisms is linked to the desire to develop products that can be marketed. It is also undoubtedly due to the difficulties encountered in studying complex relationships between plants and microorganisms in a complex environment, the soil. Whatever the method of biological control to be used, introducing microorganisms will not be successful unless the questions of WHERE and WHEN they act are considered. Similarly, the stimulation of resident antagonists will require more knowledge about WHAT they are and HOW they act. Baker (1990) pointed out the need to investigate the "what," "how," "where," and the "when" and to integrate these into a truly comprehensive whole. But these questions are not independent and their interaction must also be taken into account for successful biological control. This is probably the element missing from the present approaches, research is all too often analytical. If, as concluded by Cook (1990), the emphasis for the next few years must be on maximizing all biological controls (both introduced and resident microorganisms) we need to know how to integrate compatible biological systems in the plant-soil environment where they must work.

The lack of tools for the study of soil-borne microorganisms at the populations level, interactions within populations, and interactions between the plant and soil environments remain a major limitation. Modeling complex biological processes should help bring together approaches that look quite isolated such as genetic regulation of antibiotic production the and impact of environment, for example. Molecular biology should also help developing new tools, such as reporter genes, for investigating biological phenomena *in situ.*

B. The Nature of Current Agents Used in the Biological Control of Soil-Borne Pathogens

Disease suppression in soils has been the basis of most research on biological control agents. Soil suppressiveness may be constitutive, being an inherent property of the soil whatever its cropping history (e.g., soil suppressiveness to fusarium wilt of melons found in the Châteaurenard region of France (Louvet *et al.,* 1976)). Alternatively, it may be adaptive when soil suppression is only achieved after specific cultural practices are adopted such as monocropping (e.g., soil suppressiveness to take-all (Shipton, 1972) or rhizoctonia root rot (Lucas *et al.,* 1993) of wheat). Both forms have been extensively studied. The first provides a strong, stable model whereas the second provides the possibility of studying the way a susceptible soil can develop the capacity to suppress disease.

The first question addressed was that of origin of soil suppression and studies indicated that the basis of the phenomena was mainly biological (Baker, 1990). Microorganisms were isolated from soils, some showing evidence of biological control activity against pathogens *in vitro* and of being more or less successful *in situ.* Some of these were considered to be potential biological control agents. As far as take-all of wheat *(Gaeumannomyces graminis* var. *tritici (=Ggt))* was concerned, hypotheses flourished from the mid-1970s suggesting causal agents for observed soil suppression or plant protection: amoebae (Homma *et al.,* 1979; Chakraborly *et al.,* 1983); fungi, i.e., hypovirulent strains of *Gaeumannomyces graminis tritici (=Ggt)* (Tivoli *et al., 1974), Gaeumannomyces graminis graminis (=Ggg)* (Wong, 1975), *Phialophora* sp. (Deacon, 1976), and *Trichoderma* sp. (Simon and Sivasithamparam, 1989); bacteria, particularly *Pseudomonas fluorescens* (Cook and Rovira, 1976), *Bacillus* sp. (Capper and Campbell, 1986).

Thus, several microorganisms can be involved in soil suppression to a single disease. These explanations have not been all investigated in the same detail. Some, such as hypovirulent *Ggt*, amoebae, and *Phialophora* receive little or no attention today. These microorganisms were identified in studies conducted in different countries (France for hypovirulent *Ggt*, U.S.A. for amoebae and *Pseudomonas* spp., U.K. for *Phialophora* and *Bacillus*, and Australia for *Ggg* and *Trichoderma),* but only pseudomonads (fluorescent and nonfluorescent) have been intensively studied both within and outside the countries where they were first demonstrated (Weller and Cook, 1983; Sarniguet and Lucas, 1992; Sarniguet *et al.,* 1992a,b; Ryder and Rovira, 1993). Although the control obtained by *Pseudomonas* spp. in experiments was not reliable and below the level of control obtained in natural soil suppression, several observations justify further study. *Pseudomonas* spp. are important members of the wheat rhizosphere, producing antibiotics and thus acting as major biological components in soil suppression. These microorganisms are more rapidly and easily amplified than any other microorganisms which are too dependent on the plant (e.g., *Phialophora, Ggg,* and hypovirulent *Ggt*) or the water status of the soil (e.g., amoebae (Cook and Homma, 1979)).

C. The Mode of Action of Biological Control Agents of Soil-Borne Pathogens

Having identified what these agents are, we can examine how they act. Early studies showed that different mechanisms were involved. These include pathogen suppression for *Trichoderma* spp. or amoebae; disease suppression through direct antagonistic activity for *Pseudomonas* spp. and *Bacillus* spp.; and disease suppression through cross protection of the plant. An example of the latter for *Ggg*, hypovirulent *Ggt,* and *Phialophora* is competition for potential infection site and/or induced resistance. This diversity of mechanisms might have been seen as evidence that soil suppression was complex and that all facets must to be taken into account to reproduce experimentally significant disease control. In fact, focusing on their hypothesis, most research groups seemed to be eager to demonstrate that biological control was possible and ready for commercial release. So far, three have been field tested with that objective.

In France, experiments were conducted between 1980 and 1982 throughout the country to assess the efficacy of the control provided by coating seeds with a hypovirulent isolate of *Ggt*. The experiments were done by farmers and consisted of comparing two 0.5 ha plots in the same field; one sown with coated seeds and one sown with uncoated seeds. About 60 experiments were conducted each year. Yield comparisons between treated and untreated plots expressed as (yield treated/yield untreated)*100 ranged from 82 to 123 for year 1980-1981 and 90 to 115 for year 1981-1982 (Lucas *et al.*, 1984). When yield increases were obtained, there was evidence that control of take-all was partial and that the benefit obtained by the treatment in heavily infested soils did not make the wheat crop economically competitive. Furthermore, the variability observed in responses to the treatment was unpredictable.

To examine the effect of fluorescent pseudomonads, Cook and collaborators set up experiments in commercial fields naturally infested with the take-all fungus in the U.S.A. Over a 14-year period, thanks to the application of fluorescent pseudomonads they claimed an average of 10 to 15% greater yield; with one increase of 33% (from 5 t/ha to 6.7 t/ha) (Cook, 1994). Yield increases of 13 to 28% also were reported for field tests in China where take-all was the main yield-limiting factor (Peng and Ellingboe, 1990). So far, no method of control based on the use of fluorescent pseudomonads has reached the market. Cook (1994) recognized that although the increases in yield due to biological control were remarkable, the best yield in their test were still 50 to 60% of the yield from the same areas in response to crop rotation.

The third microorganism that has been the subject of attempts at commercial development is *Ggg*. Wong *et al.* (1993) claimed increases in yield of 27 to 45% in 1991 and 1992 with isolates of *Ggg* grown on sterilized moist oat and millet grains and inoculated into soils. Yield increases of only 10 to 30% were sometimes obtained with fluorescent pseudomonads. These *Ggg* isolates have been patented and are apparently in the process of being marketed (Wong *et al.*, 1993). The question remains as to whether this method will be really more successful than either of the two previous ones.

While none of these developments may provide a useful method of control for farmers. The studies on these biological phenomena and especially on fluorescent pseudomonads have led to a considerable increase in the knowledge of the compounds responsible for antagonistic activity (Tomashow and Weller, 1988) and the genetics of their production (Cook *et al.*, 1995). There appear to be different levels of genetic regulation of antibiotic production, the primary one being dependent on the environment of the bacteria. In some way, investigating fundamental determinism at a molecular level emphasizes the interaction with the soil-plant environment and other microbial populations of the rhizosphere. Thus, there will probably be still more focus on fluorescent pseudomonads and future progress on this group will add to our knowledge of new areas of this complex puzzle which is soil suppression to take-all. The question remains as to whether fluorescent pseudomonads must only be considered as a model for studying direct antagonism responsible for some disease suppression or as realistic candidates for a commercial biological method providing enough control when used alone. Focusing on a single phenomenon has so far been unsuccessful. Several mechanisms probably occur together in natural soils. This diversity should undoubtedly be taken into account. This raises the question of the advantages it might provide and the problems that could arise.

III. THE COMPLEXITY OF THE ENVIRONMENT AND INTERACTIONS THEREIN

We have shown that a whole variety of microorganisms and mechanisms are involved in soil suppression of take-all. Despite a great deal of work screening candidates and improving methods of application economically acceptable control has not yet been obtained. There is probably still potential for improving the antibiotic production capacity of fluorescent pseudomonads but focusing on a single vehicle of suppression may lead to problems. As shown by Mazzola *et al.* (1994) there is, within *Ggt* populations, important variation in sensitivity to phenazine-1-carboxylic acid and 2,4-diacetylphluoroglucinol; 2 major antibiotics produced by two strains of *Pseudomonas fluorescens.* There is therefore a risk of providing selective pressure in favor of resistant strains of *Ggt,* leading to the failure of the control method. This further justifies the use of biological control systems that employ multiple strains or multiple mechanisms. Diversifying mechanisms of action (and candidates for biological control) is thus a reasonable strategy for durable control. But achieving more efficiency in the control requires cumulative effects of the different mechanisms to be combined.

On a theoretical basis this should be possible for take-all, considering the different stages and niches of the fungus on which some of the biological control agents act. For example, *Trichoderma* spp. and amoebae are pathogen suppressive (they affect the saprophytic growth or the survival of the pathogen in bulk soil). Antagonistic bacteria are disease suppressive (they limit the extent of the root lesions and the spread of the fungus to secondary infections in the rhizosphere or on the rhizoplan). Cross-protecting fungi are also disease suppressive: acting by enhancing the plant host's resistance to the pathogen and resulting in slower disease progression in plant tissues.

From a practical point of view, the accurate and effective timing (when) and mode of application (where) of these potential biological control agents at present appears unrealistic, even without considering the cost of this strategy of biological control. There have been some attempts to combine fluorescent pseudomonads and hypovirulent *Ggt* on the assumption that limited necroses caused by the hypovirulent fungus would enhance the establishment of the bacteria on the roots while at the same time providing some increased resistance for the plant. Unfortunately disease control was not significantly improved (Lucas and Sarniguet, unpublished data).

So, it seems that even cumulative effects of different biological control mechanisms will be of poor value without management of the soil-plant environment. Thus, identifying cultural practices that optimize both pathogen and disease suppression will be necessary but knowing the mechanisms on which they have an impact would also help in defining a strategy that combines practices to give efficient disease control. Furthermore, hierarchy and interactions between cultural practices will have to be taken into account. Modeling disease development could provide answers to several of these important questions.

IV.   A WAY TO SORT OUT AND UNDERSTAND MULTIVARIATE COMPLEXITY: THE USE OF MATHEMATICAL MODELS

Some work has been done on modeling field disease progress for take-all (Brasset and Gilligan, 1989). Colbach *et al.* (1997) simplified these models and used them to assess the impact of crop management on primary and secondary infection cycles of take-all epidemics. Central to the model are origin of inoculum and infection rates. Inocula can be found in soils on plant debris or on roots of the living plant. Each inoculum is associated with an infection rate. Rate (c_1) is a measure of the capacity of the soil reservoir inoculum to cause infection and disease. The rate of secondary infection (c_2) is a measure of the capacity of infected roots to spread disease to other roots. The percentage of diseased plants Is given by the following equation, where time t is expressed as cumulative degree- days (basis 0°C since sowing:

$$y = \frac{1 - e^{-(c_1 + c_2)t}}{1 + \dfrac{c_2}{c_1} e^{-(c_1 + c_2)t}}$$

This equation was first successfully tested (r^2=0.99) on a plot assessed every two weeks after growth stage 30 (Zadoks *et al.,* 1974). It was then fitted to the buildup of take-all for each experimental treatment on three sites (i.e., three regions of France), where different cultural practices (sowing date, sowing density, total nitrogen dose, nitrogen fertilizer form, and burial or removal of preceding crop residue) were tested. Curve fitting estimated the parameters c_1 and c_2 for each experimental treatment of each site.

A linear model was tested to interpret the parameters c_1 and c_2 for each set of estimates at each site as a function of the factors analyzed and covariables measured. The analysis showed that sowing date always affected c_1 (i.e., primary infections) whereas c_2 (i.e., secondary infections) was only influenced on the most favorable sites (i.e., the highest infection rates under favorable climatic conditions). The parameter c_1 was always increased by early sowing. This is consistent with previous results (Hornby *et al.*, 1990) and the fact that early sowing provides a longer period for infection before winter. The effect of early sowing on c_2 was variable, positive for one experimental site and negative for another.

There was a positive correlation between plants per m^2 and parameter c_1, but only in the most favorable sites. As for the sowing date, the influence of this factor on c_2 varied. A high plant density at early stages, when the roots are still few and short, probably increases the chance of contact between the soil inoculum and living roots whereas it has a more inconsistent effect when the root system is well developed. The high nitrogen dose increased parameter c_1 and decreased c_2 but both were reduced when the nitrogen was applied as ammonium. As reported by Sarniguet *et al.* (1992a), nitrogen can stimulate both pathogen and antagonistic microflora. An increase in early infection of seminal roots allows the development of fluorescent pseudomonads on necroses, which interfere with pathogen expansion later on, especially when ammonium nitrogen forms are applied.

The hierarchy and interaction between the various factors were shown to be important. Factors other than sowing date were usually significant only when the sowing date was also significant. Sowing date may therefore be considered to be the dominant factor and its interactions with the other factors as the most important. The type of interaction therefore strongly resembles the one with site: several factors had a stronger influence or were only significant when the site was favorable to disease development. Thus, each factor seemed to amplify the risk due to the other effects and low effect factors could only influence disease if high effect factors also were favorable to its expression.

We postulate that parameter c_1 is partly dependent on pathogen suppression whereas parameter c_2 is mostly related to disease suppression. These studies therefore provide an initial approach to analyses of the way in which cultural practices act and interact and therefore how biological controls of soil pathogen might be conserved. Models can be improved to take into account other aspects of pathogenic fungus behavior or host-plant development. Some of the models developed by Brasset and Gilligan (1989), for example, include root development and inoculum decay.

However, model development does not give any information on which biological phenomena are responsible for enhancing pathogen suppression or disease suppression. The development of the microbial populations involved in these suppressions must also be analyzed. The problem is that an increase in the numbers of these populations is not necessarily important for soil suppressiveness but rather changes in population structure (Sarniguet *et al.*, 1992b). Therefore, the diversity of these populations must be well characterized according to their antagonistic activity (*sensu latu*). It seems unrealistic to engage in such studies for each of the biological phenomena described above but this should be done at least for those that appear to be representative of pathogen and disease suppression. *Trichoderma* populations would probably be a good candidate for this first phase and fluorescent pseudomonads for the second. Some important results have been obtained on fluorescent pseudomonads using analytical approaches. These attempts to obtain biological control with efficient strains of these bacteria have not been commercially successful as yet. But the modes of action and population diversity is now well documented and should help to define the soil resident population structure. Wider approaches using fluorescent pseudomonads populations as biological indicators have also attempted to link crop management to the specific enhancement of soil microbial activity responsible for disease reduction.

V. FLUORESCENT PSEUDOMONADS AND BIOLOGICAL CONTROL; INUNDATIVE RELEASE OR MANIPULATION OF THE ENVIRONMENT: THE NEED FOR CONVERGENCE OF DIFFERENT APPROACHES

A. Inundative Release: Use of Single Antagonistic Strains

The evidence that take-all decline is related to specific antagonism in the rhizosphere led to a search for the narrowest microorganism group playing an active role in soil suppressiveness. Initial *in vitro* studies indicated that fluorescent pseudomonad strains were always well represented among all microorganisms showing antagonistic activity against *Ggt*. The definition of the fluorescent pseudomonad group came from the phenotypic observation of the production of fluorescent siderophores (= microbial iron transport cofactors) when grown on iron-poor medium. This group is quite large and includes several species and subspecies of the genus *Pseudomonas* (Palleroni, 1984; Barett *et al.*, 1986). Some very active strains were isolated and tested successfully in small scale experiments (Weller, 1988). Despite the widespread distribution of these rhizobacteria and these initial successes the jump to field application was premature and failed to demonstrate effectiveness. The inundative incorporation of these bacteria into soils imitated the spreading of fungicide. The failure was attributed to poor root colonization (due to a poor active growth along the roots), low survival in the rhizosphere (due to a great sensitivity to environmental stresses), and poor competitiveness (due to an inability to compete for nutrients in the rhizosphere).

Instead of studying what changed when increasing the scale of observation (from pots to field, climatic chambers to seasonal changes, etc.) most research teams chose to describe and explain how it worked and when it worked. The best results came with the use of molecular biological techniques although limited because the studies started from *a priori* hypotheses on involved mechanisms. The best approach would probably have been to create random mutants and to keep only those mutations which were lacking or which improved efficiency in biological control tests.

Considering the size of *Pseudomonas* spp. genome over 10,000 mutants would have been necessary for such a screening. This practice is easy to realize with *Rhizobium* spp. when the result is the presence or absence of nodules on roots (Rolfe *et al.,* 1982) or with pathogenic bacteria when the result is a hypersensitive response (HR) on the leaf surface in less than 24 hours (Boucher *et al.,* 1985; Lingren *et al.,* 1986). But from a practical point of view this approach is difficult with a soil-borne pathogen like *Ggt* because of the long development time of the disease (even in pots) and because of the great variability in response due to variability in host plants, bacterial application, soil water and nutrient, and soil structure. This kind of experiment requires extensive replication as well as a lot of space, soil volume, and time. The choice of soil could also influence the data obtained. The use of sterile soil may help to standardize the biological test but it prevents any assessment of competitiveness against other microorganisms.

An alternative way is to test factors such as antibiosis, nutritive competition, root agglutination, or survival in the rhizosphere under defined conditions and then determine if the factor is involved in antagonism by evaluating the importance of the phenomena in more complex conditions. This approach requires that many teams work on various aspects of biological control and has the risk of producing large amounts of data hardly transposable to biological control. The siderophore hypothesis is a good illustration. Fluorescent pseudomonads were found to produce this high affinity iron chelate which was thought to be involved in nutritive competition for iron between a pathogen *(Fusarium* spp.) and the fluorescent pseudomonads (Kloepper *et al.,* 1980). The iron trapped by bacterial siderophores is no longer available for the pathogenic fungus, so its pathogenicity will decrease. The first attempts to demonstrate this were conducted with artificial iron chelates without any concern about the side effects on plant and total microbial physiology (Scher and Baker, 1982). The complex genetics of siderophore production prevented finding the best mutant in which to investigate iron competition. This hypothesis was not verified for take-all (Hamdan *et al.*, 1991) perhaps because of the need for a saprophytic food basis for *Ggt* that would impair effective iron deficiency.

The demonstration of antibiotic production followed quite the same process. Toxic secondary metabolites from pseudomonads have been known for a long time (Leisinger and Margraff, 1979) and were recovered from *in vitro* antibiosis tests. Deficient mutants unable to produce some antifungal products were tested for their lack of *in situ* antagonistic activity. Biological activity was restored by transforming the mutants with a cosmid harboring the same functional DNA region. Phenazine (Tomashow and Weller, 1988) and 2,4-diacetylphloroglucinol (Keel *et al.,* 1992) were shown to be active. This procedure is fine when antibiotic biosynthesis genes are directly affected but can be controversial when a pleiotropic gene is mutated or when the production of other antibiotics masks the role of a single gene. Many of the mechanisms that have been investigated like agglutination and resistance to oxidative stress to explain root colonization or plant- induced resistance by pseudomonad metabolites follow the same rule. But, the possible involvement of these mechanisms must not be rejected. They need to be included as parts of the overall activity of biological control that may lead to another level of gene regulation that is more linked to environmental conditions.

The PhzR gene that is sensitive to cell density enhances phenazine production in P. aureofaciens when it is activated in high cell density environments (Pierson et al., 1994). GacA (Laville et al., 1992) and ApdA (Corbell and Loper, 1995) are thought to be the genes of a two-component gene system that regulates all antibiotic production in P. fluorescens and which is a probable intracellular relay of external signals that influence overall cell physiology. The relationship between the physiological state, antibiotic production, and stress resistance via a general regulator of the stationary phase (the plateau that follows the exponential phase of growth of bacteria) clearly establishes interdependence with the environment (Sarniguet et al., 1995). A mutation in this regulator (rposS gene) leads to greater cell sensitivity to stress, thus to lower survival in the rhizosphere when in the stationary phase; but a higher production of some antibiotics and so to a better control of the damping-off of cucumber caused by Pythium. These data indicate that exponentially growing cells may not be the most active for antagonism and that looking for ways to enhance antibiotic production and searching for better cell multiplication on roots may not be independent, as the two phenomena may be negatively correlated. Thus, potential antagonistic microorganisms are effective provided they encounter conditions that are suitable for the expression of their antagonistic activity. Increasing their numbers may make no sense without managing the environment towards more favorable conditions for the expression of their antagonistic activity.

One could suggest that manipulating microorganisms instead of managing the environment might be easier and that genetically modified microorganisms (GEMMOs) with, for example, enhanced antibiotic production could be used. But the introduction of GEMMOs into the soil is not without problems as there is a risk of spreading their modified DNA to the genomes of other rhizosphere inhabitants. This type of practice is not accepted everywhere because of local laws and rules about the use of GEMMOs.

More complex but probably more promising would be to take into account the diversity of antagonism mechanisms. One way would be to establish a hierarchy of the different modes of antagonism, not only in terms of their importance but also in terms of dependence on external management. Although a population of fluorescent pseudomonads can adapt to diverse environments it would be hard finding a single strain that is antagonistic under all the conditions along the roots. Associated strains having complementary or synergetic activities in different ecosystems would have to be used. But more knowledge about this kind of relationship is needed and can be obtained, perhaps with the help of reporter genes for assessing in situ activities, before such mixes could be successfully managed. The association of these markers with genes involved in biological control is perhaps the best route for getting into the rhizosphere black box and discovering their real activity and also the exact microsites at which microorganisms interact (De Weger *et al.,* 1994; Loper and Lindow, 1994; Meikle *et al.,* 1994; Kraus and Loper, 1995). But why create such artificial associations? There must be enough diversity in the rhizosphere to take advantage of antagonistic actions of resident microorganisms. This type of global approach has been used with a focus on one group, pseudomonad populations.

B. Manipulating the Environment: Pseudomonads as a Population and Multifactorial Analysis

A global approach is sustained by the prospect of increasing soil suppressiveness through managed microbial activity. The importance of nutrients and site competition is illustrated by the greater severity of take-all in sterile soil than in a living soil with an active auxiliary microflora or by reduced severity of the disease when the global microbial activity is increased by slightly increasing the temperature (Cook and Baker, 1983). Of course these conditions cannot be directly applied in the field.

The first issue is how to study the whole soil microflora. Each species and subspecies cannot be routinely described. An indirect way is to assess global microbial antagonism by measuring soil suppressiveness. The development of disease on susceptible plants is assessed in different soils with different amounts of introduced pathogen inoculum (Alabouvette *et al.,* 1982; Lucas *et al.,* 1989). instead of analyzing the whole microflora, fluorescent pseudomonads have been used as biological indicators of soil microbial activity and to relate soil suppressiveness to biotic phenomena. This choice is supported by the major contribution of these rhizobacteria to antagonism by their great diversity and by the experiments whose results are summarized below. The latter indicate the direct relationship between field disease, soil suppressiveness, and the structural diversity of fluorescent pseudomonad populations.

Different agronomic practices generate different amounts of disease in the same soil: monocropping of wheat compared to rotations leads to take-all decline, using ammonium rather than nitrate nitrogen fertilization reduces the disease and a disease decline is observed in the center part of a take-all patch on turf grass (Sarniguet *et al.,* 1992a,b; Sarniguet and Lucas, 1992, 1993). The relationship between induction of disease suppression and changes in pseudomonad populations have been established in all these cases. Despite the variety of these situations, a concordance of events is necessary for the build-up of microflora that are antagonistic to take-all. The plant, the pathogenic fungus, and the soil microflora must all be simultaneously and durably associated to generate soil suppressiveness. An initial severe attack is always necessary for the development of antagonistic microflora such as fluorescent pseudomonads (Sarniguet and Lucas, 1993). Soil suppressiveness develops very early on soils taken from the field at early stages of the crop; the limiting of the disease incidence or severity is only seen later on plants in the field.

The form of nitrogen applied may act on different phases of the fungus pathogenic cycle. When there is little native inoculum in the field the ammonium form of nitrogen reduces the frequency of attacked plants (i.e., disease incidence), thus acting on early infection by the soil inoculum. When the level of *Ggt* inoculum is high, ammonium nitrogen influences only disease severity, thus acting on the preparasitic and parasitic phase through the rhizosphere microflora (Sarniguet *et al.*, 1992a).

This population approach also illustrates interactions between microorganisms. Disease reduction is not only attributed to the additive activity of all the antagonistic strains but is the resulting activity of two subgroups of microorganisms: antagonistic ones that reduce disease and deleterious ones that increase disease severity (Sarniguet *et al.*, 1992b). The coexistence of such subgroups in fluorescent pseudomonads confirms the importance of this group as biological indicators.

These results reveal that biological control is really based on microbial dynamics and complex interactions between microorganisms that depend greatly on their environment. Biological control is produced by diverse microorganisms, even if it is defined as specific. It may involve complementary activities and there may be a balance of opposite activities. An external input can help to structure these populations toward more antagonism, as shown with nitrogen fertilization. Greater success in this and other forms of conservation biological control can only be achieved if we have more knowledge of the impact of other soil-plant environment characteristics that are important for the activity of useful and deleterious microorganisms. As proposed by Huber and McCay-Buis (1993), manganese in soils might be an important one.

The role of the plant in orienting the population structure of microorganisms in the rhizosphere has yet to be explored. Early take-all necroses of the seminal root system of wheat have been shown to lead to the establishment of useful microorganisms in the rhizosphere and consequently to better subsequent protection of the nodal root system against take-all (Sarniguet *et al.*, 1992a). But the capacity of plant cultivars to sustain antagonistic microbial activity has not been explored because of the lack of suitable biological markers for such screening. Future research is now needed to find the most controllable inducers; for example, how to structure the population without too much root necrosis. Greater knowledge of the links between diversity and functionality of the bacteria in the rhizosphere within pseudomonads populations would be useful.

VI. CONCLUSION

Disease development is the result of a succession of events in which the two main actors are the plant and the pathogen. It is classically represented as a cycle from inoculum source to development of symptoms on the plant, with intermediate phases, such as inoculum dispersion, plant contamination, plant infection, and disease progression in plant tissues. Application of fungicides or plant resistance aim at breaking the cycle at some point: plant infection with preventive fungicide or disease progression with curative fungicide and plant resistance. Attempts to achieve the same results with introduced antagonistic microorganisms have not been widely successful so far. Common hypotheses are that they do not interrupt the cycle with the same efficacy as fungicides or plant resistance do and that they do not have a sufficient durable efficacy. The second point can be neglected when the diseases to be controlled are diseases which develop in a short-term period such as damping off but it is important for pathogens that can infect the plant at different stages of its growth.

Considering the first point, the control obtained in the field is not the sum of the effects of the applied biological control agents. It results from interactions between microorganisms applied as biological control agents and resident deleterious or antagonistic microorganisms as well as from interactions with the plant-soil environment which can regulate the antagonistic activity of introduced biological control agents. Furthermore, survival of introduced biological control agents is also dependent on interactions with other microorganisms involving competition for nutrients and dependent on the physico-chemical characteristics of the plant-soil environment.

Soil suppression of disease is well known for several diseases caused by soil-borne plant pathogens. The natural level of suppression (by most soil-borne pathogens) is often insufficient, may occur only in some soils (e.g., soil suppressive to fusarium wilt (Alabouvette *et al.,* 1982)), or may be achieved only in particular cropping conditions after an important development of the disease (as in take-all decline and monocropping of wheat (Shipton, 1972)).

Conservation of natural soil suppression is important but may not be enough and it needs to be enhanced, extended, or managed in an agronomically acceptable way. There are different approaches to studying the complex phenomenon, soil suppression of diseases: (1) the epidemiological approach which assesses the impact of cultural practices on phases of disease development, (2) the approach which tries to relate modifications of the plant-soil environment to beneficial modifications in the structure of resident populations of microorganisms, and (3) the single biological control agent approach which aims at identifying microorganisms that might be good candidates for biological method of control based on inundative application of these biological control agents. None of these approaches will probably be sufficient to achieve success if taken alone. Linking all of them is the only way to propose integrated method of control based on the use and conservation of biological agents.

It is important for the epidemiologist to know on which component of soil suppression he or she is acting when maintaining or enhancing soil suppression by managing the environment. This can be done through use of biological indicators to provide insights into the involved groups of microorganisms but also on the mode of action that are expressed by these microorganisms. Such tools can be elaborated by microbiologists with an expertise in microorganism physiology and genetics (e.g., the use of reporter genes for example) but they will have to be adapted to field studies. Furthermore, as discussed in this chapter, groups of microorganisms and modes of action are multiple and space and time of action may be different; all of which results in interactions with hierarchies, synergism, or antagonism. There is a need for modeling these complex interactions. This cannot be done without close collaborations between microbiologists and epidemiologists. Progress has been made in both conservation and augmentation approaches. They barely have achieved success in terms of efficacy in controlling disease at an economically acceptable level. It is time for both approaches to converge.

REFERENCES

Anon. (1995). Index Phytosanitaire. ACTA, Paris. 566p.
Alabouvette, C., Couteaudier, Y., and Louvet, J. (1982). Comparaison de la réceptivité de différents sols et substrats de culture aux fusarioses vasculaires. *Agronomie* 6, 243-284.
Baker, R. (1990). An overview of current and future strategies and models for biological control. *In* "Biological Control of Soil-borne Pathogens." (D. Hornby, ed.), pp. 375-388. C.A.B. International. Wallingford, U.K.
Barett, E. L., Solanes, R. E., Tang, J., and Palleroni N. J. *(1986)*. *Pseudomonas fluorescens* biovar V: its resolution into distinct component groups and the relationship of these groups to other *P. fluorescens* biovars, to *P. putida*, and to psychotropic pseudomonads associated with food spoilage. J. Gen. *Microbiol.* 132, 2709-2721.
Boucher, C. A., Barberis, P. A., Trigalet, A., and Demery, D. A. (1985). Transposon mutagenesis of *Pseudomonas solanacearum:* Isolation of Tn-5 induced avirulent mutants. *J. Gen. Microbiol.* 131, 2449-2457.
Brasset , P. R., and Gilligan, C. A. (1989). Fitting of simple models for field disease progress data for the take-all fungus. *Plant Pathol.* 38, 397-407.
Capper, A. L., and Campbell, R. (1986). The effect of artificially inoculated antagonistic bacteria on the prevalence of take-all disease of wheat in field experiments. *J. Appl. Bacteriol.* 60, 155-160.
Chakraborty, S., Old, K. M., and Warcup, J. H. (1983). Amoebae from a take-all suppressive soil which feed on *Gaeumannomyces graminis tritici* and other soil fungi. *Soil Biol. Biochem.* 15, 17-24.
Colbach, N., Lucas, P., and Meynard, J. M. (1997). Influence of crop management on take-all development and disease cycles of winter wheat. Phytopathol. 87, 26-32.
Cook, R. J. (1990). Twenty-five years of progress towards biological control. *In* "Biological Control of Soil-borne Pathogens." (D. Hornby, ed.), pp. 1-14. C.A.B. International. Wallingford, U.K.
Cook, R. J. (1993). Making greater use of introduced microorganisms for biological control of plant pathogens. *Annu. Rev. Phytopathol.* 31, 53-80.
Cook, R. J. (1994). Problems and progress in the biological control of wheat take-all. *Plant Pathol.* 4, 429-437.

Cook R. J., and Baker K. F. (1983). "The Nature and Practice of Biological Control of Plant Pathogens." Amer. Phytopathol. Soc. St Paul, MN.

Cook, R. J., and Rovira, A. D. (1976). The role of bacteria in the biological control of *Gaeumannomyces graminis* by suppressive soils. *Soil Biol. Biochem. 8,* 269-273.

Cook, R. J., and Homma, Y. (1979). Influence of water potential on activity of amoebae responsible for perforation of fungal spores. *Phytopathol.* 69, 914. (abstract).

Cook, R. J., Tomashow, L. S., Weller, D. M., Fujimoto, D., Mazzola, M., Bangera, G., and Kim, D-S. (1995). Molecular mechanisms of defense by rhizobacteria against root disease. *Proc. Natl. Acad. Sci. USA* 92, 4197-4201.

Corbell N., and Loper, J. E. (1995). A global regulator of secondary metabolites production in *Pseudomonas fluorescens* Pf-5. *J. Bacteriol.* 177, 6230-6236.

Deacon, J. W. (1976). Biological control of the take-all fungus *Gaeumannomyces graminis*, by *Phialophora* and similar fungi. *Soil Biol. Biochem.* 8, 275-283.

De Weger, L. A., Dekkers, L. C., van der Bij, A. J., and Lugtenberg, B. J. J. (1994). Use of phosphate-reporter bacteria to study phosphate limitation in the rhizosphere and in bulk soil. *Mol. Plant-Microbe Interact.* 7, 32-38.

Doussinault, G., Delibes, A., Sanchez-Monge, R., and Garcia-Olmedo, F. (1983). Transfer of a dominant gene for resistance to eyespot disease from a wild grass to hexaploid wheat. *Nature* 303, 698-700.

Hamdan, H., Weller, D. M., and Tomashow L. S. (1991). Relative importance of fluorescent siderophores and other factors in biological control of *Gaeumannomyces graminis* var. *tritici* by *Pseudomonas fluorescens* 2-79 and M4-80R. *Appl. Environ. Microbiol.* 57, 320-327.

Homma, Y., Sitton, J. W., Cook, R. J., and Old, K. M. (1979). Perforation and destruction of pigmented hyphae of *Gaeumannomyces graminis* by vampyrellid amoebae from Pacific Northwest wheat field soils. Phytopathol. 69, 1118-1122.

Hornby, D., Bateman, G. L., Gutteridge, R. J., Lucas, P., Montfort, F., and Cavelier, A. (1990). Experiments in England and France on fertilizers, fungicides and agronomic practices to decrease take-all. *Proc. Brighton Crop Prot Conf.* - Pests and Dis. 2, 771-776.

Huber, D. M., and McCay-Buis, T. S. (1993). A multiple component analysis of the take-all disease of cereals. *Plant Dis.* 77, 437-447.

Keel, C., Schnider, U., Maurhoffer, M., Voisard, C., Laville, J., Burger, V., Wirthner, P., Haas, D., and Défago, G. (1992). Suppression of root diseases by *Pseudomonas fluorescens* CHAO: importance of the bacterial secondary metabolite 2,4 diacetylphloroglucinol. *Mol. Plant-Microbe Interact.* 5, 4-13.

Kloepper, J. W., Leong, J., Teintze, M., and Schroth, M. N. (1980). Pseudomonas siderophores: a mechanism explaining disease suppressive soils. *Curr. Microbiol.* 4, 317-320.

Kraus, J., and Loper, J. E. (1995). Characterization of a genomic region required for production of the antibiotic pyoluteorin by the biological agent *Pseudomonas fluorescens* Pf-5. *Appl. Environ. Microbiol.* 61, 849-854.

Laville, J., Voisard, C., Keel, C., Maurhofer, M., Défago, G., and Haas, D. (1992). Global control in *Pseudomonas fluorescens* mediating antibiotic synthesis and suppression of black root rot of tobacco. *Proc. Natl. Acad. Sci. USA* 89, 1562-1566.

Leisinger T., and Margraff, R. (1979). Secondary metabolites of fluorescent pseudomonads. *Microbiol. Rev.* 43, 422-442.

Lingren, P. B., Peet, R. C., and Panopoulos, N. J. (1986). Gene cluster of *Pseudomonas syringae* pv. *phaseolicola* controls pathogenicity on bean plants and hypersensitivity on non-host plants. *J. Bacteriol.* 168, 512-522.

Loper, J. E., and Lindow, S. E. (1994). A biological sensor for iron available to bacteria in their habitats on plant surfaces. *Appl. Environ. Microbiol.* 60, 1934-1941.

Louvet, J., Rouxel, F., and Alabouvette, C. (1976). Mise en évidence de la nature microbiologique d'un sol au développement de la fusariose vasculaire du melon. *Ann. Phytopathol.* 8, 425-436.

Lucas, P., and Sarniguet, A. (1990). Soil receptivity to take-all: Influence of some cultural practices and soil chemical characteristics. *Symbiosis* 9, 51-57.

Lucas, P., Lemaire, J. M., and Doussinault G. (1984). Le piétin-échaudage des céréales dû à *Gaeumannomyces graminis* var. *tritici* Problèmes posés par la mise au point d'une méthode biologique basée sur l'utilisation d'une souche hypoagressive du parasite. *In* "EEC Program on Integrated and Biological Control-Final Report 1979-1983." (R. Cavalloro, and A. Paviaux, eds.), pp. 367-377. CEE, Bruxelles.

Lucas, P., Sarniguet, A., Collet, J. M., and Lucas, M. (1989). Réceptivité des sols au piétin-échaudage: influence de certaines pratiques culturales. *Soil Biol. Biochem.* 21, 1073-1078.

Lucas, P., Smiley, R. W., and Collins, H. P. (1993). Decline of rhizoctonia root rot on wheat in soils infested with *Rhizoctonia solani AG-8. Phytopathol.* 83, 260-265.

Mazzola, M., Fujimoto, D. K., and Cook, R. J. (1994). Differential sensitivity of *Gaeumannomyces graminis* populations to antibiotics produced by biocontrol fluorescent pseudomonads. *Phytopathol.* 84, 1091.

Meikle, A., Glover, L. A., Killham, K., and Prosser, J. I. (1994). Potential luminescence as an indicator of activation of genetically-modified *Pseudomonas fluorescens* in liquid culture and in soil. *Soil Biol. Biochem.* 26, 747-755.

Palleroni, N. J. (1984). Genus I. *Pseudomonas. In* "Bergey's Manual of Systematic Bacteriology." (N. R. Krieg, and J.G. Holt, ed.), pp. 141-199. 3rd ed., Vol 1, Williams and Wilkins Co. Baltimore, MD.

Peng, Y., and Ellingboe, A. H. (1990). Mutations in *Pseudomonas fluorescens* improve antibiosis and biocontrol of take-all of wheat. *Phytopathol.* 90, 1016.

Pierson III, L. S., Keppenne, V. D., and Wood, D. W. (1994). Phenazine antibiotic biosynthesis in *Pseudomonas aureofaciens* 30-84 is regulated by *phzR in* response to cell density. *J. Bacteriol.* 176, 3966-3974.

Rolfe, B. G., Gresshoff, P.M., and Shine J. (1982). Rapid screening method for symbiotic mutants of *Rhizobium leguminosarum* biovar *trifolii* and white clover plant. *Plant Sci. Lett.* 19, 227-284.

Ryder, M. H., and Rovira, A. D. (1993). Biological control of take-all of glasshouse-grown wheat using strains of *Pseudomonas corrugata* isolated from wheat field soil. *Soil Biol. Biochem.* 25, 311-320.

Sarniguet, A., and Lucas, P. (1992). Evaluation of populations of fluorescent pseudomonads related to decline of take-all patch on turf grass. *Plant and Soil* 145, 11-15.

Sarniguet, A., and Lucas, P. (1993). Induction de l'antagonisme microbien vis-à-vis du piétin-échaudage sur blé *(Gaeumannomyces graminis* var. *tritici*) et sur *gazon (Gaeumannomyces graminis* var. *avenae). In* "Abstracts of the 6th International Congress of Plant Pathology." p. 268. National Research Council Canada, Ottawa, Canada.

Sarniguet, A., Lucas, P., and Lucas, M. (1992a). Relationships between take-all, soil conduciveness to the disease, populations of fluorescent pseudomonads and nitrogen fertilizers. *Plant and Soil* 145, 17-27.

Sarniguet, A., Lucas, P., Lucas, M., and Samson, R. (1992b). Soil conduciveness to take-all of wheat: influence of the nitrogen fertilizers on the structure of populations of fluorescent pseudomonads. *Plant and Soil* 145, 29-36.

Sarniguet, A., Kraus, J., Henkels, M. D., Muehlchen, A. M., and Loper, J. E. (1995). The sigma factor σ^s affects antibiotic production and biological control activity of *Pseudomonas fluorescens* Pf-5. *Proc. Natl. Acad. Sci. USA* 92, 12255-12259.

Scher M. F., and Baker R. (1982). Effect of *Pseudomonas putida* and a synthetic iron chelator on induction of soil suppressiveness to fusarium wilt pathogens. *Phytopathol.* 72, 1567-1573.

Scott, P. R., Hollins, T. H., and Summers, R. W. (1989). Breeding for resistance to two soil-borne diseases of cereals. *Vortr. Pflanzenzüchtg.* 16, 217-230.

Shipton, P. J. (1972). Take-all in spring-sown cereals under continuous cultivation: disease progress and decline in relation to crop succession and nitrogen. *Ann.* Appl. Biol. 71, 33-46.

Simon , A., and Sivasithamparam, K. (1989). Pathogen-suppression: a case study in biological suppression of *Gaeumannomyces graminis* var. *tritici* in soil. *Soil Biol. Biochem.* 21, 331-337.

Smiley, R. W. (1978). Antagonists of *Gaeumannomyces graminis* from the rhizoplane of wheat in soils fertilized with ammonium or nitrate nitrogen. Soil *Biol. Biochem.* 10, 169-174.

Thomashow, L. S., and Weller, D. M. (1988). Role of a phenazine antibiotic from *Pseudomonas fluorescens* in biological control of *Gaeumannomyces graminis* var. *tritici. J. Bacteriol.* 170, 3499-3508.

Tivoli, B., Lemaire, J. M., and Jouan, B. (1974). Prémunition du blé contre *Ophiobolus graminis* Sacc. par des souches peu agressives du même parasite. *Ann. Phytopathol.* 6, 395-406.

Weller, D. M. (1988). Biological control of soil-borne plant pathogens in the rhizosphere with bacteria. *Annu. Rev. Phytopathol.* 26, 379-407.

Weller, D. M., and Cook, R. J. (1983). Suppression of take-all of wheat by seed treatment with fluorescent pseudomonads. *Phytopathol.* 73, 463-469.

Wong, P. T. W. (1975). Cross-protection against the wheat and oat take-all fungi by *Gaeumannomyces graminis* var. *graminis. Soil Biol. Biochem.* 7, Wong, P. T. W., Mead, J. A., and Holley, M. P. (1993). Improved field control of take-all in wheat by new isolates of *Gaeumannomyces graminis* var. *graminis. In* "Abstracts of the 6th International Congress of Plant Pathology." p. 265. National Research Council Canada, Ottawa, Canada.

Zadoks, J. C., Chang, T. T., and Konzac, C. F. (1974). A decimal code for the growth stages of cereals. *Weed Res.* 14, 415-421.

CHAPTER
20

CONSERVATION STRATEGIES FOR THE
BIOLOGICAL CONTROL OF WEEDS

R. M. Newman, D. C. Thompson, and D. B. Richman

I. INTRODUCTION

Success of the biological control of weeds is largely dependent on the establishment and maintenance of adequate populations of biological control agents. Conservation techniques involve the identification and manipulation of factors that limit or enhance the abundance and effectiveness of control agents. Although conservation is thus essential for effective biological control of weeds relatively little attention has been given to conservation strategies to enhance weed control. Conservation strategies for the enhancement of insect natural predators and parasitoids are relatively well developed (e.g., other chapters in this volume; DeBach and Rosen, 1991; Whitcomb, 1994). Although conservation strategies are often mentioned in most reviews of the biological control of weeds (e.g., Wapshere et al., 1989; DeBach and Rosen, 1991; DeLoach, 1991; Harris, 1991) coverage is limited, with little documentation. Reviews of insect conservation give little attention to their use for the control of weeds (e.g., Collins and Thomas, 1991; Gaston et al., 1993). The recent text by van Driesche and Bellows (1996) considers conservation strategies. However, it provides few examples of successful conservation techniques for the control of weeds. Harris (1991) noted that conservation for weed biological control was largely a theoretical concept. The historical lack of scientific information on conservation strategies for the biological control of weeds may in part be due to a general lack of rigorous evaluation of weed biological control projects (McClay, 1995); reasons for success or lack of success are not evaluated or reported. More research on factors that influence the success of weed biological control agents, essentially the testing of conservation strategies, is now being reported. The topics of papers in the two recent proceedings of international symposia on the biological control of weeds are evidence of this change (Delfosse and Scott, 1995; Moran and Hoffmann, 1996).

Interest in the use of native agents for the control of weeds is also increasing (DeBach and Rosen, 1991; Buckingham, 1994; Sheldon and Creed, 1995). This approach is controversial, with skeptics arguing that native control agents are constrained by their own natural enemies and will thus be ineffective weed control agents. This objection is largely undocumented. Rigorous testing with native agents is needed to provide true tests of efficacy. Furthermore, with native weeds the use of native agents may be preferable to the use of exotic species (the traditional approach to biological control) that may attack nontarget species (Simberloff and Stiling, 1996; Louda *et al.*, 1997). Given the increasing concern about the potential negative effects of introduced classical control agents (e.g., Louda and Masters, 1993; Randall, 1996; Simberloff and Stiling, 1996), rigorous testing of native control agents appears justified. The effective use of native biological control agents will be dependent upon the development of effective conservation strategies.

In this chapter, we first review the factors that limit the success of insects for weed biological control, including factors that regulate control agent populations and factors that influence weed response to insects. These factors will be considered in the context of conservation strategies that have been applied with classical biological control. We will then address the use of conservation biological control strategies with native agents and we will summarize our experiences with rangeland weeds and the aquatic weed Eurasian watermilfoil. Lastly, we will summarize conservation strategies and suggest that conservation will become an increasingly important strategy for weed control but that this will require better evaluation of control projects.

II. FACTORS THAT LIMIT THE SUCCESS OF WEED BIOLOGICAL CONTROL AGENTS

Numerous factors can affect the success of weed biological control agents by regulating their abundance or by determining the weed's response to a population of agents. Adequate densities of control agents are required if the agent is to effectively control the target species. Thus, conservation strategies are needed to ameliorate factors that control or reduce agent densities. However, thriving populations of control agents do not always result in effective control. Factors that influence the response of the weed and the rest of the plant community must be considered to provide lasting control. Effective conservation strategies will also enhance the weed's susceptibility to control. Relatively few studies have systematically evaluated the effects of biological control agents and many have failed to measure population response of the weeds (McClay, 1995). Determination of weed response is central to determining the success of a control agent (Cullen, 1995).

To determine what factors influence biological control of weeds, Cullen (1995) examined the causes of failure for 25 different weed-agent associations that had shown significant success in at least one place. His analysis indicated that about half of the failures were associated with temperature or moisture regime and more than half of these were due to negative effects on the agent. Predation and parasitism of agents were important in about 17% of the cases. Host plant resistance, habitat suitability, and weed response (including competition with other weeds) were each important in 10% of the cases; the causes of several regional failures could not be determined.

A. Factors That Regulate Control Agent Populations

1. Climate and weather

Climate and weather are clearly important in regulating insect populations. Crawley (1986) found that climate was important in 44% of the failures of weed control agents to establish or control weeds. Cullen's (1995) analysis produced comparable results. Proper climate matching is a major concern in classical biological control; however, it can be difficult to predict how agents will perform in a new environment. Room *et al.* (1989) demonstrated the obvious importance of proper climatic matching for control of *Salvinia*. For example, the weevil *Cyrtobagous salviniae* Calder and Sands will not lay eggs at temperatures below 21°C and populations cannot thrive in temperate climates. Conversely, the moth *Samea multiplicalis* (Guenee) cannot survive at high temperatures and will not be effective in tropical climates. The flea beetle *Agasicles hygrophila* Selman and Vogt is unable to maintain population densities sufficient to control alligatorweed *Alternanthera philoxeroides* (C. Martius) in both the tropical and the cooler regions of the weed's distribution (Julien *et al.*, 1995). In cooler regions, *A. hygrophila* may be replaced by *Vogtia malloi* Pastrana which becomes more abundant and effective at controlling alligatorweed in its northern range (Vogt *et al.*, 1992).

Within a suitable climatic region outcomes of annual variation in weather, such as droughts or floods, may affect agent densities (DeLoach, 1995; Hight *et al.*, 1995). Similarly, microclimate can be important in herbivore distribution and damage (e.g., Collinge and Louda, 1988; Louda and Rodman, 1996). The classic case of control of Klamath weed *(Hypericum perforatum* L.) by leaf beetles (primarily *Chrysolina quadrigemina* (Suffrian) but also *C. hyperici* (Foster)) illustrates the importance of both regional and local scale climatic factors. The degree of control of Klamath weed varies both within and among countries, with high success in most western U.S.A. states (Huffaker and Kennett, 1959; Campbell and McCaffrey, 1991). However, mixed results have been obtained in other western states, British Columbia, Canada (Williams, 1985; McCaffrey *et al.,* 1995), New Zealand, and Australia (Briese, 1991; Syrett *et al.*, 1996). In general, control by *C. quadrigemina* has been most successful in dryer sites, in part due to better survival of the beetle under these conditions (Huffaker *et al.*, 1984; Williams, 1985; Myers, 1987; Briese, 1991). Performance of two other agents, *Chrysolina hyperici and Agrilus hyperici* (Creutzer), appears better in wetter sites or sites with colder winters where these agents may become more important than *C. quadrigemina* (Williams, 1985; Briese, 1991; Campbell and McCaffrey, 1991). The rainfall regime affects insect phenology, survival, and their effect on the plant (Huffaker *et al.*, 1984; Williams, 1985).

Microclimate variation can determine local variation in herbivore damage and plant control. For example, herbivores of *Cardamine cordifolia* A. Gray are more common at drier, sunnier sites where the plants also may be more affected by herbivores (Collinge and Louda, 1988; Louda and Rodman, 1996). The presence of shade and other vegetation can enhance populations of the cactus moth borer *Melitara dentata* (Grote) and the stem boring weevil *Gerstaeckeria* sp., resulting in a greater effect on the cactus at these sites (Burger and Louda, 1994, 1995). Large-scale climate will determine the suitability of control agents but small-scale effects may be amenable to manipulation. Conservation strategies such as promoting increases or decreases in shade by manipulation of other vegetation such as trees or shrubs or providing windbreaks (plantings or snow fences) to alter the microclimate should be investigated.

2. Habitat manipulation and pesticides

Fires or controlled burns used to manage natural habitat can affect biological control agent populations. Populations of the grasshopper *Hesperotettix viridis* (Thomas) can be dramatically reduced due to poorly timed prescribed burns in southwestern U.S.A. rangelands (D.C. Thompson, unpublished data). Prescribed burns in an Australian *Eucalyptus* forest resulted in major decreases in *Chrysolina quadrigemina* density and a resurgence of St. John's wort *(H. perforatum) (Briese*, 1996). However, the released nutrients may have resulted in higher quality plants which, in turn, promoted a subsequent increase in control agent density. Thus, it may be possible to adjust the timing, frequency, and distribution of prescribed burns to promote rather than suppress biological control (Briese, 1996) by burning only when resistant stages (e.g., eggs or belowground larvae) are present or by leaving adequate patches of unburned habitat.

Cultivation, crop rotation, physical disturbance, and grazing can also reduce populations of control agents. Annual disturbance such as cultivation or grazing can greatly reduce control agent densities and increase weed densities. However, provision of nearby refuges can sometimes result in weed control even in the disturbed areas (Peschken and McClay, 1995). Crop rotation appears to limit the ability of the ragweed leaf beetle *(Zygogramma suturalis* (F.)) to control ragweed *(Ambrosia artemisiifolia* L.) in Russia, due to the poor dispersal ability of the beetle (Reznik, 1996). Mowing at the wrong time can have severe effects on biological control agents. Mowing the thistle *Carduus thoermeri* Weinmann at the bud or bloom stage can greatly reduce or eliminate the weevil *Rhinocyllus conicus* (Frölich). However, mowing later in the season to eliminate lateral flowers after primary inflorescences have senesced can enhance *C. thoermeri* control by *R. conicus* because the lateral inflorescences usually escape attack by the weevil (Tipping, 1991). Grazing by livestock can also affect populations; however as with mowing, timing can be important. Early season grazing of knapweed when the plants are acceptable to cattle may help suppress the plant but summer grazing, when only flowers are acceptable, would reduce populations of agents that attack the capitula (Harris, 1991).

In aquatic systems, weed harvesting can dramatically reduce herbivore density (Sheldon and O'Bryan, 1996). Water levels and flooding also can be important for aquatic and semi-aquatic control agents. Prolonged spring flooding can affect the establishment and maintenance of purple loosestrife *(Lythrum salicaria* L.) control agents (Hight *et al.,* 1995), but would not adversely affect the plant. Water levels are also important in the establishment of *Bagous affinis* Hustache, a weevil that feeds on the tubers of the submersed macrophyte *Hydrilla verticillata* (L.f.). The weevil cannot withstand long periods of submergence and it has failed to establish at most sites in Florida due to lack of a cyclical drought period (Buckingham, 1994). In contrast, the weevil established in release cages at several sites in northern California that undergo annual water level reductions, despite low winter temperatures (Godfrey *et al.,* 1994). Fluctuating water levels and sedimentation also can be important. Silt on the roots of waterhyacinth can greatly reduce the successful pupation of the waterhyacinth weevil *Neochetina eichhorniae* Warner (Visalakshy and Jayanth, 1996). Thus, populations of the weevil and its degree of control appear to be limited in situations where the plant is not free floating or is rooted in the sediment.

Chemical control can have major effects on biological control agents. DeBach and Rosen (1991) stress the importance of avoiding insecticides. However, there appear to be few documented cases of insecticides negatively affecting weed biological control agents (Hoffmann and Moran, 1995). Often there would be no reason to apply insecticides to areas with weed control agents. However, when weeds are in close proximity to managed crops or in rangeland situations the potential for conflict arises. For example, Pomerinke *et al.* (1995) note that extensive insecticide use to control grasshoppers and caterpillars on rangelands may have disrupted the natural control of purple locoweed *(Astragalus mollissimus* Torr.) by the native weevil *Cleonidius trivittatus* (Say). The weevil *Trichapion lativentre* (Béguin-Billecocq), a successful control agent of the legume *Sesbania punicea* (Cavanille) in South Africa can be severely affected by insecticide use in citrus orchards (Hoffmann and Moran, 1995). Drift of insecticides from the orchards can reduce *T. lativentre* summer populations and results in much higher densities of the weed up to several hundred meters from the orchards.

In contrast to insecticides, herbicides are more likely to be used in areas with weed biological control agents. Herbicide use can affect the success of biological control agents positively or negatively. When herbicides are used as a planned IPM strategy with biological control agents they can enhance the effect of the agents (Wapshere *et al.,* 1989; Harris, 1991; Messersmith and Adkins, 1995). However, when herbicide use is not coordinated with biological control efforts it can reduce the abundance and effectiveness of the agents (e.g., Zimmermann, 1979; Center, 1994; Messersmith and Adkins, 1995). Adverse effects of herbicides are usually not via direct toxicity but either by a reduction of food supply and habitat or changes in plant acceptance, nutritional quality, or habitat quality (Messersmith and Adkins, 1995). Direct toxicity, however, does occur (Messersmith and Adkins, 1995) and susceptibility can vary among taxa and life stage for the same herbicide (Haag and Buckingham, 1991).

Most herbicide effects are indirect and as with other cultural practices the timing and extent of treatment or the availability of refugia can be important in maintaining adequate biological control agent densities. Repeated herbicide treatments that eliminate or greatly reduce host plants can eliminate weed control agents (Center, 1994); however, untreated refugia can maintain populations. Thus, rather than repeated and long-term herbicide treatments of an entire area the provision of refugia can permit development of long-term weed suppression by biological control agents while allowing for immediate or high-intensity management of priority areas by herbicides. For example, by leaving untreated areas adjacent to herbicide treatments, high populations of the control agent *Neochetina eichhorniae* were maintained and control of waterhyacinth, both within the untreated refugia and the remainder of the lake, was achieved (Haag and Habeck, 1991). Julien and Storrs (1996) found a short-term reduction in *Cyrtobagous salviniae* populations associated with herbicide reduction of *Salvinia molesta* Mitchell. However, weevil populations appeared to be able to rebound from early herbicide treatments and also via colonization from uncontrolled areas. Herbicides applied to the thistle *Carduus thoermeri* can reduce the survival of the weevil *Rhinocyllus conicus* primarily by eliminating its seed source. However, spraying at later plant life stages has less effect on the weevil while reducing seed production and viability (Tipping, 1991). Herbicides, for high priority treatments, can thus be compatible with biological control agents if properly timed and if local refugia are provided.

There are several examples of integration of herbicide timing and spatial distribution with biological control agents or the application of growth inhibitors to enhance agent success. Integration of herbicides with the gall midge *Spurgia esulae* Gagne can be effective for control of leafy spurge. Although herbicides can reduce the number of galls they have little impact on gall midge populations (Lym and Carlson, 1994). They suggest that leaving 15-25% of an area untreated, especially less accessible areas, should provide good integrated control. Plant growth retardants (EL-509 and paclobutrazol) have been shown to enhance the effectiveness of the waterhyacinth weevil for controlling waterhyacinth (Van and Center, 1994). Paclobutrazol appeared to be particularly effective; it did not alter consumption by the weevil and acted synergistically to enhance control by preventing the plants from outgrowing the damage produced by the weevil (Van and Center, 1994). However, it does not appear that these approaches are routinely used in an integrated weed management approach.

Population protection includes the informed and appropriate use of pesticides to maintain native and/or exotic biological control agents. Insecticides must be used with caution to insure that biological control agents are not directly influenced. This may be as simple as timing treatments such that nontarget biological control agents are protected (e.g., before adult root boring beetles emerge (Knight and Thompson, 1996)) or choosing not to use an insecticide when biological control agents make up a significant portion of the community (Lockwood, 1993; Thompson *et al.*, 1996). Properly timed mowing, grazing, burning, and water level management allow biological control agents to survive and supplement other management techniques. Leaving undisturbed critical habitat or refugia for biological control agents is important. Populations of weed control agents can be maintained by leaving strips or islands of untreated weed populations (Haag and Habeck, 1991; Tipping, 1991; Turcotte, 1993; Lym and Carlson, 1994). The optimal size of the strips or islands is dependent on the mobility and phenology of the biological control agents at implementation of weed control (Thomas *et al.*, 1991). Unfortunately this information is unknown for most species. For example, herbicide control of large healthy patches of *Opuntia* should be avoided as these are the most suitable for maintaining control agent populations (Zimmermann, 1979). For successful implementation of conservation biological control all efforts should be melded into an integrated weed management strategy (Watson and Wymore, 1990; Johnson and Wilson, 1995; Messersmith and Adkins, 1995).

3. Predators and parasites

Historically, one of the rationales for successful classical biological control of weeds was that the newly released agents would be freed from their normal natural enemies and thus reach higher densities to effect control. However, concern remains about the role of predators and parasitoids in failures of agents to control weeds. Goeden and Louda (1976) presented a detailed analysis of the role of predators, parasitoids, and pathogens in the success of classical weed control. They found that natural enemies could have been important in up to half the weed control projects they examined. However, they noted that experimental evidence was lacking in all but two cases and separating natural enemies from other factors was difficult. It did appear that generalist predators were much more important limiting factors than parasitism and disease (Goeden and Louda, 1976). These findings suggest that imported herbivores are less vulnerable to attack by specialist natural enemies but can be susceptible to generalist predators. Price (1987) summarized the conclusions of several authors who had extensively studied ten populations. Natural enemies were identified as major factors regulating population density in 70% of these populations; the tenuous nature of these conclusions was stressed. However, additional experimental studies also supported an important role for natural enemies (Price, 1987). In contrast, other studies have shown lesser effects of natural enemies on herbivores of weeds. Crawley (1986) found that predators were the cause of reduced effectiveness of weed control agents in 22% of the cases reviewed, parasitoids in 11 % and diseases in 8%. Cullen (1995) showed that predation and parasitism were important in about 17% of the cases he examined. Because he only examined systems where control was successful at least once he may have missed taxa that are particularly susceptible to control by natural enemies.

Some recent studies support the contention that parasitoids and predators are often not a major factor in limiting weed control agents and herbivores in general, including studies with native herbivores (Harrison and Cappuccino, 1995; Price *et al.,* 1995). Several careful studies of classical biological control agents and their natural enemies have also failed to demonstrate a strong influence of parasitoids even though these may be common and thus presumed to be important. For example, although an introduced seed feeding bruchid acquired ten parasitoids in three years the parasitoids had little effect on establishment or population density (Hoffmann *et al.,* 1993). Similarly, a review of parasitoids found on classical biological control agents in South Africa showed that although 40% of the introduced weed control agents acquired native parasitoids, the parasitoids were not an important factor in establishment or control success. Only one instance of a failure to establish was due to a parasitoid (Hill and Hulley, 1995).

However, predators and parasitoids can be important at least for some taxa, in some years (Goeden and Louda, 1976; Price, 1987; Belovsky and Joern, 1995; Olckers, 1995; Roland and Taylor, 1995; Gardner and Thompson, 1997). Reductions in predator abundance via the application of an insecticide resulted in increased cochineal densities and improved control of prickly pear cactus (Annecke *et al.,* 1969). Although Müller *et al.* (1990) found that poor host plant synchronization and generalist predators were usually the most important factors regulating the density and success of the moth *Coleophora parthenica* Meyrick, a Russian thistle control agent. Parasitoid attack was variable among years and in one year was the main source of mortality. The caterpillar *Samea multiplicalis* is generally an ineffective control agent for *Salvinia,* in part due to population regulation by parasitoids and pathogens (Room, 1990). In contrast, no parasitoids or pathogens have been found for the weevil *Cyrtobagous singularis* Hustache a less successful congener of the highly successful *C. salviniae.* Nechols *et al.* (1996) suggested that parasitism was not important but predation by generalist arthropod predators, particularly on eggs, could limit populations of purple loosestrife control agents. Predation may also be dependent on habitat. Predation of *Galerucella nymphaeae* (L.) a native congener of introduced control agents was lower in the spring when the adjacent vegetation was submerged than in late summer when the sites were dryer (Nechols *et al.,* 1996). Habitat fragmentation can also alter the efficacy of parasitoids and predators and thus their effects on herbivores (Roland and Taylor, 1995).

Generalizations about the extent or importance of natural enemies are difficult, and for the same taxa the importance of natural enemies may vary by habitat or year (Müller et al., 1990; Belovsky and Joern, 1995; Nechols et al., 1996). Certainly, release of new natural enemies to control other insects must be carefully evaluated to ensure that desirable weed control agents are not affected. For example, there was considerable concern that the use of classical control agents for nuisance grasshoppers could severely affect desirable species that are suppressing weeds (Lockwood, 1993). Insecticides that selectively kill predators may be used to increase control agent densities (Annecke et al., 1969) but this approach requires testing for each insecticide-herbivore-predator combination. Temporary removal of predators might be feasible to build initial populations; however, longer-term strategies will likely rely on habitat manipulation and protection. If natural enemies are suspected to be important, manipulative or experimental studies that consider the effects of environmental variables along with natural enemies will be required to determine the importance of predation and parasitism and to suggest conservation strategies to ameliorate the effects of natural enemies.

4. Plant quality

Plant quality can be a major determinant of the density and effectiveness of weed control agents (Myers, 1987). These influences may result from variation in host plant quality (Slansky, 1992), differential resistance to biological control agents (i.e., plant genotype effects) (Maddox and Root, 1987; Fritz and Price, 1988; Strauss, 1990; Stiling and Rossi, 1995), and differential attack by the insects' natural enemies (Karban, 1989, 1992; Fritz 1995). Secondary plant chemicals can contribute significantly to the defense of plants against herbivorous insects (Roitberg and Isman, 1992; Rosenthal and Berenbaum, 1992). Plant nitrogen content and defensive chemicals can affect the palatability of plants as well as the growth and survival of the herbivores (e.g., Slansky, 1992, Newman et al., 1996b). Similarly, levels of defense and nutrient content can vary by habitat, affecting levels of weed control by herbivores (Louda and Rodman, 1996).

The importance of plant quality in biological control is well illustrated by the now classic case of the floating fern Salvinia molesta, where it was demonstrated that plant nutrient status was critical in successful control by the weevil Cyrtobagous salviniae (Room et al., 1989; Room, 1990; Room and Fernando, 1992). Salvinia in low nutrient waters did not provide adequate nutrition for the weevil to increase. Addition of fertilizer increased Salvinia nitrogen content and increased its acceptability to the caterpillar Samea multiplicalis and its nutritional quality for both insects, which resulted in higher populations of Cyrtobagous salviniae. Weevil populations will increase sufficiently on the higher nitrogen plants to effect control. Damage by the weevil will maintain higher N content in the remaining plants, eliminating the need for further fertilization (Room *et al.,* 1989; Room, 1990).

Similar responses to plant quality have been seen with control agents for *Opuntia spp., Hydrilla verticillata,* and waterhyacinth *(Eichhornia crassipes (*Mart.*)).* Increased nitrogen increased the suitability of *Opuntia* to *Cactoblastis* resulting in better control (Wilson, 1960; Andres, 1982). *Hydrilla verticillata* nitrogen content significantly influences the growth and survival of *Hydrellia pakistanae* Deonier and may affect its ability to control the plant (Wheeler and Center, 1996). The waterhyacinth weevil *N. eichhorniae,* prefers higher nitrogen tissue for feeding and oviposition and performs better on less damaged and higher nitrogen plants (Center and Wright, 1991; Center and Dray, 1992; Center, 1994). However, extensive weevil damage can reduce plant nitrogen content (Center and Van, 1989) and thus suitability, resulting in dispersal of adults and reduced populations once the weed is damaged (Center and Dray, 1992; Julien *et al.,* 1996).

Management of weeds via biological control agents can be very sensitive to variability in weed genotypes (Sheppard, 1992). Plant resistance and acceptability can vary among genotypes and biotypes within a species. The success of biological control agents (e.g., *Spurgia esula)* imported to control leafy spurge *(Euphorbia esula* L.) has been correlated to weed genotype variability (Nissen *et al.,* 1995), whereas success was not correlated to variability among populations of musk thistle *Carduus nutans* L. (Zwölfer and Harris, 1984). The secondary metabolite profiles (terpenoids) differ among biotypes of leafy spurge which also differ in their acceptability to several biological control agents (Spencer, 1995). These differences in plant chemistry may explain differences in the feeding and survival of these agents. A similar influence of plant quality has been found for the gall midge *Spurgia esulae* (Lym *et al.,* 1996).

Host plant quality can be an important determinant of successful control. Previous work has shown that habitat manipulation or fertilization can be used to substantially increase host plant quality and the degree of control. Although these manipulations have not been tested with many weeds their high success suggests that these strategies could work in other systems and practitioners should determine if plant quality is a limiting factor for successful control.

5. Competition

Although biological control researchers have long been concerned about possible competition among control agents there is relatively little evidence of competitive interactions being the cause of biological control failures (e.g., Crawley, 1989; Harris, 1991). Until recently, competition was not thought to be important in the structure and abundance of insect herbivore communities. However, evidence now suggests that competition is more important for certain herbivores than previously thought (Denno *et al.,* 1995; Stewart, 1996) and is an equal or greater source of mortality than natural enemies or host plant resistance (Denno *et al.*, 1995). Competition was most intense between congeners, introduced taxa, and relatively immobile endophytic taxa.

In spite of these concerns, there appear to be relatively few instances of weed control failures that can be attributed to competition. Competition either is not important or in the several documented cases of competitive replacement an inferior agent may have been replaced with a better agent (Blossey, 1995). For example, Saner *et al.*, (1994) found little evidence of competition between two control agents for Dalmatian toadflax. Similarly, McEvoy *et al.*, (1993) found no competition between tansy ragwort *(Senecio jacobaea* L.) control agents. However, adding an inferior control agent to a superior control agent did not enhance control (McEvoy *et al.*, 1993). There are a few examples of competition and competitive displacement among biological control agents. Jordan (1995) indicates that the weevil *Larinus minutus* Gyll., a potential control agent for knapweeds, may be displaced by the tephritid *Urophora affinis* (Frauenfeld) which apparently has not been a successful control agent (Julien, 1992). Competition among control agents *of Hypericum* also has been suggested based on some evidence that *Chrysolina quadrigemina* can limit the abundance and distribution of *Chrysolina hyperici and Agrilus hyperici* (Campbell and McCaffrey, 1991; Briese, 1991). In these instance, *C. quadrigemina* appears to be the superior control agent.

Harris (1991) suggested that better control may be obtained by the cumulative stress imposed by several control agents that either attack the plant in a temporal sequence or attack different plant parts. This was the basis for the release of multiple control agents of purple loosestrife, several which are congeners and attack the same plant parts (Blossey, 1995; Hight *et al.*, 1995). Intensive evaluation of this system should further our understanding of whether competition is important to the survival of these agents or to the successful control of purple loosestrife and whether the strategy of multiple agents is advisable. Current evidence suggests that competition will generally not inhibit control and that superior competitors will likely be better control agents.

B.   Factors That Determine Effectiveness of Agent Populations

1.   Agent density

The density or abundance of a control agent will obviously affect its ability to control the plant (e.g., Campbell and McCaffrey, 1991; McEvoy *et al.*, 1991; Blossey, 1995). In classical biological control of weeds, the failure to establish populations of agents accounts for about 40% of the control failures (Greathead, 1995); which clearly illustrates the importance of factors limiting agent populations. However, as will be pointed out in the following two sections abundance or high density of the control agent alone is often not sufficient for successful control. Other factors such as plant competition, disturbance regimes, and microclimate may be quite important.

2. Plant response to herbivores

Because most weed biological control agents are specialists and thought to be adapted to overcome their host's defensive systems, the role of resistance has rarely been considered in weed biological control. However, specialists can be affected by the defense system of their host plants (e.g., Louda and Rodman, 1996; examples in Rosenthal and Berenbaum, 1992) and native generalist agents will likely be affected by plant defensive responses (Bernays and Chapman, 1994). Furthermore, differential acceptability of, and performance on, different genotypes and biotypes of target weeds (e.g., Spencer, 1995; Lym *et al.,* 1996) suggests that resistance may be more important than previously thought. Factors that influence resistance may affect control. Considerable variation in defense levels and deterrence can also occur among habitats or vary with degree of plant stress (Louda and Collinge, 1992; Louda and Rodman, 1996). In addition to direct defensive responses, reductions in nitrogen content due to herbivore damage (Center and Van, 1989) can make plants less acceptable to control agents (Center and Wright, 1991). Similarly, fertilization of *Opuntia and Salvinia* increased their suitability and therefore the degree of control by *Cactoblastis* and *Cyrtobagous,* respectively (Wilson, 1960; Room *et al.,* 1989; see Section II,A,4).

Plant tolerance and escape will also influence the impact of herbivores (Myers *et al.,* 1990; Rosenthal and Kotanen, 1994). Plants with a high tolerance to herbivore attack may be difficult to control (McEvoy *et al.,* 1993). The above-mentioned differential control of *Hypericum* by *Chrysolina* leaf beetles in dry and sunny vs. wet and cool habitats may be as much due to the effects of these environments on plant tolerance as on control agent population dynamics. Control may be effective at the dryer sites because *Hypericum* is less able to withstand and recover from herbivore damage than at wetter sites where soil water availability is adequate for regrowth (Rosenthal and Kotanen, 1994). Likewise, control of *Opuntia* is affected by microclimate differences that affect its ability to withstand herbivory as well as regulate cochineal densities (Zimmermann *et al.,* 1986). Similar variation in the ability of weeds to recover from damage has been reported for other systems (Cullen, 1995).

Herbivore effects on plants may emerge after multiple attacks or persist after herbivory is stopped (Karban and Strauss, 1993), an observation that indicates the need for longer-term monitoring of plants at individual and population levels. Herbivores and defoliation can reduce plant carbohydrate stocks (Lacey *et al.,* 1994; Ang *et al.,* 1995; Newman *et al.,* 1996a) and may reduce plant overwintering survival or competitive ability. Factors that influence the plant's ability to recover from the stresses of herbivory can have an important effect on the success of control and more attention to plant response is needed to develop effective control strategies (McClay,

1995). Conservation strategies that enhance plant quality for herbivores, such as fertilization, or decrease plant tolerance, such as the enhancement of plant community response should enhance control.

3. Plant community response

Increasing evidence suggests that plant community response may often be as important for successful biological control as herbivore performance (e.g., Crawley, 1990; McEvoy *et al.*, 1991, 1993; Groves, 1995; Sheppard, 1996). Herbivory can alter plant competitive outcomes (Anderson and Briske, 1995), however, without a strong response from native or desirable plants biological control can result in the replacement of the target weed with another weed (Randall, 1996; Sheppard, 1996). For example, Randall (1996) notes that successful biological control of St. Johnswort *(Hypericum perforatum)* in one county in Oregon was followed by an increase of tansy ragwort *(Senecio jacobaea)*. After successful biological control of tansy ragwort, Italian thistle *(Carduus pycnocephalus* L.) invaded and attained nuisance levels. Without effective changes in cultural practices an endless series of new weed outbreak and biological control introductions may be perpetuated (Randall, 1996).

Sheppard (1996) investigated the relative impact of biological control agents and plant competition in the control of pasture weeds. In 80% of the studies examined, competitive interactions were the dominant factor regulating weed performance (i.e., biomass, seed production, or survival). However, for most systems the effects were additive; control agents and competition combined to increase control (Sheppard, 1996). Synergistic effects were less common as were antagonistic interactions (i.e., negative interaction between control agent and competitor effects).

Specific case studies illustrate the importance of competition (see Section II,A,5). During successful biological control of tansy ragwort *(Senecio jacobaea)*, other species become as, or more, abundant and replaced ragwort (McEvoy *et al.*, 1991). Competition with these other plants is quite important (McEvoy *et al.*, 1990; 1993) and in undisturbed areas with no herbivory on any species ragwort may be outcompeted by other plants. The biological control agents alone could suppress tansy ragwort but agents were more effective when plant competition was present. The effects of herbivory were additive to plant competition and biological control in conjunction with plant competition yielded the best control (McEvoy *et al.*, 1993). These studies provide direct evidence that plant competition can enhance the biological control of weeds. Control of Canada thistle *(Cirsium arvense* (L.)) and musk thistle by the leaf beetle *Cassida rubiginosa* Müller may also be enhanced by competition with other plants (Harris, 1991; Ang *et al.*, 1995). Similar results have been seen with native plant-herbivore interactions. Competition with grasses can reduce growth and abundance of *Opuntia fragilis* Nutt. Competition combined with enhanced herbivory are both important determinants of the distribution and abundance of the cactus (Burger and Louda, 1994, 1995).

These studies indicate that successful control is often dependent on factors other than the density of the agent. Other factors such as disturbance regime also can affect plant community response and resurgence of weeds (McEvoy and Rudd, 1993; Anderson and Briske, 1995; Burke and Grime, 1996). In some instances, control agents can respond quickly to outbreaks in metapopulations of weeds (McEvoy and Rudd, 1993). However, in other instances control agents may go locally extinct or may be unable to keep up with weed outbreaks. Effective conservation strategies will not only ensure adequate populations of control agents but a reduction in factors that result in weed outbreaks. Protection and enhancement of native plant communities by managing disturbances such as cultivation or grazing should be effective. Nutrient loads that favor weeds should be reduced. These approaches along with plantings of native plants should hasten recovery of native plant communities and enhance control.

III. CONSERVATION AND USE OF NATIVE BIOLOGICAL CONTROL AGENTS

Although native insects commonly cause considerable damage to weed species their use in biological control is often overlooked. However, there is increasing experimental as well as observational evidence that native insect herbivores can control plant abundance and distribution (e.g., Blossey, 1995; Creed and Sheldon, 1995; Louda and Potvin, 1995). Despite skepticism that native agents will be effective at controlling either native or exotic weeds there are numerous examples of native agents controlling weeds (see DeBach and Rosen, 1991; Julien, 1992). It is clear that for many of these cases, contrary to prevailing dogma, herbivore populations are not so limited by natural enemies as to render them ineffective. The successful use of native agents for biological control of weeds requires the development of effective conservation strategies.

Research to determine the potential of using conservation strategies to manage native insects on native rangeland weeds in the southwestern U.S.A. has been initiated in ecosystems influenced by the perennial snakeweeds (broom snakeweed, *Gutierrezia sarothrae* (Pursh) and threadleaf snakeweed, *G. microcephala* D.C.)) and locoweeds *(Astragalus mollissimus* and *Oxytropis sericea* (Nutt.)). The damage potential and basic biology of several native biological control agents on the snakeweeds are being quantified (Richman and Huddleston, 1981; Parker, 1985; Wisdom *et al.,* 1989; Richman *et al.,* 1992; Thompson *et al.,* 1995; Thompson *et al.,* 1996). Population dynamics of the perennial poisonous locoweed *A. mollissimus* are driven by feeding of the root boring weevils *Cleonidius trivittatus* (Pomerinke and Thompson, 1995) and *Sitona californicus (F.)* (Pomerinke, 1993). Almost 100% mortality of *A. mollissimus* occurs when populations of the native weevil *C. trivittatus* exceed two weevils per plant (Pomerinke *et al.,* 1995). Understanding the biology of the biological control agent is essential before attempting conservation biological control techniques.

Properly timed range management practices can protect populations of native biological control agents, increasing their effectiveness. Populations of the grasshopper *Hesperotettix viridis* can be dramatically increased by altering the dates of prescribed burns. Grasshopper densities in the 50m area surrounding burned plots averaged 16.1±6.3 per plant when plots were burned before grasshopper egg hatch in mid-April and 2.8±0.6 per plant when plots were burned after egg hatch (Thompson, unpublished data). In field plots where *H. viridis has* attacked snakeweeds, the standing crop of grasses increased 23% during the year of grasshopper herbivory and 44% one year after herbivory compared to plots with no grasshoppers (Thompson *et al.,* 1996). Herbicides are usually slow acting enough that mobile native insects will disperse in search of better host plants. Turcotte (1993) showed that although populations of the highly immobile (larvae live in leaf ties and adult females do not fly) snakeweed leaf-tiers decreased in plots where 10% of the weed population was left as refugia, *H. viridis* maintained or increased population densities in these plots. Pomerinke *et al.* (1995) question whether insecticides sprayed annually to control rangeland grasshoppers and range caterpillars *(Hemileuca olivacea* (Cockerell)) have disrupted the interrelationships of the native root borer *Cleonidius trivittatus* on its native host *A. mollissimus.* However, densities of a snakeweed root boring beetle, *Crossidius pulchellus* LeConte, were unaffected by insecticides used for range pests if sprayed prior to adult emergence (Knight and Thompson, 1996).

Techniques to enhance native aquatic insects are also being investigated for the control of both native and exotic weeds (e.g., Oraze and Grigarick, 1992; Buckingham, 1994; Creed and Sheldon, 1995; McGregor *et al.,* 1996). Three insects, an indigenous weevil, an indigenous midge, and a naturalized European moth are being investigated for control of Eurasian watermilfoil *(Myriophyllum spicatum* L.) in North America (MacRae *et al.,* 1990; Sheldon and Creed, 1995). The weevil *Euhrychiopsis lecontei* (Dietz) appears to be the most promising control agent, occurring more commonly and in greater abundance than the moth and midge (Creed and Sheldon, 1995; Newman and Maher, 1995). The weevil is host specific to watermilfoils for feeding and oviposition and prefers the exotic Eurasian watermilfoil over its native host *M. sibiricum* Komarov (Sheldon and Creed, 1995; Solarz and Newman, 1996). Weevil development and survival is at least as good on the exotic as on the native milfoil (Newman *et al.,* 1997) and host plant resistance by the exotic species does not appear to be a limiting factor.

Newman *et al.,* (1996a) postulated that a weevil density of 200-300/m^2 (i.e., 1-2 weevils per stem) should result in control of Eurasian watermilfoil but noted that densities at Minnesota sites generally were not this high. We are currently investigating factors that limit the effectiveness of the weevils. Although the adults can fly in the fall and spring overwintering habitats appear restricted to dry litter areas close to lakes. It is unclear if overwintering habitat or survival are limiting but spring shoreline densities do appear to influence spring in-lake densities (D. W. Ragsdale and R. M. Newman, unpublished data). Populations in some lakes have failed to increase during the summer suggesting that natural enemies may be limiting in these lakes. Weed harvesting can reduce weevil populations (Sheldon and O'Bryan, 1996); however, no harvesting occurs at our sites. Parasitism appears nonexistent. Fish predation appears inadequate to affect weevil populations in lakes with moderate to high weevil density but sunfish could have a substantial impact on low-density weevil populations (Sutter and Newman, 1997). Augmentations of adults to small open plots have failed in several lakes, possibly due to predation by bluegills.

In addition to weevil density, plant community response appears to be important. Even with moderate densities of weevils (20-50/m^2), declines of Eurasian watermilfoil have persisted where biomass of native plants increased (Newman, unpublished data). At sites with no persistent decreases in Eurasian watermilfoil, the native plant community has failed to increase in patches where milfoil was heavily damaged by the weevils. Competition from native plants may be essential to provide sustained biological control.

Native herbivores can control Eurasian watermilfoil, but factors that limit success need to be identified and manipulated for control to become predictably effective. Factors that need further investigation include: fish predation, overwintering habitat, transition from water to shore and back to water, host plant resistance among lakes, plant community response, and competition with other plants. Control may be more effective in lakes which lack bluegills (e.g., Brownington Pond, VT (Creed and Sheldon, 1995)). In lakes with high bluegill populations, refugia (such as dense plant beds) may be needed to increase weevil populations. Shoreline overwintering habitat should be protected from development. Management practices that help retain a native plant community such as maintenance of good water clarity and lack of disturbance may enhance the effects of weevils.

We suspect that native control agents will not work in all situations nor will they show the spectacular successes seen with the several prime examples of classical biological control such as the *Opuntia-Cactoblastis, Salvinia-Cyrtobagous,* or tansy ragwort systems. These spectacular successes are relatively rare and only 10% of the attempted introductions, or 16% of the established introductions have given good control (Greathead, 1995). However, in many natural systems (e.g., lakes, wetlands, rangelands), other methods of control are equally problematic, being either too expensive or causing unacceptable levels of nontarget impacts. In many of these systems selective herbicide control is ineffective, unpredictable, or prohibited. Even moderate success rates with native control agents would greatly enhance our ability to manage these systems. The challenge is to determine when, where, and how native agents can effect acceptable levels of conservation biological control.

IV. CONCLUSIONS AND RECOMMENDATIONS

Defining specific conservation strategies in the biological control of weeds is difficult. The success of any conservation strategy is dependent on a thorough knowledge of the biology of the potential biological control agent, its host and their interactions. Lack of resources limit exploration into the complex biologies and interactions of most of the herbivores associated with target weeds. We suggest most conservation strategies can be divided into three general areas: (1) population protection or the informed and appropriate use of pesticides to maintain native or exotic biological control agents; (2) habitat protection to preserve critical habitat or refugia; and (3) plant community management to maintain and enhance the effectiveness of existing biological control agents. Other strategies to protect populations from natural enemies, climate, weather or competitors are less clearly defined.

Harris (1991) noted that conservation for weed biological control was largely a theoretical concept. This is unfortunate because conservation strategies should play an important role in the management of weeds with both exotic and native biological control agents. The importance of establishing the factors responsible for the success or failure of weed biological control projects is becoming more and more apparent. The strategies outlined in this chapter have come largely from researchers who have done a thorough job of evaluating biological control projects. Virtually none of the examples we provided in this chapter were the result of a conscious effort to test conservation strategies. Integration of conservation strategies with classical biological control along with other weed management techniques, should be implemented and carefully evaluated. The resulting integrated weed management programs would be more likely to succeed. Finally, we contend that conservation strategies should not be considered as seldom used, theoretical concepts rather they should be considered integral parts of any weed biological control effort.

Acknowledgments

We thank Drs. Rob Creed, Svata Louda, and David Ragsdale for providing literature or access to unpublished data. Comments on the manuscript and discussion provided by Drs. Pedro Barbosa, Bernd Blossey, Gary R. Buckingham, and C. Jack DeLoach were most helpful and greatly appreciated. D. C. Thompson and D. B. Richman were supported in part by the USDA/CSREES and the New Mexico Agricultural Experiment Station, New Mexico State University, Las Cruces. R. M. Newman was supported in part by a grant from the Minnesota Department of Natural Resources, based on funds appropriated by the Minnesota Legislature as recommended by the Legislative Commission on Minnesota Resources from the Minnesota Future Resources Fund and funding from the Minnesota Agricultural Experiment Station. This paper is published as Paper Number 974740002 of the contribution series of the Minnesota Agricultural Experiment Station based on research conducted under Project 74.

REFERENCES

Anderson, V. J., and Briske, D. D. (1995). Herbivore-induced species replacement in grasslands: is it driven by herbivory tolerance or avoidance? *Ecol. Appl.* 5, 1014-1024.

Andres, L. A. (1982). Integrating weed biological control agents into a pest-management program. *Weed Sci.* 30 (Supplement), 25-30.

Ang, B. N., Kok, L. T., Holtzman, G. I., and Wolf, D. D. (1995). Canada thistle *(Cirsium arvense* (I.) Scop.) response to density of *Cassida rubiginosa* Müller (Coleoptera: Chrysomelidae) and plant competition. *Biol. Cont.* 5, 31-38.

Annecke, D. P., Karny, M., and Burger, W. A. (1969). Improved biological control of the prickly pear, *Opuntia megacantha* Salm-Dyk, in South Africa through the use of an insecticide. *Phytophylactica* 1, 9-13.

Belovsky, G. E., and Joern, A. (1995). The dominance of different regulating factors for rangeland grasshoppers. *In* "Population Dynamics: New Approaches and Synthesis." (N. Cappuccino, and P. W. Price, eds.), pp. 359-386. Academic Press. San Diego, CA.

Bernays, E. A., and Chapman, R. F. (1994). "Host-plant Selection by Phytophagous Insects." Chapman and Hall. New York, NY.

Blossey, B. (1995). Impact of *Galerucella pusilla and G. calmariensis* (Coleoptera: Chrysomelidae) on field populations of purple loosestrife *(Lythrum salicaria). In* "Proceedings of the VIII International Symposium on Biological Control of Weeds." (E. S. Delfosse, and R. R. Scott, eds.), pp. 27-31. CSIRO, Melbourne, Australia.

Briese, D. T. (1991). Current status of *Agrilus hyperici* (Coleoptera: Buprestidae) released in Australia in 1940 for the control of St. John's wort: lessons for insect introductions. *Biocont. Sci. Technol.* 1, 207-215.

Briese, D. T. (1996). Biological control of weeds and fire management in protected natural areas: are they compatible strategies? Biol. Cons. 77, 135-141.

Buckingham, G. R. (1994). Biological control of aquatic weeds. *In* "Pest Management in the Subtropics: Biological Control - a Florida Perspective." (D. Rosen, F. D. Bennett, and J. L. Capinera, eds.), pp. 413-480. Intercept Ltd. Andover, U.K.

Burger, J. C., and Louda, S. M. (1994). Indirect versus direct effects of grasses on growth of a cactus *(Opuntia fragilis):* insect herbivory versus competition. Oecologia 99, 79-87.

Burger, J. C., and Louda, S. M. (1995). Interaction of diffuse competition and insect herbivory in limiting brittle prickly pear cactus, *Opuntia fragilis* (Cactaceae). *Amer. J. Bot.* 82, 1558-1566.

Burke, M. J. W., and Grime, J. P. (1996). An experimental study of plant community invasibility. *Ecology* 77, 776-790.

Campbell, C. L., and McCaffrey, J. P. (1991). Population trends, seasonal phenology, and impact of *Chrysolina quadrigemina, C. hyperici* (Coleoptera: Chrysomelidae), and *Agrilus hyperici* (Coleoptera: Buprestidae) associated with *Hypericum perforatum* in Northern Idaho. *Environ. Entomol.* 20, 303-315.

Center, T. D. (1994). Biological control of weeds: waterhyacinth and waterlettuce. *In* "Pest Management in the Subtropics: Biological Control - a Florida Perspective." (D.

Rosen, F. D. Bennett, and J. L. Capinera, eds.), pp. 481-521. Intercept Ltd. Andover, U.K.

Center, T. D., and Dray, F. A. (1992). Associations between waterhyacinth weevils *(Neochetina eichhorniae and N. bruchi)* and phenological stages of *Eichhorniae crassipes* in southern Florida. *Fla. Entomol.* 75, 196-211.

Center, T. D., and Van, T. K. (1989). Alteration of water hyacinth *(Eichhornia* crassipes (Mart.) Solms) leaf dynamics and phytochemistry by insect damage and plant density. *Aquat. Bot.* 35, 181-195.

Center, T. D., and Wright, A. D. (1991). Age and phytochemical composition of waterhyacinth (Pontederiaceae) leaves determine their acceptability to *Neochetina eichhorniae* (Coleoptera, Curculionidae). *Environ. Entomol.* 20, 323-334.

Collinge, S. K., and Louda, S. M. (1988). Herbivory by leaf miners in response to experimental shading of a native crucifer. *Oecologia* 75, 559-566.

Collins, N. M., and Thomas, J. A. (1991). "The Conservation of Insects and Their Habitats." Academic Press. San Diego, CA.

Crawley, M. J. (1986). The population biology of invaders. *Philos. Trans. R. Soc.* London B 314, 711-731

Crawley, M. J. (1989). The successes and failures of weed biocontrol using insects. *Biocont. News Inform.* 10, 213-223.

Crawley, M. J. (1990). Plant life-history and the success of weed biological control projects. *In* "Proceedings of the VII International Symposium on Biological Control of Weeds." (E. S. Delfosse, ed.), pp. 17-26. Istituto Sperimentale per la Patologia Vegetale, Ministero dell'Agricoltura e delle Foreste. Rome, Italy.

Creed, R. P., and Sheldon, S. P. (1995). Weevils and watermilfoil: did a North American herbivore cause the decline of an exotic plant? *Ecol. Appl. 5,* 1113-1121.

Cullen, J. M. (1995). Predicting effectiveness: fact and fantasy. *In* "Proceedings of the VIII International Symposium on Biological Control of Weeds." (E. S. Delfosse, and R. R. Scott, eds.), pp. 103-109. CSIRO. Melbourne, Australia.

DeBach, P., and Rosen, D. (1991). "Biological Control by Natural Enemies." 2nd ed. Cambridge University Press. New York, NY.

Delfosse, E. S., and Scott, R. R. (1995). "Proceedings of the VIII International Symposium on Biological Control of Weeds." CSIRO, Melbourne, Australia.

DeLoach, C. J. (1 991). Past successes and current prospects in biological control of weeds in the United States and Canada. *Nat. Areas J.* 11, 129-142.

DeLoach, C. J. (1995). Progress and problems in introductory biological control of native weeds in the United States. *In* "Proceedings of the VIII International Symposium on Biological Control of Weeds." (E. S. Delfosse, and R. R. Scott, eds.), pp. 111-122. CSIRO. Melbourne, Australia.

Denno, R. F., McClure, M. S., and Ott, J. R. (1995). Interspecific interactions in phytophagous insects: competition reexamined and resurrected. *Annu. Rev. Entomol.* 40, 297-331.

Fritz, R. S. (1995). Direct and indirect effects of plant genetic variation on enemy impact. *Ecol. Entomol.* 20, 18-26.

Fritz, R. S., and Price, P. W. (1988). Genetic variation among plants and insect community structure: willows and sawflies. *Ecology* 69, 845-856.

Gardner, K. T., and Thompson, D. C. (1997). Influence of avian predation on a grasshopper assemblage that feeds on threadleaf snakeweed *(Gutierrezia microcephala). Environ. Entomol.* 26, (in press).

Gaston, K. J., New, T. R., and Samways, M. J. (1993). "Perspectives on Insect Conservation." Intercept Ltd. Andover, U.K.

Godfrey, K. E., Anderson, L. W. J., Perry, S. D., and Dechoretz, N. (1994). Overwintering and establishment potential of *Bagous affinis* (Coleoptera: Curculionidae) on *Hydrilla verticillata* (Hydrocharitaceae) in northern California. *Fla. Entomol.* 77, 221-230.

Goeden, R. D., and Louda, S. M. (1976). Biotic interference with insects imported for weed control. *Annu. Rev. Entomol.* 21, 325-342.

Greathead, D. J. (1995). Benefits and risks of classical biological control. *In* "Biological Control: Benefits and Risks." (H. M. Hokkanen, and J. M. Lynch, eds.), pp. 53-63. Cambridge University Press. Cambridge, U.K.

Groves, R. H. (1995). Biological control of weeds - past, present and future. *In* "Proceedings of the VIII International Symposium on Biological Control of Weeds." (E. S. Delfosse, and R. R. Scott, eds.), pp. 7-11. CSIRO. Melbourne, Australia.

Haag, K. H., and Buckingham, G. R. (1991). Effects of herbicides and microbial insecticides on the insects of aquatic plants. *J. Aquat. Plant Manag.* 29, 55-57.

Haag, K. H., and Habeck, D. H. (1991). Enhanced biological control of waterhyacinth following limited herbicide application. *J. Aquat. Plant Manag.* 29, 24-28.

Harris, P. (1991). Invitation Paper (C. P. Alexander Fund): Classical biocontrol of weeds - its definition, selection of effective agents, and administrative political problems. *Can. Entomol.* 123, 827-849.

Harrison, S., and Cappuccino, N. (1995). Using density-manipulation experiments to study population regulation. *In* "Population Dynamics: New Approaches and Synthesis." (N. Cappuccino, and P. W. Price, eds.), pp. 131-147. Academic Press. San Diego, CA.

Hight, S. D., Blossey, B., Laing, J., and Declerckfloate, R. (1995). Establishment of insect biological control agents from Europe against *Lythrum salicaria* in North America. *Environ. Entomol.* 24, 967-977.

Hill, M. P., and Hulley, P. E. (1995). Host-range extension by native parasitoids to weed biocontrol agents introduced to South Africa. *Biol. Cont.* 5, 297-302.

Hoffmann, J. H., and Moran, V. C. (1995). Localized failure of a weed biological control agent attributed to insecticide drift, *Agric., Ecosys., Environ.* 52, 197-203.

Hoffmann, J. H., Impson, F. A. C., and Moran, V. C. (1993). Biological control of mesquite weeds in South Africa using a seed-feeding bruchid, *Algarobius prosopis* - initial levels of interference by native parasitoids. *Biol. Cont.* 3, 17-21.

Huffaker, C. B., and Kennett, C. E. (1959). A ten year study of vegetational changes associated with biological control of Klamath weed. *J. Range Manag.* 12, 69-82.

Huffaker, C. B., Dahlsten, D. L., Janzen, D. H., and Kennedy, G. G. (1984). Insect influences in the regulation of plant populations and communities. *In* "Ecological Entomology." (C. B. Huffaker, and R. L. Rabb, eds.), pp. 659-691. John Wiley and Sons. New York, NY.

Johnson, M. W., and Wilson, L. T. (1995). Integrated pest management: Contributions of biological control to its implementation. *In* "Biological Control in the Western United States: Accomplishments and Benefits of Regional Research Project W-84, 1964-1989." (J. R. Nechols, L. A. Andres, J. W. Beardsley, R. D. Goeden, and C. G. Jackson, eds.), pp. 7-24. University of California, Div. Agric. Nat. Res., Publ. 3361. Oakland, CA.

Jordan, K. (1995). Host specificity of *Larinus minutus* Gyll. (Col., Curculionidae), an agent introduced for the biological control of diffuse and spotted knapweed in North America. *J. Appl. Entomol.* 119, 689-693.

Julien, M. H. (1992). "Biological Control of Weeds: a World Catalogue of Agents and Their Target Weeds," 3rd ed. CAB International. Wallingford, U.K.

Julien, M. H., and Storrs, M. J. (1996). Integrating biological and herbicidal controls to manage *Salvinia* in Kakadu National Park, northern Australia. *In* "Proceedings of the IX International Symposium on Biological Control of Weeds." (V. C. Moran, and J. H. Hoffmann, eds.), pp. 445-449. University of Capetown. Rondebosch, South Africa.

Julien, M. H., Skarratt, B., and Maywald, G. F. (1995). Potential geographical distribution of alligator weed and its biological control by *Agasicles hygrophila. J. Aquat. Plant Manag.* 33, 55-60.

Julien, M. H., Harley, K. L. S., Wright, A. D., Cilliers, C. J., Hill, M. P., Center, T. D., Cordo, H. A., and Cofrancesco, A. F. (1996). International cooperation and linkages in the management of water hyacinth with emphasis on biological control *In* "Proceedings of the IX International Symposium on Biological Control of Weeds." (V. C. Moran, and J. H. Hoffmann, eds.), pp. 273-282. University of Capetown, Rondebosch, South Africa.

Karban, R. (1989). Community organization of *Erigeron glaucus* folivores: Effects of competition, predation, and host plant. *Ecology* 70, 1028-1039.

Karban, R. (1992). Plant variation: its effects on populations of herbivorous insects. *In* "Plant Resistance to Herbivores and Pathogens." (R. S. Fritz, and E. L. Simms, eds.), pp. 195-215. University of Chicago Press. Chicago, IL.

Karban, R., and Strauss, S. Y. (1993). Effects of herbivores on growth and reproduction of their perennial host, *Erigeron glaucus. Ecology* 74, 39-46.

Knight, J. L., and Thompson, D. C. (1996). Effects of controlling rangeland insect pests on natural enemies of a perennial rangeland weed. *Proc. Western Soc. Weed Sci.* 49, 47.

Lacey, J. R., Olsonrutz, K. M., Haferkamp, M. R., and Kennett, G. A. (1994). Effects of defoliation and competition on total nonstructural carbohydrates of spotted knapweed. *J. Range Manag.* 47, 481-484.

Lockwood, J. A. (1993). Benefits and costs of controlling rangeland grasshoppers (Orthoptera: Acrididae) with exotic organisms: search for a null hypothesis and regulatory compromise. *Environ. Entomol.* 22, 904-914.

Louda, S. M., and Collinge, S. K. (1992). Plant resistance to insect herbivores: a field test of the environmental stress hypothesis. *Ecology* 73, 153-169.

Louda, S. M., and Masters, R. A. (1993). Biological control of weeds in Great Plains rangelands. *Great Plains Res.* 3, 215-247.

Louda, S. M., and Potvin, M. A. (1995). Effect of inflorescence-feeding insects on the demography and lifetime fitness of a native plant. *Ecology* 76, 229-245.

Louda, S. M., and Rodman, J. E. (1996). Insect herbivory as a major factor in the shade distribution of a native crucifer *(Cardamine cordifolia* A. Gray, bittercress). *J. Ecol.* 84, 229-237.

Louda, S. M., Kendall, D., Connor, J., and Simberloff, D. (1997). Ecological effects of an insect introduced for the biological control of weeds. *Science* 277, 1088-1090.

Lym, R. G., and Carlson, R. B. (1994). Effect of herbicide treatment on leafy spurge gall midge *(Spurgia esulae)* population. *Weed Technol.* 8, 285-288.

Lym, R. G., Nissen, S. J., Rowe, M. L., Lee, D. J., and Masters, R. A. (1996). Leafy spurge *(Euphorbia esula)* genotype affects gall midge *(Spurgia esulae)* establishment. *Weed Sci.* 44, 629-633.

MacRae, I. V., Winchester, N. N., and Ring, R. A. (1990). Feeding activity and host preference of the milfoil midge, *Cricotopus myriophylli* Oliver (Diptera: Chironomidae). *J. Aquat. Plant Manag.* 28, 89-92.

Maddox, G. D., and Root, R. B. (1987). Resistance to 16 diverse species of herbivorous insects within a population of goldenrod, *Solidago altissima:* genetic variation and heritability. *Oecologia* 72, 8-14.

McCaffrey, J. P., Campbell, C. L., and Andres, L. A. (1995). St. Johnswort. *In* "Biological control in the Western United States: Accomplishments and Benefits of Regional Research Project W-84, 1964-1989." (J. R. Nechols, L. A. Andres, J. W. Beardsley, R. D. Goeden, and C. G. Jackson, eds.), pp. 281-285. University of California, Div. Agric. Nat. Res. Public. 3361.Oakland, CA.

McClay, A. S. (1995). Beyond "before-and-after:" experimental design and evaluation in classical weed biological control. *In* "Proceedings of the VIII International Symposium on Biological Control of Weeds." (E. S. Delfosse, and R. R. Scott, eds.), pp. 213-219. CSIRO. Melbourne, Australia.

McEvoy, P. B., and Rudd, N. T. (1993). Effects of vegetation disturbances on insect biological control of tansy ragwort Senecio jacobaea. *Ecol. Appl.* 3, 682-698.

McEvoy, P. B., Cox, C. S., James, R. R., and Rudd, N. T. (1990). Ecological mechanisms underlying successful biological weed control: field experiments with ragwort *Senecio jacobaea. In* "Proceedings of the VII International Symposium on Biological Control of Weeds." (E. S. Delfosse, ed.), pp. 55-66. Istituto Sperimentale per la Patologia Vegetale, Ministero dell'Agricoltura e delle Foreste. Rome, Italy.

McEvoy, P. B., Cox, C., and Coombs, E. (1991). Successful biological control of ragwort, *Senecio jacobaea,* by introduced insects in Oregon. *Ecol. Appl.* 1, 430-442.

McEvoy, P. B., Rudd, N. T., Cox, C. S., and Huso, M. (1993). Disturbance, competition, and herbivory effects on ragwort *Senecio jacobaea* populations. *Ecol. Monogr.* 63, 55-75.

McGregor, M. A., Bayne, D. R., Steeger, J. G., Webber, E. C., and Reutebuch, E. (1996). The potential for biological control of water primrose *(Ludwigia grandiflora)* by the water primrose flea beetle *(Lysathia ludoviciana)* in the southeastern United States. *J. Aquat. Plant Manag.* 34, 74-76.

Messersmith, C. G., and Adkins, S. W. (1995). Integrating weed-feeding insects and herbicides for weed control. *Weed Technol.* 9, 199-208.

Moran, V. C., and Hoffmann, J. H. (1996). "Proceedings of the IX International Symposium on Biological Control of Weeds." University of Capetown, Rondebosch, South Africa.

Müller, H., Nuessly, G. S., and Goeden, R. D. (1990). Natural enemies and host-plant asynchrony contributing to the failure of the introduced moth, *Coleophora parthenica* Meyrick (Lepidoptera, Coleophoridae), to control Russian thistle. *Agric., Ecosys., Environ.* 32, 133-142.

Myers, J. H. (1987). Population outbreaks of introduced insects: lessons from the biological control of weeds. *In* "Insect Outbreaks." (P. Barbosa, and J. C. Schultz, eds.), pp. 287-312. Academic Press. San Diego, CA.

Myers, J. H., Risley, C., and Eng, R. (1990). The ability of plants to compensate for insect attack: why biological control of weeds with insects is so difficult. *In* "Proceedings of the VII International Symposium on Biological Control of Weeds." (E. S. Delfosse, ed.), pp. 67-73. Istituto Sperimentale per la Patologia Vegetale, Ministero dell'Agricoltura e delle Foreste. Rome, Italy.

Nechols, J. R., Obrycki, J. J., Tauber, C. A., and Tauber, M. J. (1996). Potential impact of native natural enemies on *Galerucella* spp. (Coleoptera: Chrysomelidae) imported for biological control of purple loosestrife: a field evaluation. *Biol. Cont.* 7, 60-66.

Newman, R. M., and Maher, L. M. (1995). New records and distribution of aquatic insect herbivores of watermilfoils (Haloragaceae: *Myriophyllum* spp.) in Minnesota. *Entomol. News* 106, 6-12.

Newman, R. M., Holmberg, K. L., Biesboer, D. D., and Penner, B. G. (1996a). Effects of a potential biocontrol agent, *Euhrychiopsis lecontei*, on Eurasian watermilfoil in experimental tanks. *Aquat. Bot.* 53, 131-150.

Newman, R. M., Kerfoot, W. C., and Hanscom, Z. (1996b). Watercress allelochemical defends high nitrogen foliage against consumption: effects on freshwater invertebrate herbivores. *Ecology* 77, 2312-2323.

Newman, R. M., Borman, M. E., and Castro, S. W. 1997. Developmental performance of the weevil Euhrychiopsis lecontei on native and exotic watermilfoil hostplants. *J. No. Amer. Benthol. Soc.* 16: 627-634.

Nissen, S. J., Masters, R. A., Lee, D. J., and Rowe, M. L. (1995). DNA-based marker systems to determine genetic diversity of weedy species and their application to biocontrol. *Weed Sci.* 43, 504-513.

Olckers, T. (1995). Indigenous parasitoids inhibit the establishment of a gall forming moth imported for the biological control of *Solanum elaeagnifolium* Cav (Solanaceae) in South Africa. *African Entomol.* 3, 85-87.

Oraze, M. J., and Grigarick, A. A. (1992). Biological control of ducksalad (*Heteranthera limosa*) by the waterlily aphid (*Rhopalosiphum nymphaeae*) in rice (*Oryza sativa*). *Weed Sci.* 40, 333-336.

Parker, M. A. (1985). Size-dependent herbivore attack and the demography of an arid grassland shrub. *Ecology* 66, 850-860.

Peschken, D. P., and McClay, A. S. (1995). Picking the target: a revision of McClay's scoring system to determine the suitability of a weed for classical biological control. *In* "Proceedings of the VIII International Symposium on Biological Control of Weeds." (E. S. Delfosse, and R. R. Scott, eds.), pp. 137-143. CSIRO. Melbourne, Australia.

Pomerinke, M. A. (1993). Bionomics of two root feeding weevils (Coleoptera: Curculionidae) on *Astragalus mollissimus* (Fabaceae). M.S. Thesis, New Mexico State University. Las Cruces, NM.

Pomerinke, M. A., and Thompson, D. C. (1995). *Cleonidius trivittatus* feeds exclusively on wooly locoweed in northeastern New Mexico. *Southwest. Entomol.* 20, 107-109.

Pomerinke, M. A., Thompson, D. C., and Clason, D. L. (1995). Bionomics of the four-lined locoweed weevil (Coleoptera: Curculionidae): A native biological control of purple locoweed (Rosales: Fabaceae). *Environ. Entomol.* 24, 1696-1702.

Price, P. W. (1987). The role of natural enemies in insect populations. In "Insect Outbreaks." (P. Barbosa, and J. C. Schultz, eds.), pp. 287-312. Academic Press. San Diego, CA.

Price, P. W., Craig, T. P., and Roininen, H. (1995). Working toward theory on galling sawfly population dynamics. *In* "Population Dynamics: New Approaches and Synthesis."

(N. Cappuccino, and P. W. Price, eds.), pp. 321-338. Academic Press. San Diego, CA.

Randall, J. M. (1996). Weed control for the preservation of biological diversity. *Weed Technol.* 10, 370-383.

Reznik, S. Y. (1996). Classical biocontrol of weeds in crop rotation: a story of failure and prospects for success. *In* "Proceedings of the IX International Symposium on Biological Control of Weeds." (V. C. Moran, and J. H. Hoffmann, eds.), pp. 503-506. University of Capetown. Rondebosch, South Africa.

Richman, D. B., and Huddleston, E. W. (1981). Root feeding by the beetle, *Crossidius pulchellus* LeConte and other insects on broom snakeweed *(Gutierrezia* spp.) in eastern and central New Mexico. *Environ. Entomol.* 10, 53-57.

Richman, D. R., Thompson, D. C., and O'Mara, J. (1992). Effects of leaftiers on broom snakeweed in central New Mexico. *Southwest. Entomol.* 17, 187-189.

Roitberg, B. D., and Isman, M. B. (1992). "Insect Chemical Ecology. An Evolutionary Approach." Chapman and Hall. New York, NY.

Roland, J., and Taylor, P. D. (1995). Herbivore-natural enemy interactions in fragmented and continuous forests. *In* "Population Dynamics: New Approaches and Synthesis." (N. Cappuccino, and P. W. Price, eds.), pp. 195-208. Academic Press. San Diego, CA.

Room, P. M. (1990). Ecology of a simple plant herbivore system: biological control of *Salvinia. Trends Ecol. Evol.* 5, 74-79.

Room, P. M., and Fernando, I. V. S. (1992). Weed invasions countered by biological control: *Salvinia molesta and Eichhornia crassipes* in Sri Lanka. *Aquat. Bot.* 42, 99-107.

Room, P. M., Julien, M. H., and Forno, I. W. (1989). Vigorous plants suffer most from herbivores: latitude, nitrogen and biological control of the weed *Salvinia molesta. Oikos* 54, 92-100.

Rosenthal, G. A., and Berenbaum, M. R. (1992). "Herbivores, Their Interactions with Secondary Plant Metabolites." 2nd ed. Vol. II. Academic Press. San Diego, CA.

Rosenthal, J. P., and Kotanen, P. M. (1994). Terrestrial plant tolerance to herbivory. *Trends Ecol. Evol.* 9, 145-148.

Saner, M. A., Jeanneret, P., and Muller-Scharer, H. (1994). Interaction among two biological control agents and the developmental stage of their target weed, Dalmatian toadflax, *Linaria dalmatica* (L.) Mill. (Scrophulariaceae). *Biocont Sci. Technol.* 4, 215-222.

Sheldon, S. P., and Creed, R. P. (1995). Use of a native insect as a biological control for an introduced weed. *Ecol. Appl.* 5, 1122-1132.

Sheldon, S. P., and O'Bryan, L. M. (1996). The effects of harvesting Eurasian watermilfoil on the aquatic weevil *Euhrychiopsis lecontei. J. Aquat. Plant Manag.* 34, 76-77.

Sheppard, A. W. (1992). Predicting biological weed control. *Trends Ecol. Evol.* 7, 290-291.

Sheppard, A. W. (1996). The interaction between natural enemies and interspecific plant competition in the control of invasive pasture weeds. *In* "Proceedings of the IX International Symposium on Biological Control of Weeds." (V. C. Moran, and J. H. Hoffmann, eds.), pp. 47-53. University of Capetown, Rondebosch, South Africa.

Simberloff, D., and Stiling, P. (1996). How risky is biological control? *Ecology* 77, 1965-1974.

Slansky, F. (1992). Allelochemical-nutrient interactions in herbivore nutritional ecology. *In* "Herbivores, Their Interactions with Secondary Plant Metabolites." (G. A. Rosenthal, and M. R. Berenbaum, eds.), pp. 135-174. 2nd ed. Vol. II. Academic Press. San Diego, CA.

Solarz, S. L., and Newman, R. M. (1996). Oviposition specificity and behavior of the watermilfoil specialist *Euhrychiopsis lecontei. Oecologia* 106, 337-344.

Spencer, N. R. (1995). Biological weed control: the plant-insect interaction. *In* "Proceedings of the VIII International Symposium on Biological Control of Weeds." (E. S. Delfosse, and R. R. Scott, eds.), pp. 153-159. CSIRO. Melbourne, Australia.

Stewart, A. J. A. (1996). Interspecific competition reinstated as an important force structuring insect herbivore communities. *Trends Ecol. Evol.* 11, 233-234.

Stiling, P., and Rossi, A. M. (1995). Coastal insect herbivore communities are affected more by local environmental conditions than by plant genotype. *Ecol. Entomol.* 20, 184-190.

Strauss, S. Y. (1990). The role of plant genotype, environment and gender in resistance to a specialist chrysomelid herbivore. *Oecologia* 84, 111-116.

Sutter, T. J., and Newman, R. M. (1997). Is predation by sunfish *(Lepomis* spp.) an important source of mortality for the Eurasian watermilfoil biocontrol agent *Euhrychiopsis lecontei? J. Freshwater Ecol.* 12, 225-234

Syrett, P., Fowler, S. V., and Emberson, R. M. (1996). Are chrysomelid beetles effective agents for biological control of weeds? *In* "Proceedings of the IX International Symposium on Biological Control of Weeds." (V. C. Moran, and J. H. Hoffmann, eds.), pp. 47-53. University of Capetown. Rondebosch, South Africa.

Thomas, M. B., Wratten, S. D., and Sotherton, N. W. (1991). Creation of 'island' habitats in farmland to manipulate populations of beneficial arthropods: Predator densities and emigration. *J. Appl. Ecol.* 28, 906-917.

Thompson, D. C., McDaniel, K. C., Torell, L. A., and Richman, D. B. (1995). Damage potential of *Hesperotettix viridis* (Orthoptera: Acrididae) on a native rangeland weed, *Gutierrezia sarothrae. Environ. Entomol.* 24, 1315-1321.

Thompson, D. C., McDaniel, K. C., and Torell, L. A. (1996). Feeding by a native grasshopper reduces densities and biomass of broom snakeweed. *J. Range Manag.* 49, 407-412.

Tipping, P. W. (1991). Effects of mowing or spraying *Carduus thoermeri* on *Rhinocyllus conicus. Weed Technol.* 5, 628-631.

Turcotte, R. M. (1993). Integrated control of broom snakeweed by strip management at two selected locations in New Mexico. M.S. Thesis, New Mexico State University. Las Cruces, NM.

Van, T. K., and Center, T. D. (1994). Effect of paclobutrazol and waterhyacinth weevil *(Neochetina eichhorniae)* on plant growth and leaf dynamics of waterhyacinth *(Eichhornia crassipes). Weed Sci.* 42, 665-672.

van Driesche, R. G., and Bellows, T. S. (1996). "Biological Control." Chapman and Hall. New York, NY.

Visalakshy, P. N. G., and Jayanth, K. P. (1996). Effect of silt coverage of water hyacinth roots on pupation of *Neochetina eichhorniae* Warner and *N. bruchi* Hustache (Coleoptera: Curculionidae). *Biocont. Sci. Technol.* 6, 11-13.

Vogt, G. B., Quimby, P. D., and Kay, S. H. (1992). "Effects of weather on the biological control of alligatorweed in the lower Mississippi Valley Region, 1973-1983." USDA, Tech. Bul. 1766. Washington, DC.

Wapshere, A. J., DelFosse, E. S., and Cullen, J. M. (1989). Recent developments in biological control of weeds. *Crop Prot.* 8, 227-250.

Watson, A. K., and Wymore, L. A. (1990). Biological control, a component of integrated weed management. *In* "Proceedings of the VII International Symposium on Biological

Control of Weeds." (E. S. Delfosse, ed.), pp. 101-106. Istituto Sperimentale per la Patologia Vegetale, Ministero dell'Agricoltura e delle Foreste, Rome, Italy.

Wheeler, G. S., and Center, T. D. (1996). The influence of *Hydrilla* leaf quality on larval growth and development of the biological control agent *Hydrellia pakistanae* (Diptera: Ephydridae). *Biol. Cont.* 7, 1-9.

Whitcomb, W. H. (1994). Environment and habitat management to increase predator populations. *In* "Pest Management in the Subtropics: Biological Control - a Florida Perspective." (D. Rosen, F. D. Bennett, and J. L. Capinera, eds.), pp. 149-179. Intercept Ltd. Andover, U.K.

Williams, K. S. (1985). Climatic influences on weeds and their herbivores: biological control of St. John's wort in British Columbia. *In* "Proc. VI Intern. Symp. Biol. Contr. Weeds" (E. S. Delfosse, ed.), pp. 127-132. Agric. Can. Vancouver, Canada.

Wilson, F. (1960). "A Review of the Biological Control of Insects and Weeds in Australia and Australian New Guinea." Commonwealth Institute of Biological Control, Tech. Com. No. 1. Ottawa, Canada.

Wisdom, C. S., Crawford, C. S., and Aldon, E. F. (1989). Influence of insect herbivory on photosynthetic area and reproduction in *Gutierrezia species. J. Ecol.* 77, 685-692.

Zimmermann, H. G. (1979). Herbicidal control in relation to distribution of *Opuntia aurantiaca* Lindley and effects on cochineal populations. *Weed Res.* 19, 89-93.

Zimmermann, H. G., Moran, V. C., and Hoffmann, J. H. (1986). Insect herbivores as determinants of the present distribution and abundance of invasive cacti in South Africa. *In* "The Ecology and Management of Biological Invasions in Southern Africa." (I. A. W. MacDonald, F. J. Kruger, and A. A.Ferrar, eds.), pp. 269-274. Oxford University Press. Cape Town, South Africa.

Zwölfer, H., and Harris, P. (1984). Biology and host specificity of *Rhinocyllus conicus* (Fröl.) (Col., Curculionidae), a successful agent for biocontrol of the thistle *Carduus nutans* L. Z. *Ang. Entomol.* 97, 36-2.